The Innovation Imperative

CONTRIBUTING TO PRODUCTIVITY, GROWTH AND WELL-BEING

OECD
BETTER POLICIES FOR BETTER LIVES

Please cite this publication as:
OECD (2015), *The Innovation Imperative: Contributing to Productivity, Growth and Well-Being*, OECD Publishing, Paris.
DOI: http://dx.doi.org/10.1787/9789264239814-en

ISBN 978-92-64-23980-7 (print)
ISBN 978-92-64-23981-4 (PDF)

Photo credits: Ilhan Erogluoic

Corrigenda to OECD publications may be found on line at: www.oecd.org/publishing/corrigenda.

Preface

Innovation provides the foundation for new businesses, new jobs and productivity growth and is a key driver of economic growth and development. Innovation can also help address pressing social and global challenges, including demographic shifts, disease threats, resource scarcity and climate change. Innovative economies are more productive, more resilient, more adaptable to change and better able to support higher living standards. Strengthening innovation is therefore a fundamental challenge for countries in their quest for greater prosperity and better lives. This new OECD report on *The Innovation Imperative – Contributing to Productivity, Growth and Well-Being* provides a toolbox for governments that wish to strengthen innovation and make it more supportive of inclusive and green growth.

The list of **policies for innovation** is long and goes beyond research and technology policies. This report shows that four areas of policy are particularly important. First, innovation relies on a **skilled workforce** that can generate new ideas and technologies, implement them and bring them to the market. Addressing skills shortages and mismatching is critical in order for the workforce to adapt to technological and structural changes and promote inclusive growth. Second, innovation requires a sound **business environment** that encourages investment in technology and in knowledge-based capital, that enables innovative firms to experiment with new ideas, technologies and business models, and that helps them to grow, increase their market share and reach scale. Third, innovation needs a **strong and efficient system for knowledge creation and diffusion** that invests in the systematic pursuit of fundamental knowledge, and that diffuses knowledge throughout society. And finally, specific **innovation policies** are needed to tackle a range of barriers to innovation and entrepreneurial activity.

The impact of good innovation strategies depends on their governance and implementation, including the trust in government action and the commitment to learn from experience. Evaluation needs to be embedded into the process of policy development, and should not be an afterthought. This report extensively discusses how policies for innovation can be applied in different contexts, including for inclusive and sustainable growth, and what challenges need to be overcome in implementing such policies.

The report argues that policy makers can and should do better in marshalling the power of innovation to achieve core objectives of public policy. Policies for innovation require a strategic approach. This update of the OECD Innovation Strategy, which we have developed by bringing together expertise from across the entire Organisation, sets out how governments can deliver better innovation policies for better lives.

Angel Gurria
Secretery - General OECD

Table of contents

Tables

Figures

Acknowledgements

The OECD released its Innovation Strategy in 2010. The strategy was one of the first dedicated horizontal projects in the OECD. Since its launch, it has been well received and affected policy developments in many countries. In his 2013 Strategic Orientations, the Secretary-General pointed to the need to revisit and update the Innovation Strategy, and the 2014 OECD Ministerial Council Meeting (MCM) signalled the importance of carrying out such an update.

The resulting OECD Innovation Strategy 2015 was co-ordinated by the Directorate for Science, Technology and Innovation (STI). The OECD Committee for Scientific and Technological Policy (CSTP) and the Committee on Industry, Innovation and Entrepreneurship (CIIE) took the lead in developing this update, with other OECD committees providing substantive input in their respective areas of competence. Overall, the strategy involved contributions from 14 OECD directorates, as follows:

Chapter 1 reflects input from the Environment Directorate (ENV), the Statistics Directorate (STD) and STI. Thanks are due to Shardul Agrawala and Ivan Haščič (ENV), Paul Schreyer (STD), and Alistair Nolan (STI) for their input.

Chapter 2 draws on a number of recent and forthcoming studies of STI. Thanks are due to Sandrine Kergroach, Dominique Guellec and Alessandra Colecchia (STI) for their input.

Chapter 3 on fostering talent and skills reflects input from the Directorate for Education and Skills (EDU), the Centre for Entrepreneurship (CFE), the Economics Department (ECO), and STI. Thanks are due to Stephan Vincent-Lancrin (EDU), Jonathan Potter (CFE), Muge Adalet McGowan and Dan Andrews (ECO), and Silvia Appelt, Fernando Galindo-Rueda, Mario Cervantes and Sarah Box (STI) for their input.

Chapter 4 on the business environment for innovation reflects input from ECO, the Trade and Agriculture Directorate (TAD), the Directorate for Financial and Enterprise Affairs (DAF), CFE and STI. Thanks are due to Dan Andrews (ECO), Frank van Tongeren (TAD), Mike Gestrin (DAF), Jonathan Potter (CFE), and Chiara Criscuolo and Nick Johnstone (STI) for their input.

Chapter 5 on knowledge creation, diffusion and commercialisation reflects input from a range of colleagues in STI. Thanks are due to Christian Reimsbach-Kounatze, Rudolf van der Berg, Jeremy West, Carthage Smith, Giulia Ajmone Marsan and Mario Cervantes for their input.

Chapter 6 on effective innovation policies reflects input from CFE, the Directorate for Governance and Territorial Development (GOV), the Centre for Tax Policy Administration (CTPA) and STI. Thanks are due to Jonathan Potter (CFE), Karen Maguire (GOV), Thomas Neubig (CTPA), and Brigitte Acoca, Chiara Criscuolo, Dominique Guellec, Sandrine Kergroach, Mario Cervantes and Nick Johnstone (STI) for their input.

Chapter 7 reflects input from across the OECD. The section on the national agenda for innovation (Section 7.1) reflects input from the STI, the Development Centre (DEV) and TAD. Thanks are due to Annalisa Primi and Carl Dahlman (DEV), Frank van Tongeren (TAD), and Dominique Guellec, Gernot Hutschenreiter, Micheal Keenan, Dimitrios Pontikakis and Alistair Nolan (STI) for their input.

The section on inclusive innovation (Section 7.2) draws on a recent report prepared by Caroline Paunov and Dominique Guellec (STI).

The section on health innovation (Section 7.3) reflects input from the Directorate for Employment, Labour and Social Affairs (ELS) and STI. Thanks are due to Francesca Colombo, Isabelle Vallard and Jillian Oderkirk (ELS), and to Hermann Garden and Elettra Ronchi (STI).

The section on green innovation (Section 7.4) was prepared by STI, with inputs from ENV. The International Energy Agency (IEA) and the Nuclear Energy Agency (NEA) kindly provided material arising out of their ongoing work. Thanks are due to Nathalie Girouard, Kumi Kitamori, Tomasz Koźluk, Elisa Lanzi, Ivan Haščič, Nils-Axel Braathen and Ryan Parmentier (ENV), Nick Johnstone and Carlo Menon (STI), Cecilia Tam (IEA), and Henri Paillere (NEA).

The section on public sector innovation (Section 7.5) reflects input from GOV, EDU and STI. Thanks are due to Marco Daglio and Edwin Lau (GOV), Dirk van Damme and Stephan Vincent-Lancrin (EDU), and Alessandra Colecchia and Fernando Galindo-Rueda (STI) for their input.

Chapter 8 on governance and implementation reflects input from GOV, ENV and STI. Thanks are due to Andrew Davies and Karen Maguire (GOV); Bob Diderich and Peter Kearns (ENV), and Carthage Smith (STI) for their input.

The revision of the OECD Innovation Strategy was co-ordinated by Alistair Nolan and Dirk Pilat, under the guidance of Andrew Wyckoff, all in STI. Many thanks are due to Stella Horsin for her help in preparing the report and to Erin Crum, Janine Treves, Kate Brooks and James Arkinstall for their help in editing and publishing the report.

Extensive comments from Tera Allas and Ken Warwick on the draft are gratefully acknowledged, as are comments received at a cross-committee workshop of 5-6 February 2015, at meetings from CSTP and CIIE, as well as written comments from OECD delegations on the draft.

Executive summary

New sources of growth are urgently needed to help the world move to a stronger, more inclusive and sustainable growth path following the financial crisis. Innovation – which involves the creation and diffusion of new products, processes and methods – can be a critical part of the solution. While not a goal in itself, innovation provides the foundation for new businesses, new jobs and productivity growth and is thus an important driver of economic growth and development. Innovation can help address pressing social and global challenges, including demographic shifts, resource scarcity and the changing climate. Moreover, innovation can help address these challenges at the lowest cost. Innovative economies are more productive, more resilient, more adaptable to change and better able to support higher living standards.

Harnessing innovation requires policies that reflect the realities of innovation as it occurs today. Innovation goes beyond science and technology, and involves investments in a wide range of knowledge-based assets that extend beyond research and development (R&D). Social and organisational innovations, including new business models, are increasingly important to complement technological innovation. Innovation also involves a wide and expanding range of actors, including firms, entrepreneurs, foundations and non-profit organisations, universities, scientific institutes, public sector agencies, citizens, and consumers, often working in close collaboration. Innovation also has a strong and ever-expanding basis in the digital economy, facilitated by the growth of mobile telecommunications, the convergence of voice, video and data to the Internet, and the rapid uptake of data and sensors (the Internet of Things). Moreover, it involves a growing role of emerging economies, in particular the People's Republic of China (hereafter: China), which recently passed the European Union in becoming the second-largest funder of R&D behind the United States. This is leading to an increasingly global context, with innovation drawing on knowledge and ideas from across the world, though still often rooted in unique local and regional strengths.

Governments play a key role in fostering a sound environment for innovation, in investing in the foundations for innovation, in helping overcome certain barriers to innovation, and in ensuring that innovation contributes to key goals of public policy. Getting the policy mix right can help governments in shaping and strengthening the contribution that innovation makes to economic performance and social welfare. These *policies for innovation* are much broader than the policies that are often seen as "innovation policies" in a narrow sense – such as policies to support business R&D, financing for risk capital, etc. Such policies for innovation need to be focused on enhancing the performance of the system as a whole, as weak links in the system will hinder its performance. The priority assigned to different elements will depend on the nature and state of their system of innovation: one size clearly does not fit all. OECD analysis suggests that innovation thrives in an environment characterised by the following features, all of which are explored in detail in the OECD Innovation Strategy 2015:

- A **skilled workforce** that can generate new ideas and technologies, bring them to the market, and implement them in the workplace, and that is able to adapt to technological and structural changes across society.
- A sound **business environment** that encourages investment in technology and in knowledge-based capital, that enables innovative firms to experiment with new ideas, technologies and business models, and that helps them to grow, increase their market share and reach scale.
- A **strong and efficient system for knowledge creation and diffusion** that invests in the systematic pursuit of fundamental knowledge, and that diffuses this knowledge throughout society through a range of mechanisms, including human resources, technology transfer and the establishment of knowledge markets.
- **Policies that encourage innovation and entrepreneurial activity**. More specific innovation policies are often needed to tackle a range of barriers to innovation. Many of these actions include policies at the regional or local level. Moreover, well-informed, engaged and skilled consumers are increasingly important for innovation.
- **A strong focus on governance and implementation.** The impact of policies for innovation depends heavily on their governance and implementation, including the trust in government action and the commitment to learn from experience. Evaluation of policies needs to be embedded into the process, and should not be an afterthought.

Out of this broad toolbox for innovation policy, five priorities are particularly important and together provide the basis for a comprehensive and action-oriented approach to innovation, much of which can also be applied in the context of fiscally constrained economies. These priorities are:

1. **Strengthen investment in innovation and foster business dynamism.** Governments need to develop better policies to support investment in knowledge-based capital, which is now currently the largest area of business investment. They also should foster the growth of young and innovative small and medium-sized enterprises that remain constrained in their impact on growth and jobs, partly because some of our policies still tend to focus on incumbents.
2. **Invest in, and shape, an efficient system of knowledge creation and diffusion.** Investment in basic research remains a key priority; most of the key technologies in use today have their roots in public research. There is a risk that public investments in this area become too focused on the short term, rather than the long-term benefits. More efficient innovation policies, based on international good practice, can help strengthen the impact of government action.
3. **Seize the benefits of the digital economy.** Digital technologies continue to offer a large potential for innovation and growth. However, policy needs to preserve the open Internet, address privacy concerns, and ensure access and competition. Digitally enabled innovation also requires new infrastructure such as broadband, spectrum and new Internet addresses.

4. **Foster talent and skills and optimise their use.** Skills are a key challenge for innovation, with two out of three workers not having the skills to succeed in a technology-rich innovation. A broad and inclusive skills strategy is therefore essential for innovation.
5. **Improve the governance and implementation of policies for innovation.** Finally, the impact of good innovation strategies depends on their governance and implementation, including the trust in government action and the commitment to learn from experience. Evaluation needs to be embedded into the process, and should not be an afterthought.

Implementing an effective government strategy for innovation is particularly important as key trends – the spread of global value chains, the increasing importance and mainstreaming of knowledge-based capital, and rapid technological progress, including the rise of the digital economy – are leading to the emergence of a "next production revolution". In the current context of a weak global recovery, business and policy leaders need to take advantage of these trends to accelerate structural shifts towards a stronger, inclusive and more sustainable economic future that creates new jobs and opportunities.

The Innovation Imperative
Contributing to Productivity, Growth and Well-Being

Chapter 1

The role of innovation and the rationale for public policy

This chapter builds on the OECD's 2010 Innovation Strategy in exploring the contribution of innovation to economic growth. It extends that study by incorporating considerations about green and inclusive growth into the conceptual framework for innovation policy. This recognises the growing role that these objectives play in the overall policy agenda, but also in the context of innovation. The chapter also devotes attention to the various policy rationales for innovation and the overall framework for policies in this area.

Innovation – which involves the creation and diffusion of new products, processes and methods (Box 1.1) – is central to advanced and emerging economies; in many OECD countries, firms invest as much in the knowledge-based assets that drive innovation, such as software, databases, research and development (R&D), firm-specific skills, and organisational capital, as they do in physical capital, such as machinery, equipment or buildings. The use of information technologies has become universal in only a few decades, and new applications emerge almost daily. But while innovation is everywhere today, its impact does not appear in the productivity statistics, to paraphrase the quip by Robert Solow (1987).[1] And other important policy objectives, such as green growth, also require a stronger contribution of innovation to be addressed.

Box 1.1. **Defining and measuring innovation**

There is growing recognition that innovation encompasses a wide range of activities in addition to R&D, such as organisational changes, training, testing, marketing and design. The latest (third) edition of the *Oslo Manual* (OECD and Eurostat, 2005) defines innovation as the implementation of a new or significantly improved product (good or service) or process, a new marketing method, or a new organisational method in business practices, workplace organisation or external relations.

By definition, all innovation must contain a degree of novelty. The *Oslo Manual* distinguishes three types of novelty: an innovation can be new to the firm, new to the market or new to the world. The first concept covers the **diffusion** of an existing innovation to a firm – the innovation may have already been implemented by other firms, but it is new to the firm. Innovations are new to the market when the firm is the first to introduce the innovation on its market. An innovation is new to the world when the firm is the first to introduce the innovation for all markets and industries.

Innovation, thus defined, is clearly a much broader notion than R&D or technological change and is therefore influenced by a wide range of factors, some of which can be influenced by policy. Innovation can occur in any sector of the economy, including government services such as health or education. However, the current measurement framework applies to business innovation, even though innovation is also important for the public sector (see Section 7.5).

The broad notion of innovation also emerges from the OECD's work on knowledge-based capital (KBC) (OECD, 2013), which points to a range of investments that firms can make beyond investment in technology (e.g. in information and communications technologies [ICT] capital) or in R&D. Increasingly, firms also invest in other forms of KBC, such as data, intellectual property, firm-specific skills or organisational capital.

This report argues that policy makers can and should do better in marshalling the power of innovation to help achieve core objectives of public policy. Treating innovation as a central tool of policy making will help ensure policy coherence, since innovation policies cross government portfolios and affect a wide range of stakeholders. Integrating innovation as part of a broader toolbox can also help ensure that innovation contributes to greater overall well-being. For example, while innovation is a source of growth and many new jobs, it also contributes to job destruction, which means that complementary policies are needed to enable job reallocation and skills development for workers who have lost their jobs.

Innovation matters not only for growth, but also for health, the environment and a range of other policy objectives that are related to well-being. However, the links between innovation and these other policy objectives have been explored in less detail and require further elaboration. This chapter extends the discussion in the OECD's 2010 Innovation Strategy (OECD, 2010) by first considering the relationship between innovation and economic growth, and then explicitly incorporating considerations about green and inclusive growth into the conceptual framework for innovation policy. This recognises the growing role that these objectives play in the overall policy agenda, but also in the context of innovation. The chapter also devotes attention to the rationale for innovation policies and the overall policy framework.

1.1. Innovation for strong, green and inclusive growth

Innovation's contribution to economic growth

There is widespread agreement that innovation is an important driver of growth, especially in the long run. Despite this understanding, the conceptual and empirical links between innovation and growth are complex. Innovation is not a simple linear process, with a straightforward link between investments in innovation and economic or social outcomes. Moreover, metrics for certain aspects of innovation suffer from limitations. This has made it difficult to establish the role that policies for innovation – in a broad sense – can play in shaping or strengthening innovation performance, with most analyses focusing only on certain aspects of innovation, such as spending on R&D. Despite these challenges, our understanding of the drivers and impacts of innovation continues to improve, and this report highlights some of the new evidence and policy insights emerging from recent work.

A key interest of policy makers in innovation has long been around its potential contribution to economic growth. A long-established way to look at the relationship between innovation and economic growth is through a production function where growth in output results from the input of labour and capital (both tangible and intangible) and from increases in multifactor productivity (MFP), i.e. the part of output growth that cannot be explained by increased factor inputs. In such a framework, innovation's contribution to growth can be found in three different places (Figure 1.1):

1. A contribution resulting from technological progress embodied in physical capital; for example, investment in more advanced machinery or in new computers. The OECD's Growth Study found that between 0.2 and 0.4 percentage points of gross domestic product (GDP) growth between 1985 and 2000 was linked to such embodied technological progress (OECD, 2003). The latest OECD estimates show that about 0.35 percentage points of GDP growth between 1995 and 2013 can be attributed to investment in ICT capital alone (Figure 1.2; OECD, 2015a).
2. A contribution resulting from investment in intangible capital, or KBC, such as R&D, software, design, data, firm-specific skills or organisational capital. This type of investment has risen steadily across the OECD (OECD, 2013), and recent analysis by Corrado et al. (2012) found it accounted for around 0.5 percentage points of GDP growth

in European Union (EU) countries from 1995 to 2007, and 0.9 percentage points in the United States. This factor has not yet been incorporated in the OECD estimates shown in Figure 1.2.

3. A contribution linked to increased MFP growth, reflecting increased efficiency in the use of labour and capital, a substantial part of which can be attributed to innovation, including social and organisational innovations as well as the spillover effects of investments in technology or KBC, including at the global level. MFP accounted for over 0.7 percentage points of GDP growth between 1995 and 2013 in the countries shown in Figure 1.2, or about one-third of total GDP growth (see also OECD, 2015a).[2]

Figure 1.1. **A simplified framework to analyse economic growth**

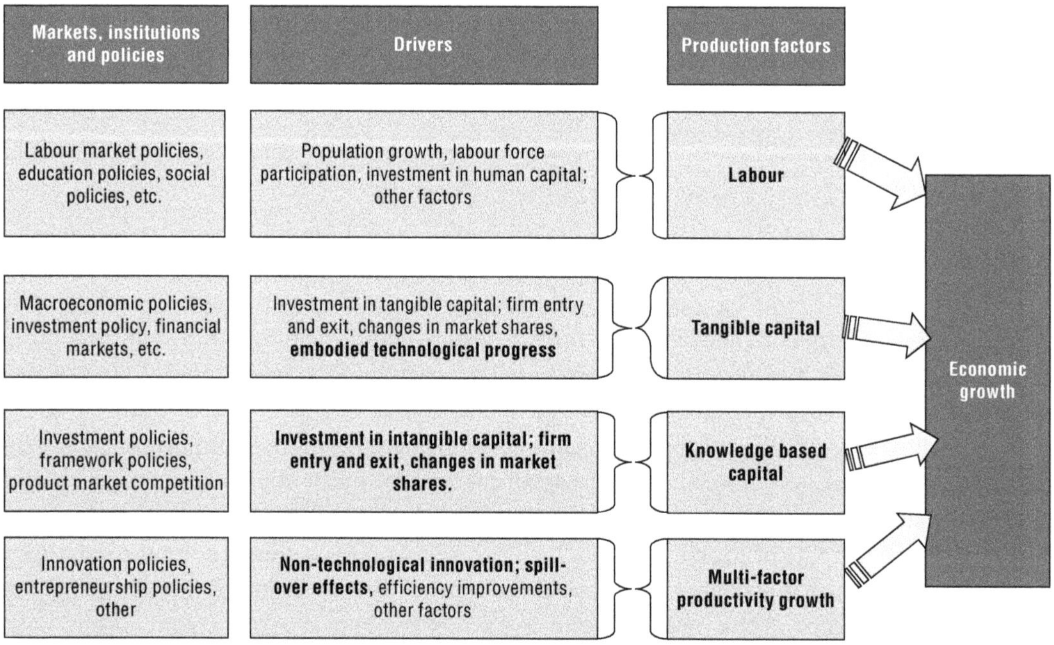

Source: Adapted from OECD (2000), *A New Economy? The Changing Role of Innovation and Information Technology in Growth*, OECD Publishing, Paris, *http://dx.doi.org/10.1787/9789264182127-en*.

The production-function framework points to important elements in the analysis of innovation, but provides a rather static and linear perspective of innovation, ignoring its dynamic nature. As suggested by Schumpeter (1942), innovation is accompanied by creative destruction as new firms enter the market, sometimes growing quickly and thus increasing their market share, replacing other firms with low productivity that are in decline or that will eventually shut down. This dynamic feature of innovation, which involves upscaling and resource allocation (OECD, 2015b), is another important element of the link between innovation and economic growth, which also has specific implications for policy.

One approach to conceptualising and quantifying this dynamic contribution was developed by Andrews and Criscuolo (2013), who distinguished three stages in the innovation process. In the first, firms invest in innovation to develop new ideas or adapt new technologies; in the second, firms implement and commercialise these ideas; and in the third phase, they realise the benefits from innovation by changes in market shares and increased profitability. This third phase emphasises the dynamic benefits from innovation that occur through changes in market shares and the reallocation of resources from declining to growing firms. The analysis of such benefits helps to modify the static view of innovation that emerges from the production-function framework. The policy implications of this more dynamic perspective of innovation will be discussed later in the report.

Figure 1.2. **Contributions to GDP growth**

Total economy, annual percentage point contribution, 1995-2013

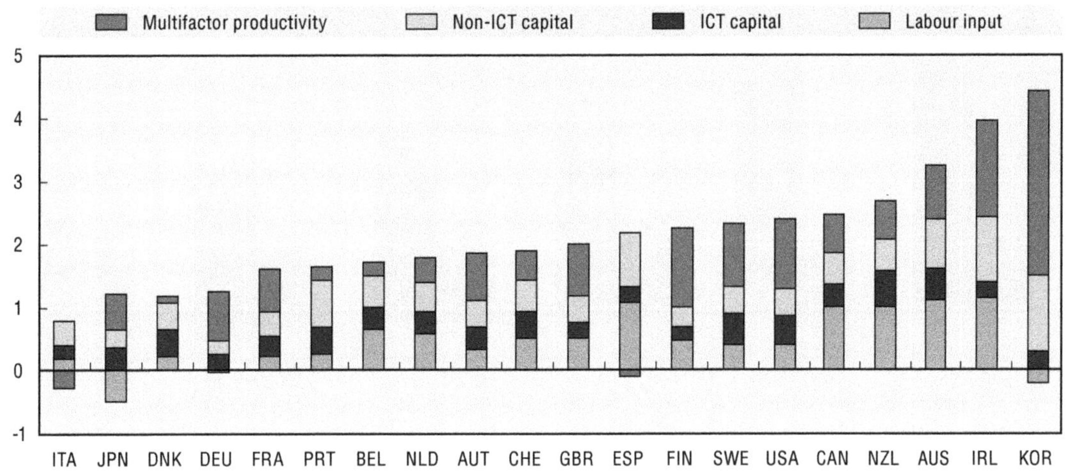

Source: OECD (2015a), *OECD Compendium of Productivity Indicators 2015*, OECD Publishing, Paris, *http://dx.doi.org/10.1787/pdtvy-2015-en.*

Together, the three elements included in the production-function framework, combined with the more dynamic perspective illustrated in Andrews and Criscuolo (2013), can account for a substantial share of economic growth, depending on the country, the level of economic development and the phase of the economic cycle. While not all elements can be fully attributed to innovation and there is likely to be some double counting, the different components of innovation together often account for at least 50% of economic growth.[3] Indeed, in the long run, it is difficult to imagine growth without innovation, as it would have to be based mostly on the accumulation of factor inputs, e.g. more labour (even if this might involve more qualified labour) or more of the same capital.

Studies of long-term economic development across countries show that MFP growth typically becomes a more important driver of growth in relative terms as countries exhaust some of the possibilities for productive investment in tangible capital. And much of the gap in income levels across countries is due to differences in MFP (OECD, 2015b). Moreover, in many OECD countries and some emerging economies, labour input has become less important as a factor of production over time, as the population has aged and the labour force has started to decline. For this reason alone, many OECD countries are increasingly emphasizing innovation-led productivity as the main source for future growth. OECD's long-term scenarios also emphasise the increasing importance of MFP growth for long-term economic growth (Figure 1.3) (Braconier, Nicoletti and Westmore, 2014).[4]

The conceptual links between innovation and economic growth set out above do not explicitly consider other important goals of public policy, such as the environment or well-being, that are also affected by innovation. However, recent OECD work has explored how to account for environmental impacts in the measurement of MFP growth (Brandt, Schreyer and Zipperer, 2014), as well as the role of environmental policy stringency on MFP (Albrizio et al., 2014). Another area of OECD work is seeking to move beyond GDP and address a broader range of measures of well-being (OECD, 2014a). The following sections discuss these issues in more detail.

Figure 1.3. **MFP as an increasingly important driver of future growth**

Contribution to growth in GDP per capita; 2000-60 (annual average)

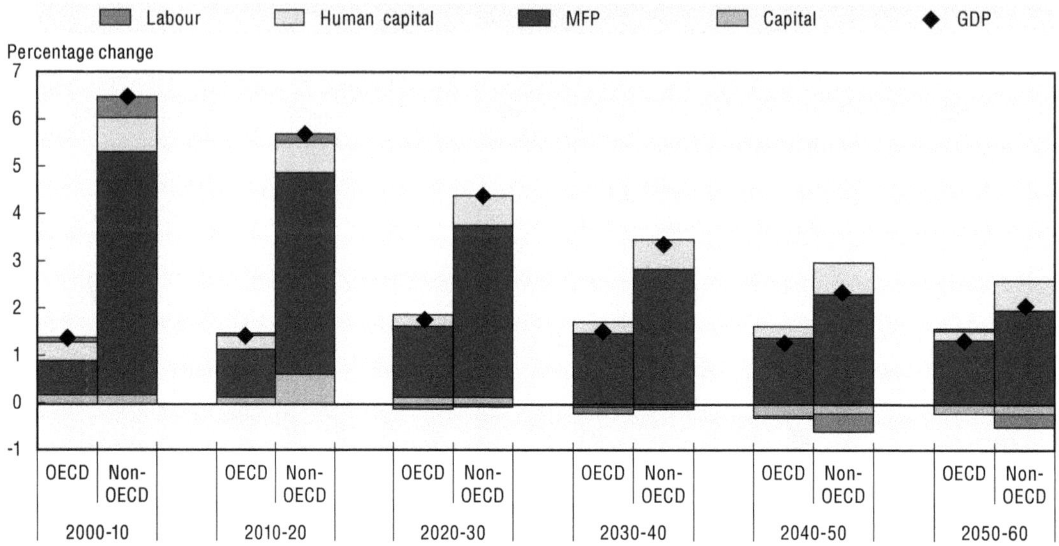

Note: Non-OECD G20 countries are Argentina, Brazil, China, India, Indonesia, Russian Federation, Saudi Arabia and South Africa.

Source: Braconier, Nicoletti and Westmore (2014), "Policy challenges for the next 50 years".

Beyond growth: Accounting for the environment

A first conceptual broadening of the framework of analysis concerns the relationship between innovation and green (or sustainable) growth. The OECD's Green Growth Strategy (OECD, 2011) noted that existing production technology and consumer behaviour can be expected to produce positive outcomes only up to a certain point, or frontier, beyond which depleting natural capital has negative consequences for overall growth. By pushing the frontier outward, innovation can contribute to decoupling growth from natural capital depletion. Innovation and the related process of creative destruction will also lead to new ideas, new entrepreneurs and new business models, thus contributing to the establishment of new markets and eventually to the creation of new jobs. Innovation is therefore a key in enabling green and growth to go hand in hand.

A first important dimension of the relationship between innovation and green growth concerns the measurement of output and productivity and the implications this has for evaluating the trade-offs and synergies between innovation and green growth (Brandt, Schreyer and Zipperer, 2014). Productivity growth that results from the production-function framework in Figure 1.1 can easily be overestimated in countries where output growth relies to a large extent on the depletion of natural capital. This can lead to an overly optimistic assessment of economic potential and growth in the long run. A similar argument applies to countries that hold production costs down by relying on heavily polluting technologies. While this can generate additional output of goods and services in the short run, it also leads to higher external costs, which can impinge on well-being and the sustainability of economic development. Conversely, the economic performance and sustainability of an economy that invests in a more efficient use of the environment in production may be underestimated, as some inputs do not serve to increase the current production of goods and services. Instead, they are aimed at reducing the associated negative externalities, improving human health, or protecting the integrity of the environment and climate stability. Therefore, when measuring productivity, it can be useful to consider as outputs not only the goods and services summarised in GDP but also the externalities, or "bad" outputs, such as air pollution and carbon dioxide (CO_2) emissions.[5]

Results presented in the paper by Brandt, Schreyer and Zipperer (2014) also suggest that considerable additional emissions reduction can be achieved at reasonable costs. The low elasticity of bad outputs presented imply that the reduction in GDP growth that would have to be accepted to achieve very considerable emissions reduction over the coming decades is limited, even in the absence of further improvements in environmental technologies. Conversely, the additional productivity growth that would be required to achieve the same emissions reduction without losses in output growth is not overly large. A combination of efficient and effective policies that price the externalities associated with bad outputs, and the promotion of R&D and new technology deployment with effective education policies, could help achieve the necessary technological progress that would allow countries to lower bad outputs without suffering income losses. Moreover, while green innovation may have some short-term implications for the (conventionally measured) rate of economic growth, these negative impacts should be outweighed by a large margin by the resulting long-term sustainability of the economy, as green innovation could contribute to avoiding the long-term impacts of diminishing environmental sinks such as potentially catastrophic climate change.

This being said, there are potential trade-offs and synergies between innovation and green growth. One key question is how a policy focus on green innovation through tighter environmental policies will affect the economy-wide rate of technological progress, as measured in the rate of MFP growth (Albrizio et al., 2014). On the one hand, proponents of the so-called Porter Hypothesis claim that environmental policies may improve incentives to innovate and lead firms to seek previously overseen efficiency improvements. On the other hand, policies that provide incentives to redirect innovation resources to reducing the cost of environmental impacts – which are not accounted for in a traditional productivity measure – may lead to an overall fall in measured productivity growth as fewer resources are channelled into "productive" innovation. Recent OECD work has tackled this question and has found that the tightening of environmental policies over the past two decades has had little effect on aggregate productivity growth (see below) (Albrizio et al., 2014).

Another argument sometimes made in favour of green innovation is that the potential spillovers arising from green innovation could well be larger than for other forms of innovation, precisely because the market is still underdeveloped and the potential for future innovation and growth may be very large. Overcoming the barriers to green innovation – such as the dominance of existing technologies and systems, a regulatory environment that may favour incumbents, or access to capital – could possibly lead to new waves of innovation comparable to those seen with other major technological revolutions. Advancing green innovation could also build on the growing interest in the private sector to use resources more efficiently. Unfortunately, this argument may be fully resolved only as green innovation expands over time and starts to have larger and more visible impacts on the economy and society.

Innovation and inclusive growth

In recent years, it has become increasingly clear that economic growth, as measured by GDP, can no longer be the overriding goal for government policy and can also not be an end in itself. Governments are increasingly focusing on inclusive growth, aiming to

Box 1.2. **Do environmental policies matter for productivity growth? Main findings**

Over the past two decades, environmental policy stringency, defined as the explicit and implicit, policy-induced price of environmental externalities, has increased significantly in the 24 OECD countries covered in new analysis. A newly developed composite indicator of environmental policy stringency, covering a range of market- and non-market-based policies, records increasingly stringent environmental policies in these 24 OECD countries, but with notable differences in stringency levels – overall, and across different policy instruments.

Countries tend to opt for similar main policy instruments but differ notably in the stringency of market- and non-market-based instruments. For example, the United Kingdom, Poland and Australia tend to show more relative stringency on market-based environmental policy instruments, e.g. environmental taxes or tradable permits. Finland, Germany and Austria tend to have relatively more stringent non-market-based policies, e.g. standards, while in the other OECD countries the relative stringency is more balanced.

There is no empirical evidence of permanent effects of environmental policy tightening on MFP, positive or negative. Analysis based on a new cross-country dataset with unprecedented time-series coverage finds that all effects tend to fade away within less than five years.

No lasting harm to productivity levels is found at the macroeconomic, industry or firm levels. On the contrary, a tightening of environmental policies is followed by a temporary increase in productivity growth, leading to an overall improvement in production efficiency for a large share of manufacturing industries.

At the macro level, the anticipation of an environmental policy tightening may also temporarily slow productivity growth – possibly due to increased investment in preparation for an expected policy change. Productivity levels subsequently rebound due to the temporary acceleration in growth rates.

The temporary effects on productivity growth are not conditional on the stringency of environmental policies already in place, but may depend on the flexibility of the environmental policy instruments. In particular, market-based instruments tend to have a more robust positive effect on productivity growth.

Advanced industries and firms generally see the largest gains in productivity levels, while less productive firms are likely to see negative effects. Highly productive firms, often the largest firms in the industry, may be best suited to profit rapidly from changing conditions – seizing new market opportunities, rapidly deploying new technologies or reaping previously overseen efficiency gains. They may also find it easier to outsource or relocate production abroad. Less advanced firms may need higher investments to comply with a new regulation, exhibiting a significant temporary fall in productivity growth.

Assuring a swift reallocation of capital and minimising barriers to entry are necessary for the efficiency gains from tightening environmental policy to be translated into economic growth. A non-negligible part of the productivity gains is likely to come from the exit of the least productive firms. To the extent that such developments are not due to increased regulatory burdens, and resources can be reallocated into fast-growing firms, they can have a positive effect on overall economic outcomes.

Designing environmental policy interventions so as to avoid generating barriers to entry and competition can help achieve both environmental and economic objectives. For example, OECD has recently measured to what extent environmental policy design unnecessarily increases fixed costs, imposes administrative burdens in permit and licensing procedures, results in the lack of coherent and consistent information, or distorts competition via vintage-differentiated regulations or tax and subsidy policies that are related to historical performance.

Cross-country evidence suggests that high barriers to entry and competition are not a necessary feature of stringent environmental policies. The new OECD indicator shows that stringent environmental policies can be pursued in a way that is more competition-friendly, with relatively low administrative burdens and little discrimination against new entrants. Countries where this has been the case include the Netherlands, Austria, Switzerland and possibly the United Kingdom. On the other hand, in Greece, Italy, Hungary and Israel environmental policies do not appear particularly stringent, but could benefit from lower distortions to entry and competition.

Source: Albrizio et al., 2014.

improve living standards and share the benefits of increased prosperity more evenly across social groups. This is particularly relevant in high-income countries and emerging market economies, where income inequality has reached levels unprecedented in the post-war period. Inequalities in other non-income outcomes, including educational attainment, health conditions and employment opportunities, are also increasingly recognised as influencing not only well-being, but also growth.

Innovation plays an important role in the debate on inclusive growth. The discussion below focuses on the three dimensions that are considered in the OECD's framework for inclusive growth (OECD, 2014a), namely income, jobs and health, although there are several other dimensions of inclusiveness that could also be affected by innovation, such as education. The OECD has recently developed a new framework for inclusive growth, which seeks to provide a clear link between individual dimensions of well-being and policies (OECD, 2014a).

A summary description of this framework is provided in Figure 1.4. The right-hand side of the diagram shows living standards and welfare as depending on both the level and the distribution of the key dimensions of well-being (e.g. income, health, jobs, education, security). The left-hand side of the diagram shows some of the policies potentially bearing on outcomes and their distribution. As the diagram indicates, there is a broad range of factors that mediate between policies and outcomes.

Figure 1.4. **OECD framework for policy analysis of inclusive growth**

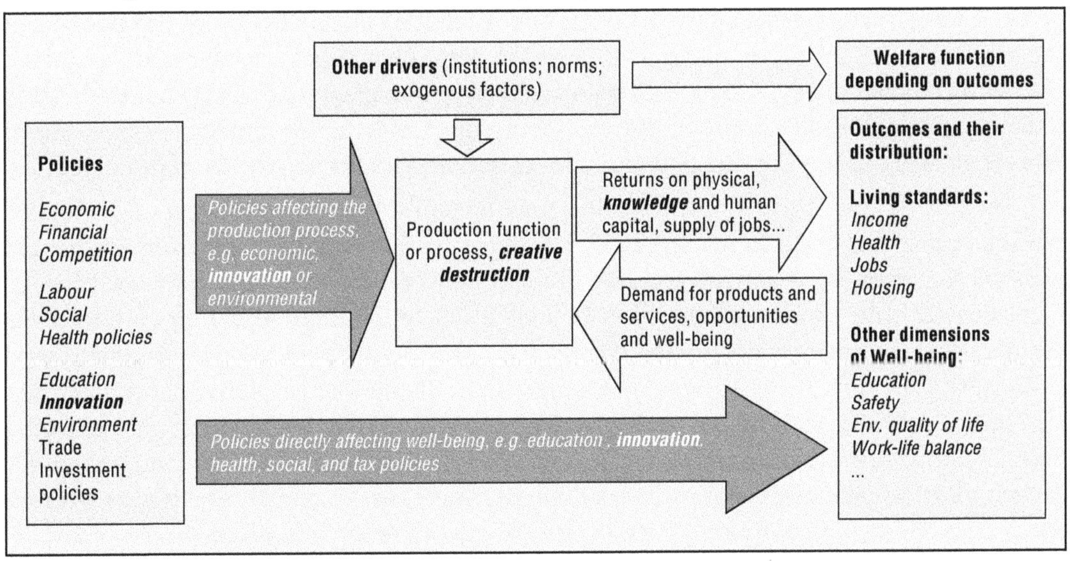

Source: Adapted from OECD, 2014a, *All on Board: Making Inclusive Growth Happen*, OECD Publishing, Paris, *http://dx.doi.org/10.1787/9789264218512-en*.

Innovation has been integrated into this framework in a number of ways. First, as discussed already above, innovation has a major impact on productivity growth through the production process, thus affecting the returns to human, physical and knowledge-based capital. It tends to increase aggregate **incomes**, which has a positive impact on living standards, but can also contribute to growing income inequality. This is partly because innovation inherently creates winners and losers in the process of creative destruction, with some firms and individuals benefiting through substantial returns to their invested labour and capital. Moreover, much technological change over the past decade has tended to be skill-biased, favouring those with the highest skills, sometimes to the detriment of those with lower or medium-level skills. Digital technologies in particular allow small differences in skill, effort or quality to yield large differences in returns, in part by increasing the size of the market that can be served by a single person or firm. For instance, average income among writers of fiction may not have changed greatly in recent decades. But a small group of writers have become multimillionaires, facilitated by the fact that digitisation allows words, images and products to be readily transmitted worldwide.[6] Innovation-related increases in inequality may not always be a concern, however, provided that those at the bottom of the income distribution also see significant increases in incomes and well-being. For example, China's growth over the past few decades can be considered inclusive as millions of people were lifted out of poverty, even if this was accompanied by an increase in measured inequality.

Policy plays an important role in shaping the relationship between innovation and inclusive growth. Many of the existing policies aimed at stimulating innovation are not neutral in their impacts on the production process. For example, in many cases, innovation policies foster excellence and tend to promote the concentration of human, financial and knowledge resources in the strongest scientific institutions, firms and regions. This concentration does tend to foster growth, but has repercussions on the distribution of the outcomes of the innovation process across the economy. Innovation policies may sometimes also end up favouring incumbents, which can also have repercussions on the inclusiveness of innovation outcomes. Recent OECD work is considering whether innovation policies can become more inclusive, including in being more focused on the needs of the poorest in society (OECD, 2015c). Policy makers will need to carefully analyse the alternative options of making innovation itself more inclusive or allowing it to create inequality but then using redistributive policies to improve the well-being outcomes for all citizens. Section 7.2 of this report will explore the link between innovation and inclusive growth in more detail.

Second, as the impacts of innovation may be skill-biased, they may tend to reduce the **jobs** prospects for some categories of workers in the economy. For example, recent technological trends, in particular related to ICT, point to a large further potential for job displacement and creative destruction linked to ICT, potentially affecting certain skills groups. Whereas this may not lead to aggregate employment losses, the resulting structural change will require complementary policies, including effective skills, labour market and social policies. Skill-biased technological change is also contributing to a growing polarisation in the labour market, with some groups with skills that are complementary to technological change benefiting through higher wages and incomes, and others without the right skills being confronted with lower wages and fewer and often more precarious jobs. The impacts of this polarisation can also reduce the ability of certain groups in the population to participate in the economy, which can contribute to (often long-term) social exclusion.[7]

Innovation is a highly disruptive force, contributing to the process of creative destruction in the economy, and thus to job displacement and the reallocation of labour and capital within the economy. This process is not new, and in many economies, about 20% of the labour force changes jobs every year. Moreover, long-term trends suggest that innovation, productivity and job creation can go hand in hand. In recent years, however, innovation and the resulting creative destruction are sometimes considered to be among the trends that may have contributed to growing income inequality in many OECD countries. At the same time, the process of creative destruction may provide new opportunities for excluded groups to be involved in the innovation process and to enhance incomes and create new job opportunities. For example, recent OECD work finds that most new job creation is due to young firms less than five years old (Criscuolo, Gal and Menon, 2014).

Third, innovation in **health** has been an important factor in improving well-being and living standards. Innovation has contributed to better care and improved diagnostics and treatments, including better medicines that have strongly contributed to increased longevity. Such innovations not only are technological, but also include social innovation, e.g. aimed at disease prevention and lifestyle changes. While these innovations have contributed to overall increases in well-being, challenges related to the access to innovations and their pricing remain, implying that some groups in society may benefit more than others. Section 7.3 of this report will explore some key dimensions of health innovation in further detail.

Innovation and the policies around it also affect other dimensions of well-being, including education and personal security. Sections 3.1 and 7.5 will include some discussion about the links between innovation and education. However, several of the interactions between innovation and inclusive growth are not yet fully understood and require further analysis.

It is also important to recognise that innovation is not always a positive force for change and can cause both harm and good. Governments therefore have a range of policies in place to manage the risks associated with innovation, e.g. in areas such as health and safety, but also related to the digital economy or financial markets. Risk management and risk governance related to innovation are therefore important in influencing the relationship between innovation and a wider set of determinants of well-being. The role of risk governance will be discussed in Chapter 8 of the report.

1.2. The rationale and role for policies for innovation

The rationale for innovation policies

There is extensive discussion among academics and policy makers on the rationale for innovation policies. The neo-classical perspective recognises only a limited set of market failures, such as externalities and information asymmetries. Other schools of thought point to a much wider range of factors and constraints that affect innovation and that can provide a rationale for policy, noting that these factors will vary from country to country and also depend on the particular area of innovation that is being considered, including the specific sector of the economy (BIS, 2014). Figure 1.5 presents a diagnostic framework for identifying the key constraints to innovation.[8] It characterises the constraints to innovation as factors that limit the returns to investment in innovation. These constraints can be divided into two categories:

- The first category refers to **low economic returns**, encapsulating factors that create inertia in economic systems (i.e. fundamental – systemic – barriers to change and innovation, e.g. linked to barriers to competition, lack of co-operation within an innovation system, prevailing norms and habits, as well as technology lock-in), and capacity constraints, or "low social returns", that are often linked to lack of skills or infrastructure, or inadequate institutions.
- The second category refers to **low appropriability of returns**. This is where market and government failures prevent firms or other innovation actors from capturing the full value of their investments in innovation, thus leading to underinvestment. Examples include

the externalities associated with investment in R&D, where a firm can never capture all the returns to its investments, due to the spillover effects associated with investment in knowledge. Another example concerns the negative externalities related to environmental damages. These damages are often not priced by the market, which adds to the difficulties faced by private investors to fully appropriate the returns from innovation.

Figure 1.5. **The potential rationale for innovation policies**

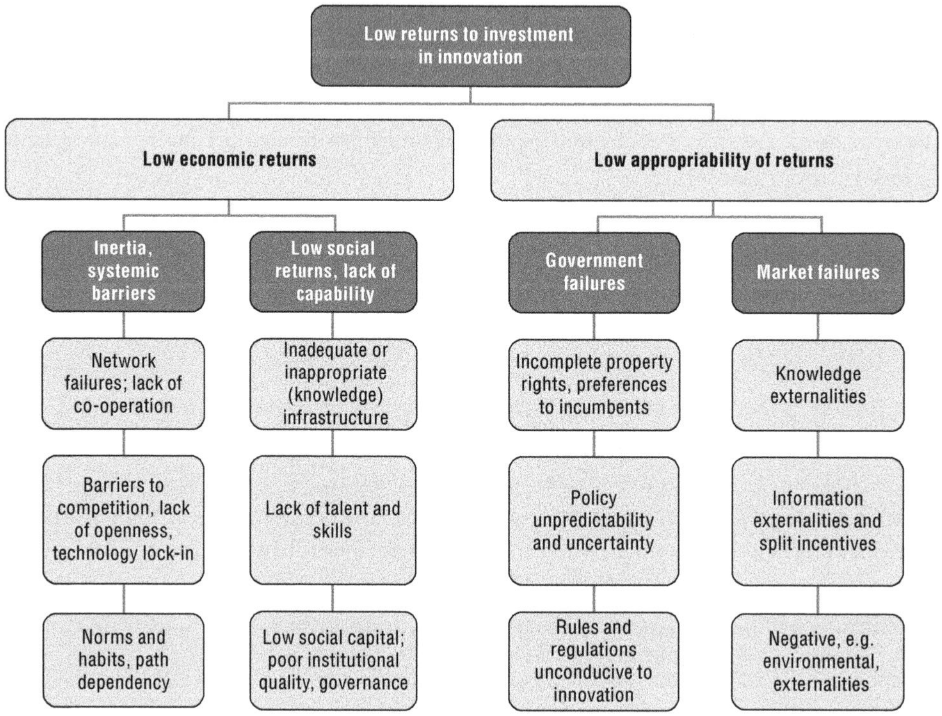

Source: Adapted from OECD (2011), *Towards Green Growth*, drawing on Hausmann, Velasco and Rodrik (2008), "Growth diagnostics".

Low economic returns due to inertia and systemic barriers can constrain the expansion of new or innovative production techniques, technologies or other forms of innovation. These constraints are often a mixture of market failures and market imperfections. For example, network effects (e.g. barriers to entry that arise from increasing returns to scale in networks) and the bias in the market towards existing technologies are examples of market imperfection. Addressing these barriers can involve a range of policies, including competition and regulatory policies, but may also require more active government intervention in the form of specific innovation policies, e.g. to overcome the lack of co-operation within an innovation system, or to overcome technology lock-in (OECD, 2015d).[9] However, government failures can arise from attempts to deal with these market failures or imperfections (e.g. governments may create regulatory barriers to competition, including in maintaining government monopolies in network industries). The policies that can help overcome these barriers are discussed in Chapters 4 (the business environment for innovation) and 6 (effective innovation policies).

"**Low social returns**" implies the absence of enabling conditions for productive investment in innovation. These constraints reduce the choices of firms and other actors to invest in innovation. For example, inadequate ICT infrastructure can limit the opportunities for firms to benefit from the network effects associated with the technology. Other barriers include insufficient human capital that may imply that firms have insufficient know-how to deploy new technologies or to adjust to them through organisational change. Lack of social capital and the poor quality of institutions are other barriers that will affect the returns to investment in innovation. Addressing these barriers typically requires capacity building, based on a mix of public and private investments in infrastructure, education and skills, as well as institutions. Some of the policies that can help overcome these barriers are discussed in Chapter 3 (talent and skills), 5 (knowledge creation and diffusion) and 8 (governance and policy coherence).

Government failures that contribute to a low appropriability of returns can affect many areas of the innovation system. Figure 1.5 points to only some of them, e.g. the preference to incumbents that can sometimes characterise policy making related to innovation (discussed in Chapters 4 and 6); the lack of policy predictability and stability that often characterises innovation policies (discussed throughout the report); and regulatory barriers that affect innovation (discussed primarily in Chapters 4 and 6). Addressing government failures may also require reform – and innovation – within the public sector, which is discussed in Section 7.5 and also in Section 8.1 of this report.

Market failures provide the main neo-classical rationale for innovation policies, and lead to well-recognised areas of policy, such as government support for business R&D (e.g. through R&D tax credits or grants, discussed in Chapter 6); government investment in basic R&D and knowledge infrastructure (discussed in Chapter 5); or policies that address negative environmental externalities, e.g. through carbon taxes, thus supporting green innovation (discussed in Section 7.4).

The categories of constraints described in Figure 1.5 are not entirely separable. There are, for example, some overlaps between market and government failures. Incomplete property rights are in many cases a market failure but are listed as a government failure as they may result from inadequate policies, e.g. in the case of an inadequate system of intellectual property rights. Similarly, the presence of regulatory uncertainty is an important impediment to private investment in many areas of innovation, such as green growth or health, even though some of these areas are also affected by market failures.

The importance of constraints to innovation will vary according to level of development, socio-economic context, and existing economic and environmental policy settings. Low human capital, inadequate infrastructure and poor institutional quality will often tend to be associated with lower levels of economic development (though not exclusively). Rectifying these constraints will be of high priority and perhaps a precondition to resolving many other constraints.

Where human capital is relatively abundant and infrastructure relatively well supplied, the focus often first shifts to resolving government and market failures. However, for innovation to have impact, attention also needs to be paid to some of the disadvantages that new firms and technologies may have relative to incumbents and existing technologies and policies that can help advance these.[10] Sequencing of reforms may be important in this context, in particular in areas where market failures are important, e.g. environmental innovation. Improving the returns to activities with low environmental impact, e.g. through carbon taxes, can help create market conditions that are conducive to the introduction of new green technologies.

Identifying which constraints are most important is not, however, entirely sequential. In particular, when institutions in some countries are not well equipped to address the prevailing barriers to innovation in a country, government failures or split incentives may need to be addressed. One constraint that is likely to be common to all countries, regardless of development, is regulatory certainty, i.e. the extent to which governments articulate and ideally legislate a clear plan for closing the gaps between private and social returns so that firms and other actors in the innovation system can plan and act without too much risk that governments will change the rules of the game. Due to the long-term character of investments in innovation, this is an important challenge for policies related to innovation, which will be further discussed in Chapter 8.

The diagnosis of key constraints will require country-specific information and data from across the economy as well as an understanding of the position and performance of a country in the global context. The OECD's indicators on innovation mentioned throughout this report provide some of the high-level measures that can be used to inform a diagnosis of constraints to innovation, as can the OECD analysis of structural policy settings, including in the area of innovation policy.

Moreover, as was already noted above, while there are many common barriers to innovation across the economy, other barriers are specific to sectors or to particular challenges (BIS, 2014), which implies they may need to be tackled through more tailored policy actions. For example, government policies to address complex societal problems, such as the development of smart cities, will need to look at the full range of system-wide barriers that affect the challenge (OECD, 2015d), with government taking an active role in supporting and managing the transition to a more sustainable system.

A further note to Figure 1.5 concerns the global context for innovation and for policy making related to innovation. As will be discussed further in Chapter 2, innovation is a global undertaking, with most innovation taking place outside national borders. The global context will affect the scope for national policy making, e.g. in attracting talent and skills, but will also provide important opportunities for governments to benefit from innovation abroad.

Finally, and crucially, while there are many barriers and obstacles to innovation, and there are many reasons governments may wish to take action to strengthen innovation, policy makers will always need to consider carefully whether they have the tools and understanding of innovation in their economy to take effective and efficient government action. This also involves a consideration of alternative policy actions where governments can best add value, and consideration of how governments can engage with other actors and encourage them to take action. Implementation is a key constraint for innovation policy, and requires an efficient and well-developed institutional framework, strong capabilities for evaluation and monitoring, and an efficient and capable government bureaucracy. This issue will be further discussed in Chapter 8 of this report.

The role of policy

Based on the conceptual thinking set out above to link innovation to key objectives of policy and the discussion on the rationale for policy making, the next question is which policies affect the various contributing factors and how government can shape and possibly strengthen the contribution that innovation makes to performance. This is the set of policies that governments need to consider when establishing policies for innovation. Obviously, this set of policies is much broader than the policies that are often seen as innovation policies in a narrow sense – policies to support business R&D, financing for risk capital, etc. These make up only a part – though an important one – of the full set of policies that affect innovation performance. Moreover, governments will need to consider how innovation and innovation policies affect other public objectives, and the complementary policies that need to be put in place to ensure that the overall goals of policy are being met, e.g. in regard to growth, jobs and income distribution, and to health and the environment.

So what are the policies that determine the contribution of innovation to economic growth? OECD analysis suggests that innovation thrives in an environment characterised by the following features, all of which will be explored further in later parts of this report:

- A **skilled workforce** that has the knowledge and skills to generate new ideas and technologies, to bring them to the market, and to adapt to technological changes across society. Reforms to education and training systems, and to skills policies more broadly, are therefore of utmost importance to innovation. They include policies aimed at science, technology, engineering and mathematics (STEM) graduates, but should go beyond this group and cover a wider set of skills. Moreover, the international mobility of talent plays an increasingly important role in meeting emerging skills needs. These policies are discussed in Chapter 3 of this report.
- A sound **business environment** that encourages investment in technology and in KBC, that also enables innovative firms to experiment with new ideas, technologies and business models, and that helps them to grow, increase their market share and reach scale. A range of empirical analysis on these issues has shown that innovation performance can be strengthened by structural reform – to product markets, encouraging competition and enabling new entry; to labour markets, enabling better resource allocation; and to financial markets, helping generate funding for risky investments. Regulatory reform is important, too, and should enable rather than stifle innovation. Openness to foreign sources of knowledge is also important for innovation, as most innovation happens outside national borders, and requires reforms to enhance the openness of an economy to trade, investment, knowledge flows and people. These policies are discussed in Chapter 4.
- A **strong and efficient system for knowledge creation and diffusion** that engages in the systematic pursuit of fundamental knowledge, and that diffuses this knowledge throughout society through a range of mechanisms, including human resources, technology transfer and the establishment of knowledge markets. Strong and well-governed universities and public research institutes and mechanisms that support and facilitate the interaction among knowledge institutions and economy and society are therefore important to strengthen innovation performance. So is investment in knowledge infrastructure, notably broadband and other digital networks that are critical tools to enable co-operation and provide new platforms for innovation to occur. Moreover, as knowledge creation and innovation are global endeavours, policies to better connect science and innovation activities around the world are crucial to the innovation policy agenda. These policies are discussed in Chapter 5.
- **Policies that encourage firms to engage in innovation and entrepreneurial activity.** More targeted innovation policies are often needed to tackle a range of barriers to innovation. The appropriate policy mix might include tax incentives for investment in R&D; direct public support through grants, subsidies and innovation competitions; and policies to facilitate

co-operation and networking, but also indirect incentives through public procurement and other so-called demand-side policies. Such policies can help to strengthen markets for innovation, and help focus it on specific challenges and opportunities, e.g. green growth. Many of these actions include policies at the regional or local level. Moreover, well-informed, dynamic engaged and skills consumers are important for innovation, and their role in innovation can be enabled by specific consumer policies. These policies are discussed in Chapter 6.

The precise application of the various policies for innovation set out in Chapters 3 to 6 will differ according to the national context. It may also be affected by the sector or technology concerned, and by the specific objectives of innovation. Chapter 7 will look at several aspects of the application of policies for innovation, in discussing the national agenda for innovation (Section 7.1); the role of innovation for inclusive growth (Section 7.2), including health innovation (Section 7.3); as well as innovation in the green growth agenda (Section 7.4). Particular attention will also be paid to innovation in the public sector (Section 7.5), which is important to improve effectiveness and efficiency within the public sector, but which can also help support innovation throughout the economy.

Chapter 8 of this report will focus on the governance and implementation of policies for innovation. Given the wide range of policies that affect innovation, it is important to ensure that the full set of government policies that affect innovation are well aligned, not only at the level of central government, but also between the central government and regional and local authorities, many of which are actively involved in innovation activities. The development and implementation of innovation policies also requires strong capabilities within the public sector, including in building trust in government action and ensuring the support of stakeholders for policy actions.

Establishing a national strategy for innovation is one thing; its implementation is often another matter. The framing of policies for innovation needs to recognise that they operate in a complex, dynamic and uncertain environment, where government action will not always get it right. A commitment to monitoring and evaluation of policies, and on learning from experience and adjusting policies over time, can help ensure that government action is efficient and reaches its objectives with the least possible cost. Moreover, the implementation of policies rests on an efficient and well-developed institutional framework, strong capabilities for evaluation and monitoring, and an efficient and capable government bureaucracy.

Finally, it is important to recognise that policies for innovation are part of a broader policy agenda that seeks to support a range of public policy objectives. In some cases, strengthening innovation may support these other objectives, e.g. in regard to growth, while in other cases, it may have some negative repercussions on specific objectives that may need to be addressed through complementary policies, e.g. to ensure that the gains of innovation are widely spread, or to address specific challenges that may come with strong innovation, such as harmful applications of specific technologies. Managing these risks and addressing them where needed is therefore also a key challenge for policies for innovation. This will also be considered in Chapter 8.

Before going into this more detailed discussion of policies for innovation, Chapter 2 briefly explores the landscape for innovation today.

Notes

1. R.M. Solow (1987).
2. Another factor is related to innovation in the way society uses its natural capital and produces undesirable outputs. This is further discussed below.
3. ICT capital and MFP growth combined account for over half of GDP growth in Figure 1.2. While MFP growth includes factors other than innovation, Figure 1.2 excludes most investment in KBC – software investment is included in the contribution of ICT capital – as well as the dynamic impacts of innovation on growth and productivity.
4. A recent OECD study (OECD, 2015b) provides a comprehensive assessment of the future of productivity growth and the various policy levers, including those related to innovation, that governments have at their disposal to strengthen innovation performance.
5. For example, recent OECD work (OECD, 2014b) suggests that the cost to society of air pollution is much larger than previously thought. Cost to society, in terms of willingness to pay to avoid the related mortalities and bad health, of outdoor air pollution in OECD countries in 2010 amounted to approximately USD 1.7 trillion. For China alone, the cost was estimated at USD 1.3 trillion, and for India, USD 0.5 trillion.
6. This "winner takes all" phenomenon is also mirrored in a widening of the distribution of profits across firms, particularly in sectors where firms invest heavily in ICT.
7. The impacts of innovation on employment also become apparent from the OECD's work on green growth (OECD, 2012). For example, this shows that shifting to a low-carbon economy will have different employment impacts across countries, with countries with highly polluting industries and rigid labour markets being more negatively impacted than countries with flexible labour markets that are able to become global green technology leaders. Overall labour market reallocation is expected to be modest with little impact on overall levels of job and skill demands.
8. This framework was already used in the context of the OECD's Green Growth Strategy and was slightly adapted to the context of innovation policies (see OECD, 2011, Annex 1).
9. OECD (2015d) discusses the barriers to system-wide innovation, where policy is aiming to solve complex societal problems, e.g. in areas such as smart cities, sustainable building, transport or healthy ageing.
10. The nature of this disadvantage will vary according to existing regulatory environments. In some cases, the regulatory environment will be such that incumbent firms enjoy an advantage over new entrants. In other cases the lack of a supporting network may prevent deployment of innovative technologies.

References

Albrizio, S., et al. (2014), "Do environmental policies matter for productivity growth? Insights from new cross-country measures of environmental policies", *OECD Economics Department Working Papers*, No. 1176, OECD Publishing, Paris, *http://dx.doi.org/10.1787/5jxrjncjrcxp-en*.

Andrews, D. and C. Criscuolo (2013), "Knowledge-Based Capital, Innovation and Resource Allocation: A Going for Growth Report", *OECD Economic Policy Papers*, No. 4, OECD Publishing, Paris, *http://dx.doi.org/10.1787/5k46bh92lr35-en*.

BIS (UK Department for Business, Innovation and Skills) (2014), *The Case for Public Support of Innovation*, Department for Business, Innovation and Skills, London.

Braconier, H., G. Nicoletti and B. Westmore (2014), "Policy challenges for the next 50 years", *OECD Economic Policy Papers*, No. 9, OECD Publishing, Paris, *http://dx.doi.org/10.1787/5jz18gs5fckf-en*.

Brandt, N., P. Schreyer and V. Zipperer (2014), "Productivity measurement with natural capital and bad outputs", *OECD Economics Department Working Papers*, No. 1154, OECD Publishing, Paris, *http://dx.doi.org/10.1787/5jz0wh5t0ztd-en*.

Corrado, C., et al. (2012), "Intangible capital and growth in advanced economies: Measurement methods and comparative results", *IZA Discussion Paper*, No. 6733, and INTAN-invest project on intangible investment, *www.INTAN-Invest.net*.

Criscuolo, C., P. Gal and C. Menon (2014), "The dynamics of employment growth: New evidence from 18 countries", *OECD Science, Technology and Industry Policy Papers*, No. 14, OECD Publishing, Paris, *http://dx.doi.org/10.1787/5jz417hj6hg6-en*.

Hausmann, R., A. Velasco and D. Rodrik (2008), "Growth diagnostics" in J. Stiglitz and N. Serra (eds.), *The Washington Consensus Reconsidered: Towards a New Global Governance*, Initiative for Global Governance Series C, Oxford University Press, Oxford.

OECD (2015a), *OECD Compendium of Productivity Indicators 2015*, OECD Publishing, Paris, *http://dx.doi.org/10.1787/pdtvy-2015-en*.

OECD (2015b), *The Future of Productivity*, OECD Publishing, Paris, *www.oecd.org/eco/growth/OECD-2015-The-future-of-productivity-book.pdf*

OECD (2015c), *Innovation Policies for Inclusive Development*, OECD Publishing, Paris, *http://dx.doi.org/10.1787/9789264229488-en*.

OECD (2015d *forthcoming*), *Synthesis Report on System Innovation*, OECD Publishing, Paris.

OECD (2014a), *All on Board: Making Inclusive Growth Happen*, OECD Publishing, Paris, *http://dx.doi.org/10.1787/9789264218512-en*.

OECD (2014b), *The Cost of Air Pollution: Health Impacts of Road Transport*, OECD Publishing, Paris, *http://dx.doi.org/10.1787/9789264210448-en*.

OECD (2013), *Supporting Investment in Knowledge Capital, Growth and Innovation*, OECD Publishing, Paris, *http://dx.doi.org/10.1787/9789264193307-en*.

OECD (2012), "The jobs potential of a shift towards a low-carbon economy", *OECD Green Growth Papers*, No. 2012/01, OECD Publishing, Paris, *http://dx.doi.org/10.1787/5k9h3630320v-en*.

OECD (2011), *Towards Green Growth*, OECD Publishing, Paris, *http://dx.doi.org/10.1787/9789264111318-en*.

OECD (2010), *The OECD Innovation Strategy: Getting a Head Start on Tomorrow*, OECD Publishing, Paris, *http://dx.doi.org/10.1787/9789264083479-en*.

OECD (2003), *The Sources of Economic Growth in OECD Countries*, OECD Publishing, Paris, *http://dx.doi.org/10.1787/9789264199460-en*.

OECD (2000), *A New Economy? The Changing Role of Innovation and Information Technology in Growth*, OECD Publishing, Paris, *http://dx.doi.org/10.1787/9789264182127-en*.

OECD and Eurostat (2005), *Oslo Manual: Guidelines for Collecting and Interpreting Innovation Data, 3rd Edition*, OECD Publishing, Paris, *http://dx.doi.org/10.1787/9789264013100-en*.

Schumpeter, J.A. (1942), *Capitalism, Socialism and Democracy*, Routledge, London.

Solow, R.M. (1987), "We'd better watch out", *The New York Times Book Review*, 12 July, p. 36.

The Innovation Imperative
Contributing to Productivity, Growth and Well-Being

Chapter 2

Innovation today

Innovation is high on the agenda of both policy makers and other actors in the innovation system. But innovation itself is changing rapidly. The growing role of the Internet, investments in knowledge-based capital (KBC) and globalisation are among the major underlying supply-side forces that affect innovation. Demand-side factors, such as challenges related to the environment, to society (ageing, inequalities) and to the slowdown in economic growth are also impacting the dynamics and orientation of innovation. This chapter provides a brief overview of these recent trends and drivers.

The statistical data for Israel are supplied by and under the responsibility of the relevant Israeli authorities. The use of such data by the OECD is without prejudice to the status of the Golan Heights, East Jerusalem and Israeli settlements in the West Bank under the terms of international law.

Innovation has never been so high on the agenda of all actors – government, businesses, universities, civil society, etc. At the same time, the conditions in which innovation is performed, as well as the very working and even definition of innovation itself, are experiencing in-depth changes in the context of broader changes in the world. The growing role of the Internet, investments in KBC and globalisation are among the major underlying supply-side forces that affect innovation. Demand-side factors, such as challenges related to the environment, to society (ageing, inequalities) and to the slowdown in economic growth are also impacting the dynamics and orientation of innovation. This chapter provides a brief overview of some recent trends and drivers of innovation, but does not aim to be exhaustive. More detail and indicators are available in several recent and forthcoming OECD reports (OECD, 2013a, 2014a, 2014b, 2015a, 2015b).

2.1. Investment in innovation has remained relatively strong

Weak demand has dampened business investments globally, including in knowledge-intensive activities and research and development (R&D) (Figure 2.1). Investments in physical assets have been slower to recover than knowledge-intensive investments, including R&D or software, reflecting both the central role of knowledge-based assets in market competition and firms' reluctance to build new production capacity. Applications to the three major patent offices in Europe, the United States and Japan give a further indication of innovation activity; they increased after 2011, but they remain low compared with earlier levels (OECD, 2014a).

Figure 2.1. **Business investment in knowledge-based assets recovered sooner**

OECD, index 2005 = 100

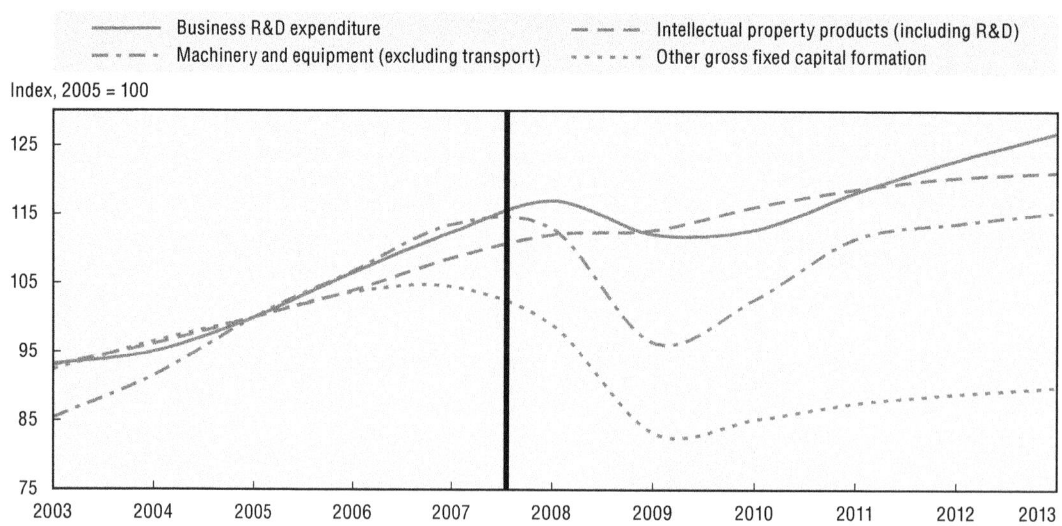

Source: OECD (2015c), *Main Science and Technology Indicators 2014-II*; OECD (2015d), *OECD National Accounts Database*, *http://stats.oecd.org/*, accessed on March 2015, based on OECD (2014a), *OECD STI Outlook 2014*, *www.oecd.org/sti/outlook*.

In a context of fiscal consolidation in many countries, fewer public resources can be mobilised, and public R&D budgets have levelled off in many countries and have started to decline in others (OECD, 2014a). In 2008-09, governments partially offset drops in business R&D spending with greater public funding, but the buffering effect of public research during the downturn has faded in the aftermath of the crisis (Figure 2.2). The average R&D intensity of OECD economies has continued to increase, however, going from just under 2% of gross domestic product (GDP) in 1995 to 2.4% of GDP in 2013. Business expenditure on R&D (BERD) accounts for about two-thirds of that spending, while the share of higher education R&D (HERD) increased slightly from just over 16% in 1995 to just below 18% in 2013. The share of government intramural R&D (GOVERD) declined slightly, from over 14% in 1995 to 11% in 2013 (Figure 2.2b).

Figure 2.2. **Gross domestic expenditure on R&D by performing sector**

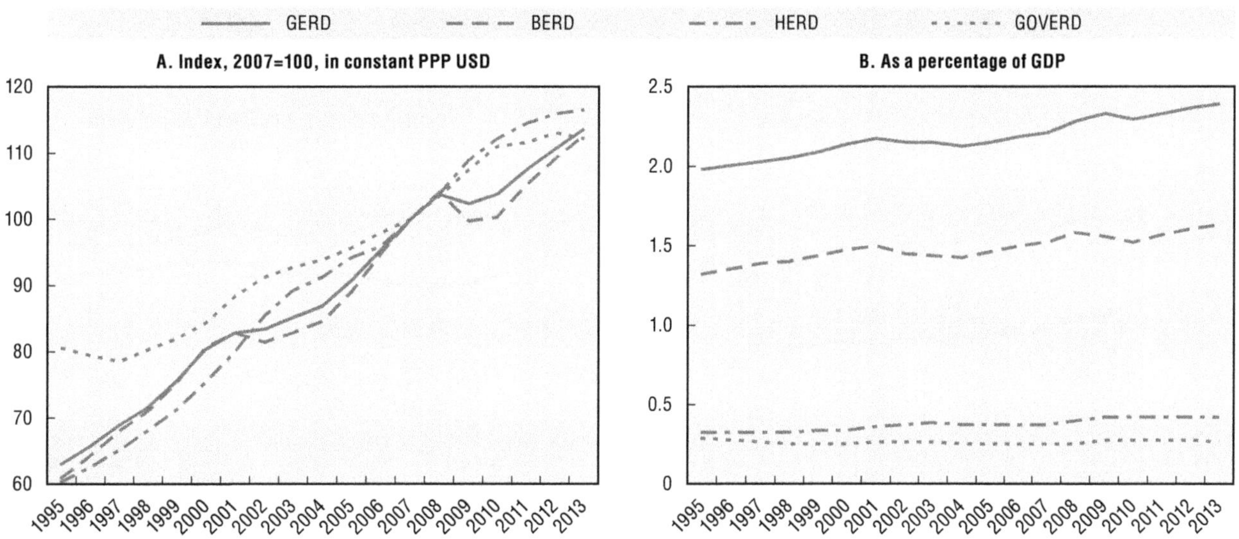

Note: PPP = purchasing power parity.

Source: OECD (2015c), *Main Science and Technology Indicators 2014_II*, *www.oecd.org/sti/msti.*

Global R&D is increasingly performed outside the OECD area. The OECD share of global R&D has slipped steadily from 90% to 70% over the past decade. The People's Republic of China (hereafter "China") is poised to become the top world R&D performer by the end of the decade if recent trends continue (Figure 2.3). Despite a slowdown in growth compared with 2001-08, its R&D expenditure doubled over 2008-12. China's R&D intensity is now on par with that of the European Union (EU28). The rise of China is driven by its economic dynamism and its long-term commitment to science, technology and innovation (STI). China's Medium and Long-Term National Plan for Science and Technology Development (2006-20) fixes a target of R&D spending of 2.5% of GDP by 2020. Korea, too, enjoyed a sizeable upward shift in R&D intensity, becoming the world's most R&D-intensive country (4.15%) in 2013. Behind Korea, Slovenia showed the sharpest increases in R&D intensity from 2008 to 2013 (+0.96%). Several other countries experienced strong increases in spending between 2008 and 2013, of over 50% in volume terms, including the Czech Republic, Poland, the Slovak Republic and Turkey. The shift in the global landscape of innovation is also apparent in patents and publications (Figure 2.3; OECD, 2014a).

Over the past decade, Asia has been the source of an unprecedented migration of talent towards the OECD area, with Asian immigrants being on average more skilled than other migrants and, for newcomers, even more skilled than OECD nationals. But according to new bibliometric indicators, China, Korea and Chinese Taipei are now also the main destinations of scientific authors from the United States and experienced a net brain gain over the period 1996-2011 (OECD, 2014a). At the same time, the BRIICS (Brazil, Russian Federation, India, Indonesia, China and South Africa) have experienced a serious slowdown in economic growth in recent years, raising fears that some may be stuck in a "middle-income trap". The BRIICS are thus attempting to switch to higher value-added activities, and shift their positions – both upstream and downstream – in global value chains (GVCs). Innovation is the key to capacity upgrading. Industrial R&D capacities have developed fast in these regions, and steady increases in R&D intensities point to growing global competition in R&D assets.

Figure 2.3. **The recent crisis has reinforced ongoing shifts in the global R&D landscape**

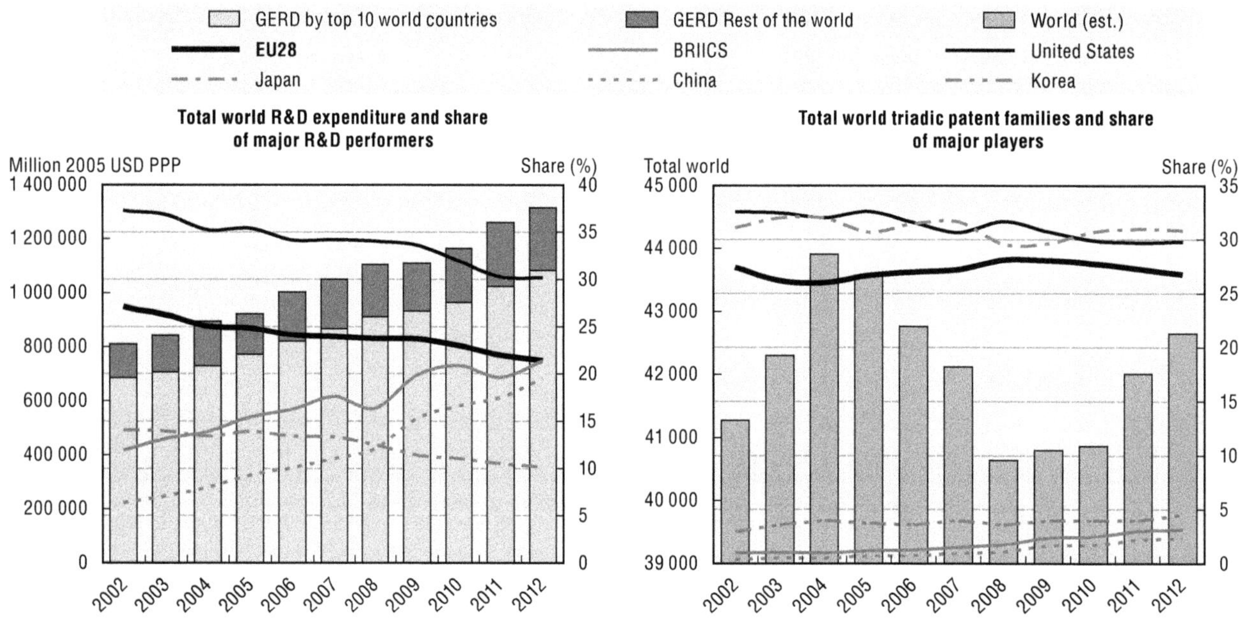

Note: Global gross expenditure on R&D (GERD) is estimated as the sum of GERD performed by OECD countries, the BRIICS, and Argentina, Colombia, Costa Rica, Egypt, Latvia, Malaysia, Romania, Singapore and Chinese Taipei. A world estimate would amount therefore to some USD 1 260 billion PPP in 2011 and USD 1 400 billion PPP in 2012.

Source: OECD (2014a), *OECD Science, Technology and Industry Outlook 2014*, OECD Publishing, Paris, *http://dx.doi.org/10.1787/sti_outlook-2014-en*.

While the share of the United States, European Union and Japan in world R&D, patents and scientific publications is on the wane, the United States still has a lead in the most advanced industries (information and communications technologies [ICTs], biotech) and benefits from world-class universities. Although Japan shows signs of renewed dynamism, firms have difficulty rebuilding R&D capacity and BERD surpassed 2007 levels only in 2013 (going from USD 116 billion to USD 121 billion PPP). Other leading R&D performers have experienced a decline in R&D intensity since 2002 – in most cases, this occurred before the crisis. Sweden, Iceland and Canada recorded the sharpest falls. Taken as a whole, EU28 business R&D intensity (1.21% in 2013) weighs on overall OECD performance. But European countries have increasingly diverged, with some moving towards meeting their R&D-to-GDP targets, while others, notably southern countries, have fallen further behind.

2.2. The digital economy continues to gain ground

A second major driver of innovation is the digital economy (OECD, 2014b). The number of Internet users in OECD countries increased from fewer than 60% of adults in 2005 to about 80% in 2013, reaching 95% among young people, although with large differences across and within countries. Fifteen-year-olds in the OECD spend about three hours on the Internet on a typical weekday, and more than 70% use the Internet at school. In OECD countries, 62% of Internet users participate in social networks and 35% use e-government services. About half of individuals in OECD countries purchase goods and services online, and almost 20% in Denmark, Korea, Sweden and the United Kingdom use a mobile device to do so.

In 2014, 76% of enterprises in the OECD area had a website or home page and 21% sold their products electronically (Figure 2.4). Over 80% of enterprises used e-government services (OECD, 2014b). Despite these figures, there are still large differences across countries in the use of ICT tools and activities within enterprises, suggesting there is much scope for further uptake and use of ICT (Figure 2.4).

Figure 2.4. **The diffusion of selected ICT tools and activities in enterprises, 2014**

Percentage of enterprises with ten or more persons employed

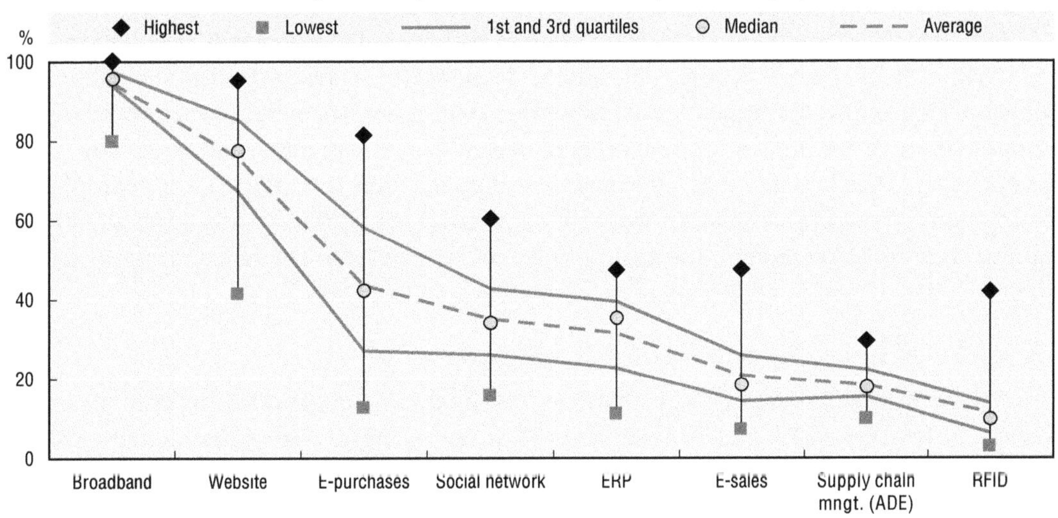

Note: ERP = enterprise resource planning; ADE = automated data exchange; RFID = radio-frequency identification.

Source: OECD (forthcoming b), *OECD Digital Economy Outlook 2015*, OECD Publishing, Paris, *http://dx.doi.org/10.1787/9789264232440-en*.

Higher-speed Internet, lower unit prices and smart devices have favoured new and more data-intensive applications. Wireless broadband subscriptions in the OECD area increased over twofold in just four years: by June 2014, more than three out of four individuals in the OECD area had a mobile wireless broadband subscription.[1] Mobile broadband is also widely available in many emerging and less developed countries (OECD, 2013a). In sub-Saharan Africa, for example, subscriptions grew from 14 million in 2010 to 117 million in 2013.

In less than two years, the number of pages viewed from mobile devices and tablets is estimated to have risen from 15% to over 30% of total. In 2013, over 75% of active Facebook users connected via a mobile device. International differences in speed and prices remain significant, however, even among OECD countries. In December 2013, the share of subscribers to high-speed broadband (above 10 megabits per second [Mbit/s]) ranged from over 70% to under 2% across OECD countries. Depending on country, smartphone users in the OECD may pay up to seven times more for a comparable basket of mobile services.

ICT-producing industries, together with publishing, digital media and content industries, accounted for about one-quarter of total OECD BERD in 2011. In 2014, patents in ICT-related technologies accounted for a third of all applications to main patent offices. In the last ten years, the share of data mining in total patents more than tripled, and the share of machine-to-machine (M2M) communication patents increased six times.

Many emerging technologies rely on innovations in ICTs. In the OECD countries, about 25% of ICT patents also belong to non-ICT areas. As a result of ICT, access to inventions and innovations is faster, cheaper and better, with technology now a part of mass culture. Widespread adoption of broadband has opened up a world of digital content to users. Cloud computing has shown great potential as a platform for new services. It has significantly reduced ICT barriers for small and medium-sized enterprises (SMEs), allowing them to expand faster and innovate. Massive open online courses are starting to change higher educa-tion and create a new field of learning informatics that could provide a new feedback for universities.

Big data could enable vast technological and non-technological innovation. The declining cost of data collection, storage and analytics, combined with the increasing deployment of smart ICT applications, generates large amounts of data, which can become a major resource for innovation and efficiency gains, on the condition that privacy issues are addressed. The benefits may also include enhanced data-driven R&D. For example, the deployment of second-generation genome sequencing techniques with embedded data-mining algorithms resulted in the cost per human-like genome sequence dropping from a million dollars to a thousand dollars in just five years (2009-14).

2.3. Technological changes are not limited to ICT

ICT is not the only technology characterised by rapid change. R&D investment by the world's largest companies is concentrated in a few sectors, with pharmaceuticals and biotechnology, technology hardware and equipment, and automobiles accounting for half of the total. The ICT-related and healthcare-related sectors remain among the most dynamic. Over the past decade, accelerated techno-logical progress (as reflected in a large burst in patenting) has been observed in (Figure 2.5):

- climate change mitigation, e.g. related to lighting, electric power, electric and hybrid vehicles, energy generation, batteries, motors, and engines
- ageing, health and food security, e.g. from chemistry and biotechnology
- information and communication management, including infrastructures for "big data" and virtual payments
- new manufacturing processes, e.g. from chemistry, nanotechnology, composite materials, new materials, 3D printing and laser technology.

The convergence of ICT and bio-, nano- and cognitive sciences has the potential to lead to "the next industrial revolution". A range of disciplines will need to be mobilised in a way that can harness multidisciplinary research.

Figure 2.5. **Acceleration in patented technologies, 1996-2001 and 2006-11**

Top 50 co-developments of international patent classes by development speed observed in the 2000s

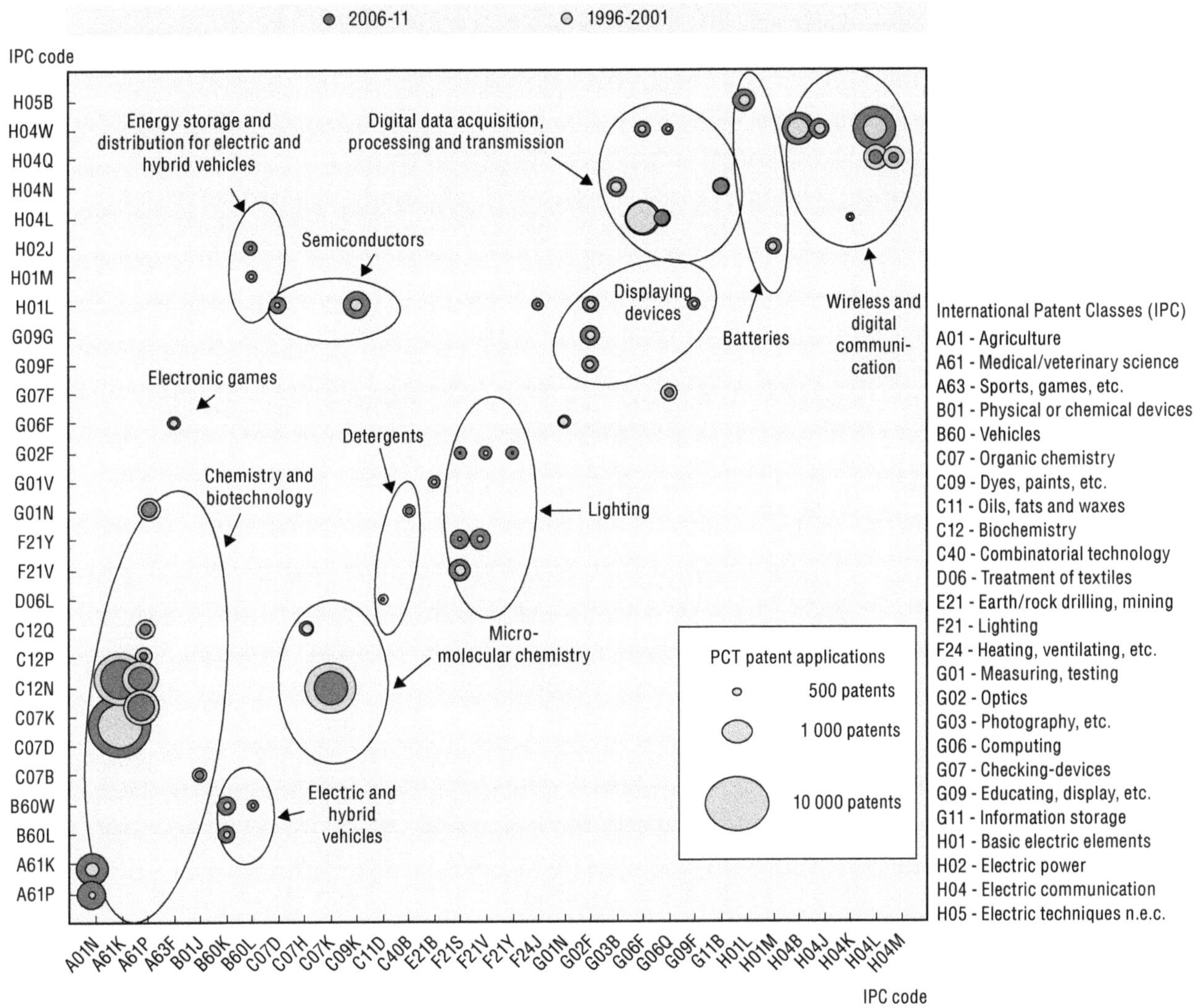

Source: OECD (2013a), *Science, Technology and Industry Scoreboard 2013*, OECD Publishing, Paris, *http://dx.doi.org/10.1787/sti_scoreboard-2013-en*.

2.4. Innovation is a very broad phenomenon, with many different features

While technological changes are a key driver of change, innovation is much broader than technological change. Firm-level data reveal innovation strategies that combine different types ("mixed modes") of innovation (Box 2.1): most innovative firms introduce new marketing or organisational methods alongside product or process innovations since these are often complementary. In fact, new organisational methods may facilitate the introduction of a new production process, or the new process may even require them. This holds true for both large firms and SMEs in both manufacturing and services.

In most countries, the shares of organisational and marketing innovating firms are relatively similar across broad sectors (Figure 2.6), although in Iceland and Portugal, the share is significantly higher in services (by 19.2 percentage points in Iceland and 13.4 percentage points in Portugal). In Israel and Brazil, more than 20% of all firms in manufacturing and 30% in services introduced organisational or marketing innovations only.

Box 2.1. **OECD definitions of innovation**

The current edition of the *Oslo Manual* identifies four types of innovation:

- **Product innovation:** The introduction of a good or service that is new or significantly improved with respect to its characteristics or intended uses. This includes significant improvements in technical specifications, components and materials, incorporated software, user friendliness, or other functional characteristics.
- **Process innovation:** The implementation of a new or significantly improved production or delivery method. This includes significant changes in techniques, equipment and/or software.
- **Marketing innovation:** The implementation of a new marketing method involving significant changes in product design or packaging, product placement, product promotion, or pricing.
- **Organisational innovation:** The implementation of a new organisational method in the firm's business practices, workplace organisation or external relations.

Source: OECD-Eurostat (2005), *Oslo Manual – Guidelines for Collecting and Interpreting Innovation Data, 3rd Edition*, OECD Publishing, Paris, *http://dx.doi.org/10.1787/9789264013100-en*.

Innovation often involves co-operation, and a large number of firms – both large and small – indicate they collaborate on innovation activities. Firms may co-develop their innovations with other companies, procure services such as R&D or design, license the rights to others' inventions, or simply imitate innovations developed and adopted elsewhere. For service innovation, more than 30% of innovating firms drew on some form of external development in the majority of countries during 2008-10 (OECD, 2013a). In terms of collaboration on innovation, large firms are far more likely to collaborate than SMEs. Among SMEs, the rate of collaboration is between 20% and 40% of innovative firms in two-thirds of the countries surveyed. For large innovative firms, collaboration rates range from more than 70% in the United Kingdom, Austria, Belgium, Finland, Denmark and Slovenia to less than a third in Brazil, Mexico and Chile.

Typically, R&D-active firms tend to collaborate more frequently on innovation than non-R&D-active firms. Moreover, collaboration with higher education or public research institutions is mainly an important source of knowledge transfer for large firms. In most countries they are usually two to three times more likely than SMEs to engage in this type of collaboration. Collaboration with other market actors, in particular suppliers and clients, is more common. Among large firms, suppliers play a key role as value chains become increasingly integrated. In Finland, the United Kingdom, Korea, South Africa and Iceland, collaboration with clients is equally as important or more important for both large firms and SMEs, an indication of the growing importance of user-driven innovation.

The co-operation on innovation is not limited to national borders. Collaboration with foreign partners can play an important role in the innovation process by allowing firms to gain access to a broader pool of resources and knowledge at lower cost and to share risks. It can take a variety of forms and levels of interaction, ranging from simple one-way information flows to highly interactive and formal arrangements. Size appears to be a strong determinant of foreign collaboration: large firms have a much higher propensity to collaborate internationally than SMEs, regardless of the overall rate of international collaboration. Among OECD countries, this is particularly true for Germany, Portugal and Italy.

Figure 2.6. **Innovation in manufacturing and services, 2008-10**

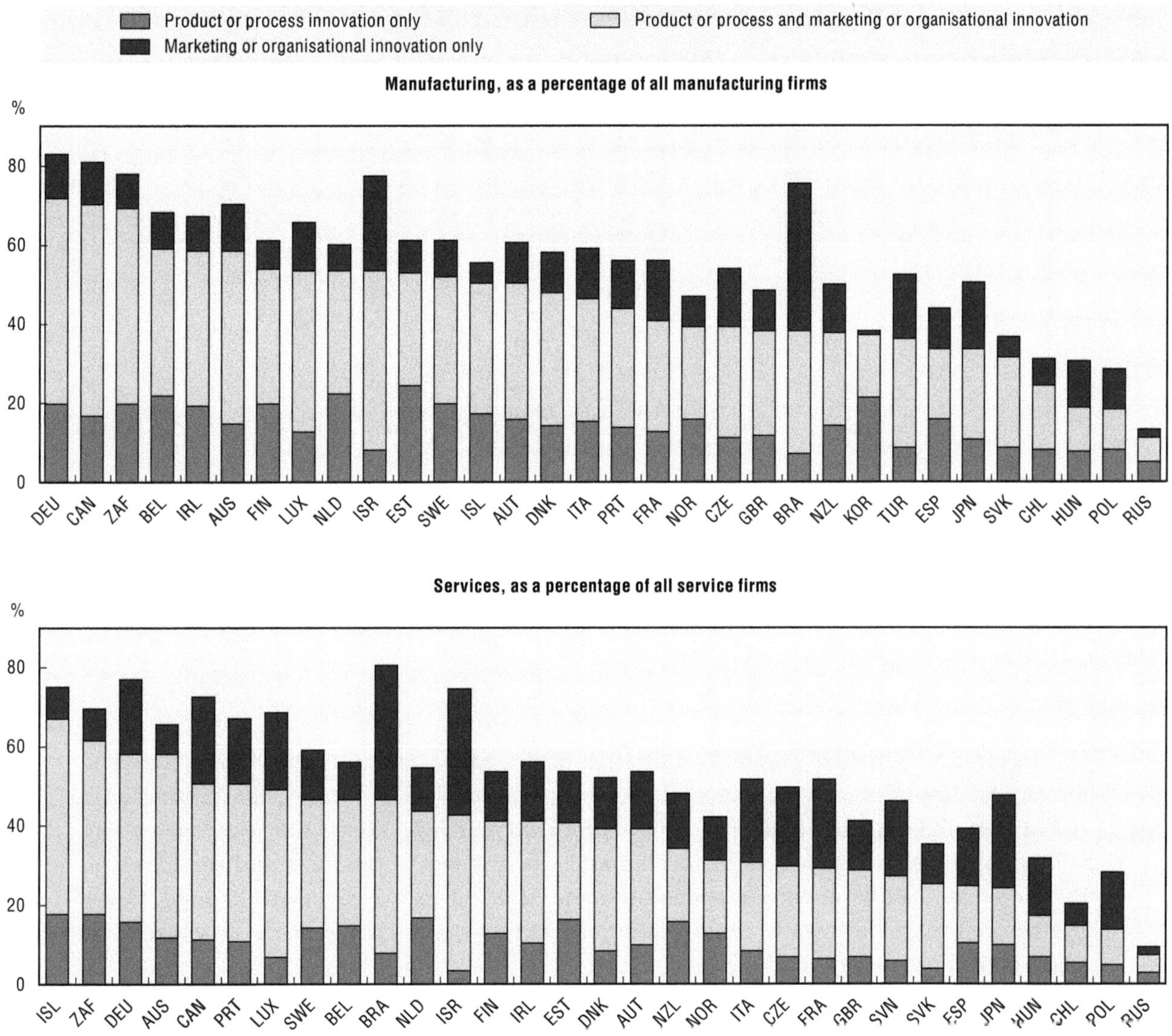

Source: OECD (2013a), *Science, Technology and Industry Scoreboard 2013*, OECD Publishing, Paris, *http://dx.doi.org/10.1787/sti_scoreboard-2013-en*.

Innovation collaboration rates vary widely across countries. In the Russian Federation and Brazil, collaboration centres on national partners, but in most countries there is more of a balance between national and international collaboration. In some small open economies – Luxembourg, the Slovak Republic, Slovenia and Estonia – collaboration is heavily skewed towards foreign partners. This may reflect factors such as sectoral specialisation, limited opportunities for domestic collaboration and, in some cases, proximity to external centres of knowledge.

Collaboration among institutions is a pervasive feature of research in, and increasingly between, countries. This is confirmed by an analysis of the affiliations and geographic locations of co-authors and co-inventors in scientific publications and patent documents (Figure 2.7). International co-authorship appears more widespread in the case of scientific publications than in that of patented inventions, except for India and Poland. Smaller countries tend to have higher rates of international collaboration. This may be partly due to the need to overcome limited domestic opportunities for collaboration and, in some cases, to the possible proximity (not only geographical) to centres of knowledge located abroad.

Figure 2.7. **International collaboration in science and innovation, 2007-11**

Co-authorship and co-invention as a percentage of scientific publications and PCT patent applications

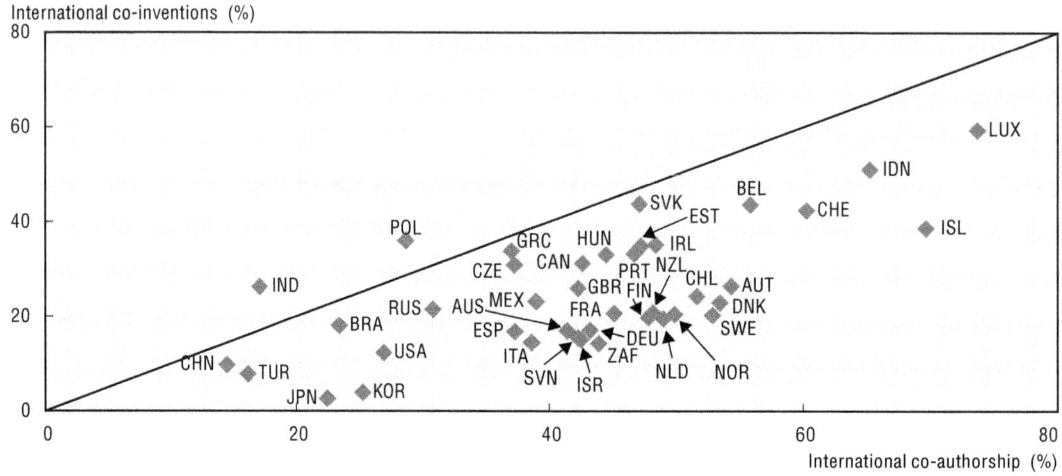

Source: OECD (2013a), *Science, Technology and Industry Scoreboard 2013*, OECD Publishing, Paris, *http://dx.doi.org/10.1787/sti_scoreboard-2013-en*.

2.5. Large and acute social and environmental challenges affect the demand for innovation

Innovation activities respond to a range of emerging demands that shape markets and act as drivers of innovative activity. As important drivers of future opportunities and challenges, these global and broad trends will increasingly affect the way innovation will be organised, where it will be located and which types of products and services will be demanded. While already present now and starting to drive current change, these transformative global trends could potentially act as game changers for innovation in OECD economies and beyond.

Changing demographics, e.g. in population size and composition, are expected to significantly impact production activities in the future, because of both supply and demand factors. Due to the decline in fertility rates and generalised gains in longevity, projections suggest that ageing over the next 50 years will result in a declining ration of the working-age population (15-74) to the total population (Figure 2.8), resulting in rising old-age dependency ratios. Apart from its macroeconomic effects – a smaller labour supply, increased fiscal pressure in some areas and an ageing workforce – this may also lead to skill mismatches and even skill shortages. An ageing population may also change consumer demand towards particular products and services (e.g. health), linked to changes in tastes, incomes, and household size and composition.

Population growth combined with rising general prosperity in emerging economies will lead to changing patterns of international demand. In addition, emerging economies will be increasingly confronted with growing flows of migration within countries (i.e. from rural areas to cities) and internationally (towards other countries, e.g. developed countries). Estimates show that the urban population in Asia will increase from 1.36 billion in 2010 to 2.64 billion by 2030 while in Africa it will rise from 294 million to 742 million, and from 394 million to over 600 million in Latin America and the Caribbean (United Nations, 2011).

This growing urbanisation will create major societal challenges to which future production will need to adjust in terms of location but also for example by providing solutions for housing, mobility, etc. (UNIDO, 2013).

Figure 2.8. **Working-age population (15-74) in relation to total population, 1990-2060**

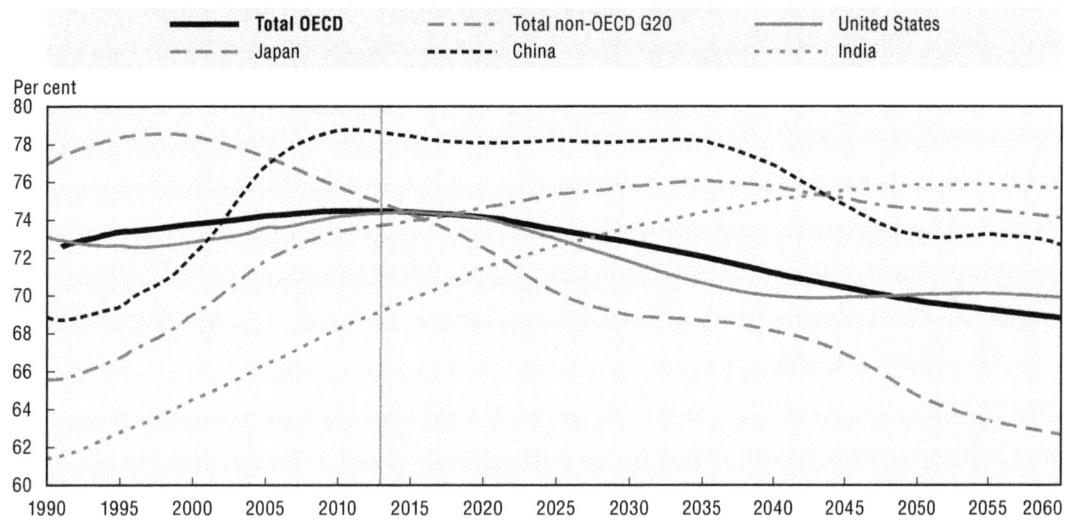

Note: The non-OECD G20 countries are Argentina, Brazil, China, India, Indonesia, Russian Federation, Saudi Arabia and South Africa.

Source: Braconier, Nicoletti and Westmore (2014), "Policy challenges for the next 50 years".

Sustainability is another main driver of change. With the demand for some natural resources outstripping available and future supplies, shortages of natural resources are predicted to emerge in many regions across the world. The issue of peak oil (i.e. the point in time at which global oil production declines) has been hotly debated in recent years because of the discovery of shale oil (and gas). However, the global resource challenge is not only about oil, since imbalances between demand and supply are also growing for energy at large, water, fish stocks, minerals (e.g. zinc, indium) and also food.

Increases in population and economic growth will result in even larger resource demands from across the globe, making the global resource challenge more acute. In addition, the challenge of climate change will have major impacts on economies and societies following the rise of average temperatures due to the growing presence of greenhouse gases in the atmosphere. There is a growing demand/pressure to become more sustainable along the value/supply/production chain and new technologies are expected to help address these challenges. As (some) production activities are energy- and resource-intensive, the development of more sustainable products and processes will become more important.

Globalisation and the emergence of GVCs have resulted in a growing interconnectedness among countries with significant flows of goods, services, capital, people and technology. Lower trade and investment barriers, falling transport costs, and advances in ICT have made it easier to offshore activities over longer distances. At the same time, GVCs have allowed countries to integrate more rapidly in the global economy. This has resulted in GVCs becoming longer and more complex over time, with production processes spanning a growing number of countries, increasingly in emerging economies. OECD work suggests that global trade integration will likely deepen over the coming 50 years (Figure 2.9), as transport costs continue to fall and trade barriers are lowered in line with already-agreed trade agreements (Johansson and Olaberría, 2014). The pace of integration is likely to slow, however, unless further agreements to lower trade barriers, transaction costs and regulatory obstacles are reached (Braconier, Nicoletti and Westmore, 2014).

Figure 2.9. **Exports as a share of GDP will continue to increase**

Exports as a percentage of GDP

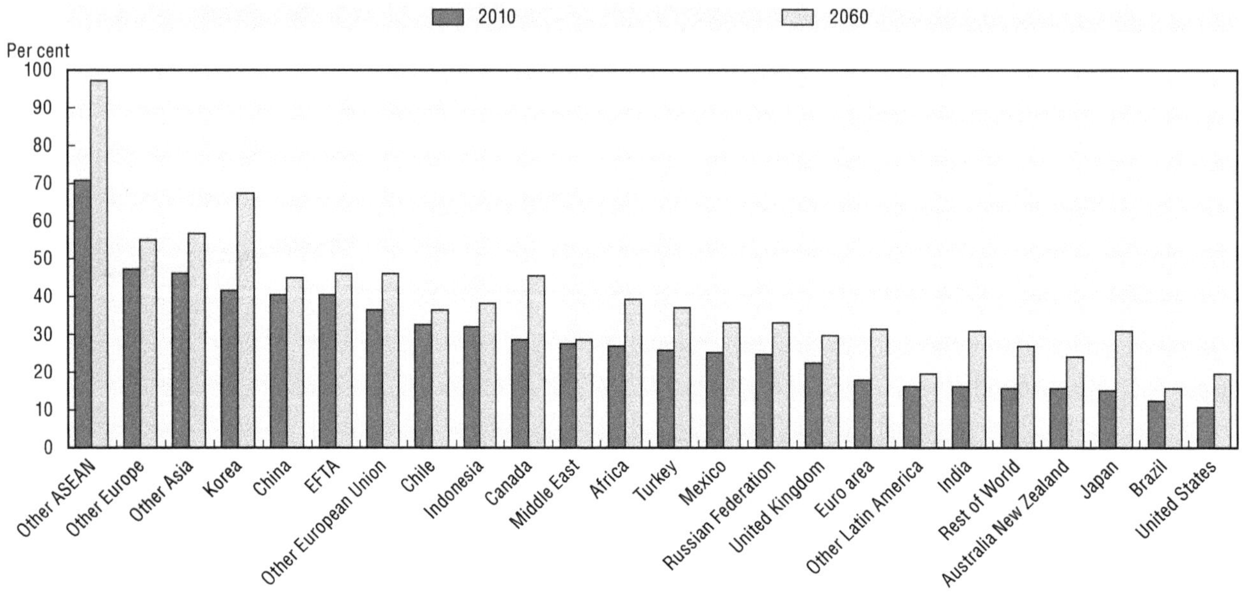

Source: Braconier, Nicoletti and Westmore (2014), "Policy challenges for the next 50 years".

Reshoring instead of offshoring of activities has increasingly attracted attention in recent years. A number of factors on the supply side may motivate companies in the near future to bring activities closer to their main markets, thereby changing the geography of GVCs in some industries (OECD, 2013b). Wage increases (for example in eastern China) are quickly eroding the labour cost advantage of emerging economies, while long and complex GVCs have exposed companies to a growing degree of supply risk in case of adverse shocks (e.g. natural disasters, political unrest, armed conflicts). In addition, management, logistical and operational problems including the protection of intellectual property rights, resulted often in significant "hidden" costs (i.e. costs that were not taken into account in the decision to offshore) and have in some cases made offshoring less/not profitable (Boston Consulting Group, 2014).

Demand factors will, however, still favour the location of production activities in emerging economies. For example, China and India are the world's most populated countries and have high GDP growth and are quickly becoming important markets for firms in many industries. While global consumer demand had previously been concentrated in (rich) OECD economies, a new middle class is emerging in China and India. While this middle class worldwide could rise from 1.8 billion to 3.2 billion by 2020 and to 4.9 billion by 2030, almost 85% of this growth is expected to come from Asia (Kharas, 2010). In 2000, Asia (excluding Japan) accounted for only 10% of global middle-class spending; this could reach 40% by 2040 and almost 60% in the long term. It is expected that income increases in emerging economies will lead to a rise in consumption of basic consumer products and other product categories.

Note

1. See *www.oecd.org/sti/broadband/oecdbroadbandportal.htm* for the latest data on broadband uptake.

References

Boston Consulting Group (2014), *The Shifting Economics of Global Manufacturing*, Boston Consulting Group, Boston.

Braconier, H., G. Nicoletti and B. Westmore (2014), "Policy challenges for the next 50 years", *OECD Economics Department Policy Papers*, No. 9, OECD Publishing, Paris.

Johansson, Å. and E. Olaberría (2014), "Long-term patterns of trade and specialisation", *OECD Economics Department Working Papers*, No. 1136, OECD Publishing, Paris, *http://dx.doi.org/10.1787/5jz158tbddbr-en*.

Kharas, H. (2010), "The emerging middle class in developing countries", *OECD Development Centre Working Papers*, No. 285, OECD Publishing, Paris, *http://dx.doi.org/10.1787/5kmmp8lncrns-en*.

OECD (forthcoming a), *Science, Technology and Industry Scoreboard 2015*, OECD Publishing, Paris.

OECD (forthcoming b), *OECD Digital Economy Outlook 2015*, OECD Publishing, Paris, *http://dx.doi.org/10.1787/9789264232440-en*.

OECD (2015c), *Main Science and Technology Indicators 2014-II*, OECD Publishing, Paris, *www.oecd.org/sti/msti.htm*.

OECD (2015d), *OECD National Accounts Database*, *http://stats.oecd.org/*, accessed on March 2015.

OECD (2014a), *Science, Technology and Industry Outlook 2014*, OECD Publishing, Paris, *http://dx.doi.org/10.1787/sti_outlook-2014-en*.

OECD (2014b), *Measuring the Digital Economy*, OECD Publishing, Paris, *http://dx.doi.org/10.1787/9789264221796-en*.

OECD (2013a), *Science, Technology and Industry Scoreboard 2013*, OECD Publishing, Paris, *http://dx.doi.org/10.1787/sti_scoreboard-2013-en*.

OECD (2013b), *Interconnected Economies: Benefiting from Global Value Chains*, OECD Publishing, Paris, *http://dx.doi.org/10.1787/9789264189560-en*.

OECD-Eurostat (2005), *Oslo Manual - Guidelines for Collecting and Interpreting Innovation Data, 3rd Edition*, OECD Publishing, Paris, *http://dx.doi.org/10.1787/9789264013100-en*.

UIS (UNESCO Institute of Statistics) (2014), *Science, Technology and Innovation Database*, *www.uis.unesco.org/datacentre/pages/default.aspx*, accessed on June 2014.

United Nations (2011), *World Population Prospects – The 2010 Revision*, United Nations, New York.

UNIDO (United Nations Industrial Development Organization) (2013), *Emerging Trends in Global Manufacturing Industries*, UNIDO, Vienna.

The Innovation Imperative
Contributing to Productivity, Growth and Well-Being

Chapter 3

Fostering talent and skills for innovation

A wide range of policies affects the various drivers of innovation. Among the most important of these policies for innovation are a skilled workforce that can generate new ideas and technologies, bring them to the market, and adapt to technological changes across society. Skilled people generate knowledge that can be used to create and implement innovations, and skills are also crucial to help absorb new innovations throughout economy and society. Human capital policy for innovation must address a wide array of skills and should help create an environment that enables individuals to choose and acquire appropriate skills and that supports the optimal use of these skills at work. This includes more incentives to institutions to improve the quality and relevance of their teaching as well as support for firm-level training. Policy makers should also assess the attractiveness of careers in academic research and improve these if necessary. Moreover, barriers to women's participation in science and entrepreneurship need to be removed. Finally, policy should facilitate the development of linkages and networks among researchers across countries.

3.1. A strategic framework for innovation policies

A range of policies affects the various drivers of innovation and can help governments in shaping and strengthening the contribution that innovation makes to economic performance and social welfare. These *policies for innovation* are much broader than the policies that are often seen as "innovation policies" in a narrow sense – such as policies to support business research and development (R&D), financing for risk capital, etc. OECD analysis suggests that innovation thrives in an environment characterised by the following features, all of which are explored in detail in the next four chapters:

- A **skilled workforce** that can generate new ideas and technologies, bring them to the market, and adapt to technological changes across society (Chapter 3). Reforms to education and training systems, and to skills policies more broadly, are therefore of utmost importance to innovation. They include policies aimed at science, technology, engineering and mathematics (STEM) graduates, but must go beyond this group and cover a wider set of skills. Moreover, the international mobility of talent plays an increasingly important role in meeting emerging skills needs and supporting knowledge creation and transfer, making supportive policies of growing importance.
- A sound **business environment** that encourages investment in technology and in knowledge-based capital (KBC); that enables innovative firms to experiment with new ideas, technologies and business models; and that helps them to grow, increase their market share and reach scale (Chapter 4). The OECD's empirical analysis shows that innovation performance can be strengthened by structural reform – to product markets, encouraging competition and enabling new entry; to labour markets, enabling better resource allocation; and to financial markets, helping generate funding for risky investments. Regulations should enable rather than stifle innovation. Economies should also enhance their openness to trade, investment, knowledge flows and people, acknowledging that innovation does not recognise borders.
- A **strong and efficient system for knowledge creation and diffusion** that engages in the systematic pursuit of fundamental knowledge, and that diffuses this knowledge throughout society through a range of mechanisms, including human resources, technology transfer and the establishment of knowledge markets (Chapter 5). Strong and well-governed universities and public research institutes and mechanisms that support and facilitate the interaction among knowledge institutions and economy and society are therefore important to strengthen innovation performance. So is investment in knowledge infrastructure, notably broadband and other digital networks that are critical tools to enable co-operation and provide new platforms for innovation to occur. Again, as knowledge creation and innovation are global endeavours, policies to better connect science and innovation activities across the world are crucial to the innovation policy agenda.
- **Policies that encourage firms to engage in innovation and entrepreneurial activity** (Chapter 6) More specific innovation policies are often needed to tackle a range of

barriers to innovation. The appropriate policy mix might include tax incentives for investment in R&D; direct public support through grants, subsidies and innovation competitions; and policies to facilitate co-operation and networking, but also indirect incentives through public procurement and other so-called demand-side policies. Such policies can help to strengthen markets for innovation, and help focus it on specific challenges and opportunities, e.g. green growth. Many of these actions include policies at the regional or local level. Moreover, well-informed, dynamic engaged and skills consumers are increasingly important for innovation.

3.2. The role of human capital for innovation

Policy statements on innovation rarely fail to emphasise the importance of human capital. One reason for this is the empirically well-established positive link between human capital – the knowledge and skills embodied in workers – and incomes, productivity and growth. Since the mid-1980s, research on macroeconomic growth has gained impetus from new theoretical insights – in particular endogenous growth theory – that highlight the role of human capital. There are many reasons to expect a positive impact of human capital on growth: more education fosters technological progress and increases the ability to absorb innovations developed abroad. (Human capital is also likely to stimulate growth through non-technological routes. For instance, if education improves health, workers might be more productive and have longer working lives.) Recently, OECD (2013a) has shown that rising business investment in a range of intangible assets – from software to designs to new forms of business organisation – is important for growth and productivity in OECD economies. Such intangible assets are often a direct manifestation of human capital: for instance, software is a translation of human expertise into code. Rising business investment in intangible assets has been enabled by rising educational attainment and investment in skills.

Policy makers in some countries are also concerned that education and training systems might not be maximising the potential for progress in science, research and innovation. For instance, fears exist in some countries that too few students choose to study science and engineering.[1] Furthermore, across the OECD area, the rapid evolution of different parts of the economy – often associated with technological change – can generate skills shortages. A recent example is the reported shortfall of managers and analysts having adequate understanding of the business uses of "big data" (McKinsey & Company, 2011). Policy makers naturally wish to ensure that technology-driven skills imbalances are quickly corrected.

A further source of policy interest in human capital relates to the effects of technology on earnings inequality. Indeed, OECD analysis finds that skill-biased technological change is the single most important driver of rising inequalities in labour income (OECD, 2011a). Human capital development is clearly central to the policy response to rising income inequality.

A further consideration relates to population ageing. Other things unchanged, population ageing in OECD countries will lead to shrinkage in the scientific workforce (relative to the total population). Such shrinkage could have numerous effects and policy implications. For instance, a decline in the availability of scientific labour could affect patterns of R&D-driven offshoring. Among other things, this would have implications for education, training and immigration policies. Questions are likely to arise in many countries regarding how much spending on education and training would need to increase to keep pace with technological change and the needs for science- and innovation-related human capital. More might also need to be done to productively engage the scientific workforce, in academia and industry, beyond today's usual ages of retirement.

Overall, this subsection emphasises that human capital facilitates innovation through numerous channels, and that many disciplines and levels of skill contribute to innovation. There is no skills-related "silver bullet" for innovation. While some generic skills such as creativity and communication skills are clearly particularly important to innovation, consensus does not yet exist on how education systems should systematically develop and test them (even if education systems increasingly include such skills in their educational objectives). A number of education and training themes loom particularly large in the innovation arena, even if some are also important for other (non-innovation) reasons. These themes include the incentives for institutions to improve the quality and relevance of teaching, support for firm-level training and lifelong learning, the attractiveness of careers in academic research, ensuring that barriers to women's participation in science and entrepreneurship are removed, and facilitating the development of enduring linkages and networks among researchers across countries.

Human capital shapes innovation in a number of ways. In particular:

- **Skilled people generate knowledge that can be used to create and implement innovations.** For instance, in American cities, a 10% increase in the share of the workforce with at least a college degree is associated with an increase in (quality adjusted) patenting per capita of about 10% (Carlino and Hunt, 2009). Locations with a high share of college graduates host more jobs that require new combinations of activities or techniques (Lin, 2009).
- **Having more skills raises the capacity to absorb innovations.** Skills that aid the adoption and adaptation of technology are beneficial across the workforce, not just within R&D teams. Innovation in firms is particularly associated with the in-house development of skills, rather than their acquisition through hiring, owing to the former's effects on absorptive capacity (Jones and Grimshaw, 2012). Educated workers also have a better foundation for further skills acquisition. And through their actions as role models, they may spur faster human capital accumulation by other workers.
- **Skills interact synergistically with other inputs to the innovation process, including capital investment.** For instance, studies show that human capital complements investment in and the use of information and communications technologies (ICT).
- **Skills enable entrepreneurship.** Entrepreneurship is often a carrier of innovation and structural change. Skills and experience are crucial to enterprise growth and survival. For example, Cressy (1999) shows that after controlling for the effects of human capital, financial capital is a relatively unimportant determinant of business longevity.
- **Skilled users and consumers of products and services often provide suppliers with valuable ideas for improvement** (Von Hippel, Ogawa and de Jong, 2011).

Human capital spurs innovation through many channels (as exemplified in the preceding paragraph). In different contexts, generic skills – such as reading, writing and problem solving – as well as technical, managerial, design and interpersonal skills, such as multicultural openness and leadership, all affect innovation. In the widest interpretation, the skills that support innovation could be any ability, proficiency or attribute that contributes to creating and implementing new products, processes, marketing methods or organisational arrangements in the workplace. Even if these skills are narrowed to only those that are teachable in the education and training system, the set remains large.

Jones and Grimshaw (2012) summarise the available assessments of how training and skills affect innovation in firms. In particular, the research shows that both tertiary and vocational education produce valuable skills, there is a positive innovation effect coming from intermediate technical skills,[2] and sectoral variation in how skills affect innovation suggests that institutions such as sector skills councils are important.

In terms of field of study, innovation policy makers often emphasise STEM. However, the importance of different fields of study varies by type of innovation and sector of activity. For example, in manufacturing, over 50% of tertiary-educated employees involved in innovation have an engineering (42.9%) or science (7.8%) degree. But in finance, the proportions are 7.0% with engineering degrees and 6.6% with science (Avvisati, Jacotin and Vincent-Lancrin, 2013).

A significant proportion of professionals with tertiary degrees from *all* fields hold highly innovative jobs. Over 45% of tertiary graduates from any field – and over 60% of science and engineering graduates – participate in at least some type of innovation. Participation varies across types of innovation. For instance, graduates in arts and engineering have the same likelihood of participating in product innovation, while engineers are significantly more likely to have a job involving technology innovation (Avvisati, Jacotin and Vincent-Lancrin, 2013).

It is useful to identify the skills that employees involved in innovation say they use in their jobs. The international REFLEX and HEGESCO surveys cover 19 European countries and Japan. These surveys show that employees who introduce innovations report using more of all types of skill in their jobs, relative to non-innovating counterparts. Among the self-reported skills used on the job that most distinguish innovative and non-innovative workers are "coming up with new ideas and solutions" (creativity), "a willingness to question ideas" (critical thinking), and "the ability to present new ideas or products to an audience" (communication) (Avvisati, Jacotin and Vincent-Lancrin, 2013).

A key principle should be the creation of an environment that enables individuals to choose and acquire appropriate skills and that supports the optimal use of these skills at work. This is the focus of the OECD Skills Strategy, the principal policy recommendations from which are set out in Box 3.1.

Educational attainment has risen; some industries have experienced large increases in workforce skills

Educational attainment, as one broad indicator of the skills available in countries, has risen steadily in OECD member countries, and around one-third of 25-34 year-olds now have a tertiary education. Graduation at the doctoral level has also grown (Box 3.2). Compared with older cohorts, young people increasingly graduate in the social sciences, business and law, and there has been a relative decline in the share of science and engineering (S&E) graduates in a number of countries. Data on wage premiums show that higher levels of study yield positive economic benefits.

Box 3.1. **Balancing the supply and demand for skills**

The OECD Skills Strategy has identified policy and institutional conditions conducive to a reasonable minimisation of skills mismatches (in any dynamic economy, such mismatches will not be eliminated entirely). The Skills Strategy has three overarching themes: developing relevant skills; activating skills supply; and putting skills to use. In a stylised manner, the key policy and institutional conditions to attain include:

- A focus on the development of strong generic skills, so that specific skills can be more easily acquired later (including through retraining).
- A focus on creating a system that is flexible, and thus responsive to economic change, rather than relying on skills forecasts as a guide to policy.
- Comprehensive information systems that allow students to understand course content, associated labour market outcomes, and the performance of education and training providers, as well as permitting employers to understand the content of qualifications.
- Arrangements allowing flexible demand-driven resource allocation across providers of education and training services, and across faculties within educational establishments. In addition, funding and financial incentives are needed that avoid distortions (for instance, inducing students to choose academic tertiary over vocational tertiary education because fees for the latter are too high) and barriers to participation (owing, for instance, to financial constraints for students from low-income backgrounds).
- The involvement of employers and other social partners in the design and delivery of skills policies. For instance, in the United Kingdom, Jaguar Land Rover has created a network from among a range of universities to deliver tailored courses in science and engineering for its staff, as part of the company's Technical Accreditation Scheme. The aim is to provide Jaguar's employees with access to "the best courses from the best sources".
- Labour market policies that help inactive workers to become active, or allow workers facing the prospect of inactivity to remain active for longer.
- Labour market policies that facilitate mobility, including mobility across local labour market areas.
- A well-developed training market for adult skills, including mechanisms that counter obstacles to training investments sometimes encountered in small and medium-sized enterprises (SMEs).
- An effective demand-driven labour migration regime. Such a regime should: identify labour market needs, considering demographic and educational changes in the non-immigrant population; establish formal recruitment channels; issue sufficient visas and process them quickly; provide efficient ways to verify residence and immigration status; and implement effective border control and workplace enforcement.
- Mechanisms to control for quality and create accountability at all levels of the system.

Source: OECD (2012a), *Better Skills, Better Jobs, Better Lives: A Strategic Approach to Skills Policies*, OECD Publishing, Paris, *http://dx.doi.org/10.1787/9789264177338-en*.

Box 3.2. **The careers of doctorate holders**

Between 2000 and 2009 there was a steady increase in the number of doctoral degrees awarded across the OECD, rising by 38% from 154 000 to 213 000. However, the association between the proportion of doctoral graduates in the labour force and a country's R&D intensity is weak.

Despite the growing supply of doctorate holders, the evidence points to a sustained labour market premium for those with a doctoral qualification. While female and younger doctorate holders fare relatively worse in terms of employment rates and earnings than their older and male counterparts, these biases are less marked for doctorate holders than for individuals with lower levels of educational attainment.

Although higher education is the main sector of employment for those with doctorates, demand for doctorates is apparent across knowledge-intensive sectors. Among doctorate holders, the take-up of jobs outside higher education is often in non-research occupations. Working as a researcher becomes less likely as careers progress and other competencies are acquired.

A wide range of monetary and non-pecuniary factors contribute to the reported attractiveness of research careers. Even when not in research, the jobs of doctorate holders in most cases relate to the subject of the doctoral degree, and doctoral graduates are generally satisfied with their employment situation.

A particular policy challenge faced by some small economies relates to the inability to provide suitable employment for researchers in certain specialised fields. This reflects a less diverse economic structure relative to many larger economies.

Source: OECD (2013b), OECD (2013a), *Science, Technology and Industry Scoreboard 2013*, OECD Publishing, Paris, *http://dx.doi.org/10.1787/sti_scoreboard-2013-en*.

3.3. Reducing skill mismatch

Higher skills do not automatically translate into higher prosperity and sustained growth. Making optimal use of existing skills and preventing the waste and attrition of skills due to mismatch or lack of use is also crucial (OECD, 2012a). Indeed, there is considerable skill mismatch in many OECD countries (Figure 3.1). New OECD research highlights that potentially significant gains to labour productivity can be achieved by a more efficient matching of workers to jobs (Adalet McGowan and Andrews, 2015a). Reducing skill mismatch expands the effective pool of labour that firms can draw workers from, enabling them to innovate and grow. Put differently, by trapping resources in relatively low productivity firms, for instance when industries have a high share of overskilled workers, skills mismatch can make it more difficult for more productive firms to attract skilled labour and gain market share at the expense of less productive firms.

Beyond the specific effects of education policies, Adalet McGowan and Andrews (2015b) show that a wider range of policies can affect skill mismatch and its consequences. Well-designed framework policies are associated with lower skill mismatch. In particular, less stringent product and labour market regulations and bankruptcy legislation that do not excessively penalise business failure are all associated with lower skill mismatch. Reforming housing market policies that inhibit residential mobility may also reduce skill mismatch. Such reforms can include lower transaction costs on buying property, less strict rent controls and less stringent building regulations. A better matching of skills and jobs is also facilitated by higher participation in lifelong learning and better managerial quality.

Figure 3.1. **Skill mismatch and productivity**

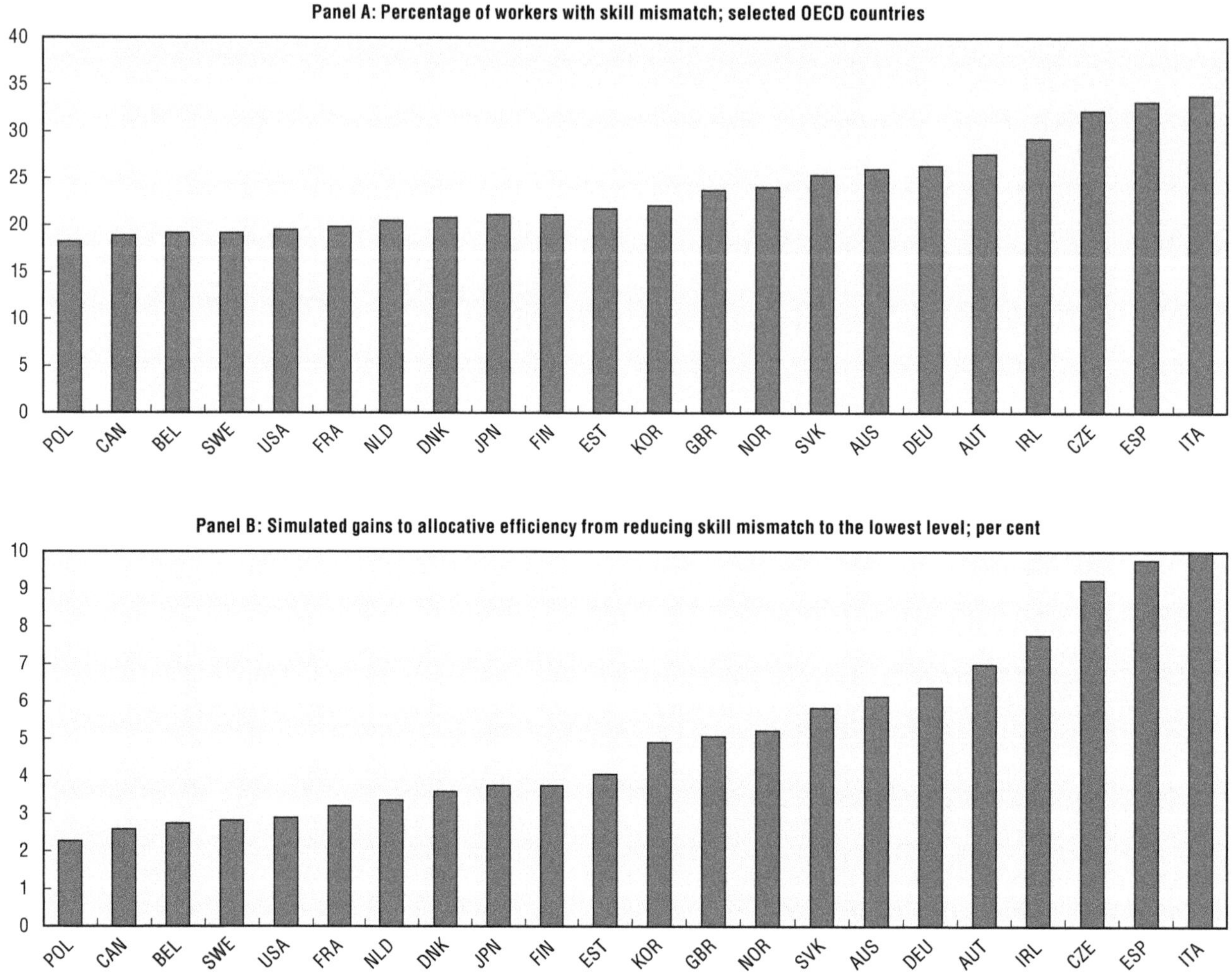

Notes: Panel A: Skill mismatch refers to the percentage of workers who are either over- or underskilled for a sample of 11 market sectors. In order to abstract from differences in industrial structures across countries, the one-digit industry level mismatch indicators are aggregated using a common set of weights based on industry employment shares for the United States. Panel B: The chart shows the difference between the actual labour productivity and a counterfactual labour productivity based on lowering the skill mismatch in each country to the best-practice level. One-digit industry level mismatch indicators are aggregated using a common set of weights based on the industry employment shares for the United States. The estimated coefficient for the impact of mismatch on productivity is based on a sample of 19 countries for which both firm level productivity and mismatch data are available.

Source: Adalet McGowan and Andrews (2015a), "Labour market mismatch and labour productivity: Evidence from PIAAC data", based on OECD (2013c), *Key Findings of the OECD-KNOWINNO Project on the Careers of Doctorate Holders*, *www.oecd.org/sti/inno/CDH%20FINAL%20 REPORT-.pdf*.

3.4. Foundation skills and innovation

Increasing students' learning outcomes in school is a key educational challenge in most countries. Every three years the OECD's Programme for International Student Assessment (PISA) tests 15-year-old students in reading, mathematics and science – as well as other domains such as problem solving and financial education. PISA gives a picture of how education systems fare in developing foundation skills.

PISA demonstrates substantial variation in learning outcomes across and within countries. Even countries that perform above the OECD average often have a long tail of poor performers. For instance, in 2012, the mathematics skills of 23% of students across

OECD countries were evaluated at Level 1 or below, suggesting that they are only able to extract relevant information from a single source and can use only basic formulae or procedures to solve problems. Eighteen per cent of students were estimated to have reading skills below the baseline level of proficiency (OECD, 2014a).

However, as already noted, some skills of particular relevance to innovation – such as creative thinking and social skills – are not traditionally tested by education systems. Education systems increasingly include such skills in their educational objectives, but there is no consensus on how to develop them. Nevertheless, a number of observations are relevant here:

- A broad curriculum exposes students to different knowledge content and ways of thinking. This could directly contribute to innovation (by enhancing the ability to make connections between different bodies of knowledge).
- Revisiting pedagogies in traditional subjects could be valuable. To take one example, in mathematics education, metacognitive pedagogies that integrate an explicit reflection about students' learning and thinking, generally by using self-questioning, have been shown to lead to better learning outcomes. Not only do students improve their mathematical reasoning, they also develop stronger skills for solving complex, unfamiliar and non-routine problems (Mevarech and Kramarski, 2014). Metacognitive pedagogies are also effective in disciplines other than mathematics.
- While countries have changed curricula to broaden the skills that they want students to acquire, many of these skills are still not explicitly assessed, at either the school or the system level. The development of new tools to assess such skills, or at least to ensure that teachers pay explicit attention to them, is critical to ensuring that students develop innovative dispositions (OECD, 2014b; Lucas, Claxton and Spencer, 2013).

As noted above, beyond subject-specific expertise, tertiary education institutions should also aim to develop students' creativity, critical thinking and communication skills. Doing so ultimately depends on pedagogy and curricula. Approaches used include problem-based learning, other pedagogical practices, and fostering inter-disciplinarity.

Problem-based learning is increasingly used to foster innovation. Problem-based learning is often characterised as the learning and teaching of theoretical material within real-world contexts. Some institutions, such as the University of Maastricht (Netherlands), integrate problem-based learning in all their educational programmes. Problem-based learning is common in medical science and increasingly in other fields such as engineering and business administration. In her review of teaching approaches in science and technology, Sagar (2014) underlines that learners often do not understand why they need to learn the specified content. Students often feel overloaded with isolated concepts that lack authentic context. A recent review of evidence on the impact of problem-based learning in higher education shows that problem-based learning fosters certain innovation-related skills compared with lecture-based forms of university instruction (Hoidn and Kärkkäinen, 2014). The literature points to benefits of problem-based learning in students' long-term retention and knowledge application as well as in social and behavioural skills such as team work. By contrast, problem-based learning appears to be inferior to traditional teaching for short-term retention of knowledge and consequently appears to have an insignificant or slightly negative impact on academic performance in tests.

Pedagogic models other than those mentioned above can also foster skills for innovation. For example, the metacognitive pedagogies presented in the previous section are also effective in higher education, even though their effect tends to be smaller than in school education (Mevarech and Kramarski, 2014). And collaborative learning, game-based learning, real-time formative assessment and the use of online laboratories have been shown to improve students' understanding, reasoning and creativity in science education (Kärkkäinen and Vincent-Lancrin, 2013). This suggests that tertiary education institutions could enhance innovation-related skills through a variety of pedagogic models.

Interdisciplinary curricula and multidisciplinary education are at the heart of the curriculum strategies by which many higher education institutions seek to train future innovators. For example, since 2006, Harvard University has integrated biological, social, behavioural and clinical sciences under its New Pathway medical programme. The Biodesign programme of Stanford University has brought together students from engineering, management, genetics, biology, medicine and business since 2003 to train medical technology innovators. And in 2014 Stanford launched a new joint programme combining computer science with either English or music. In Japan, the Shonan Fujisawa Campus of Keio University offers interdisciplinary programmes in policy management, environmental information, nursing and medical care (Hoidn and Kärkkäinen, 2013).

Recent policy reforms in higher education have often modified the governance of institutions by giving them more autonomy, increasingly opening them to the labour market and making them more accountable. Because international rankings put much emphasis on research, countries should try to give more incentives to institutions to improve the quality and relevance of their teaching.

Beyond core research competencies, researchers also need skills that apply in a broad variety of work situations. Such transferable skills include communication, business and management skills. The literature identifies several benefits that can come from formal training to develop transferable skills. PhD candidates, for example, benefit from acquiring transferable skills during their studies. These skills help in carrying out their projects and in later employment. Learning by doing in employment is of course important. However, formal skills training can add value.

Government is not typically the key player in transferable skills training for researchers (OECD, 2012b). While training activity for transferable skills is considerable, mostly through universities and research institutions, there is usually no overall national strategy. The available information does not allow detailed comparison of transferable skills training across countries, but does indicate some international differences. For instance, in certain countries the emphasis on transferable skills is relatively new (e.g. Luxembourg), while in others, activity in this area has taken place for some time (e.g. the United Kingdom).

Debate exists over the best way to learn transferable skills – whether through interaction with supervisors and peers, formal courses, or workplace-based learning (e.g. during an internship) – as well as the respective roles of governments and research institutions. There may be merit in a more systematic approach to training for transferable skills and to more thoroughly embedding such training in existing education and research structures. Making research funding conditional on transferable skills training is another possibility, notably for doctoral studies.

The attractiveness of careers in academic research

Several issues appear to reduce the attractiveness of academic research careers, including: low starting pay; limited material rewards at senior levels compared with other professions; little wage differentiation among cohorts; and difficulties in moving between institutions, and moving internationally, because of tenure arrangements, pension rights, and attitudes to movement and job changes (HLG, 2004). OECD (2007) also highlighted drawbacks linked to the use of temporary contracts, slow access to tenure and a decline in the linear career track for academics.

Developing skills for entrepreneurship

Close conceptual links exist between innovation-specific skills and entrepreneurship skills (OECD, 2014c). Moreover, as discussed in various parts of this report, entrepreneurship is a critical vehicle for the introduction of innovations. During the past decade, most OECD countries have started to promote entrepreneurship skills in all levels of education (Hytti and O'Gorman, 2004). In particular, the number of higher education institutions providing entrepreneurship support for their students, graduates, researchers and professors is growing rapidly worldwide.

Entrepreneurship support in higher education generally has two strands. One aims at developing entrepreneurial mindsets. It stresses the development of self-efficacy, creativity, risk awareness, building and managing relationships, etc. A second strand aims to build the attitudes, skills and knowledge needed to successfully launch and grow a new business.

In recent years, the frequent use of business plans to teach entrepreneurship courses has been complemented by greater involvement of entrepreneurs in the teaching process, as well as an increasing use of social media and massive open online courses. It is increasingly common to find classrooms in which students are challenged to identify and use a wider range of knowledge sources to find novel solutions (Box 3.3).

Today, more than ever, universities are expected to respond to the social and economic needs of society, such as facilitating graduate employability, contributing to economic growth and local development, assisting innovation, and stimulating the birth of new enterprises. In this connection, HEInnovate (*www.heinnovate.eu*) – a joint initiative of the OECD and the European Commission – is a tool to help higher education institutions identify and act on opportunities for capacity development, including in teaching and research to enhance innovation and entrepreneurship.

Skills development beyond initial education

Lifelong learning is an essential part of both reacting to and fostering innovation. Learning and replenishing skills is necessary to respond to economic and technological change. For those who leave formal education with relatively poor skills, formal and informal learning in adulthood provides an opportunity to enhance their skills and increase their ability to contribute to innovation.

Giving companies and individuals sufficient incentives to participate in work-related or other kinds of training over their life span is a key challenge for any lifelong learning policy. Across OECD countries more than 40% of adults are estimated to participate in job-related and other training in a given year, though patterns vary substantially by country. Training is more common among younger workers and those with higher educational attainment.

Box 3.3. **Curricula-embedded entrepreneurship learning**

The University of Twente (UT) is located in Enschede, a town with approximately 170 000 inhabitants in eastern Netherlands. Established in 1961, with the aim to enhance and revive the regional economy after a major collapse of the regional textile industry, performing research that is useful for society has been UT's main goal from the beginning. All UT students should acquire entrepreneurship competencies by the end of their studies. The educational model has an emphasis on project-based and active learning, with a core emphasis on challenging students to identify and use many sources of knowledge to find novel solutions. A new interdisciplinary programme – the Academy of Technology and Liberal Arts & Sciences (ATLAS) – was recently launched for students who want to combine social and technical perspectives in engineering studies. During the three-year programme, students make use of the latest technologies in areas such as nano-robotics, tracers for personal safety, 3D printing and renewable energy. The curriculum includes a "personal pursuit" element in which students focus on their personal interests in music, sports or a second language.

Founded in 1971, the Munich University of Applied Sciences is the second-largest university of applied sciences in Germany. In 2011 a new course format was developed, building a triangle among entrepreneurship education, knowledge exchange and start-up support. REAL (Responsibility, Entrepreneurship, Action- and Leadership-Based) projects involve teams of five to six students in a one-semester project. Each REAL project course has multiple teams working on different aspects or solutions of a central innovation challenge. The course is team-taught, by a professor and an expert on entrepreneurship. Professors and students work together to define the specific challenge. One of the first REAL project courses, on urban farming, involved four faculties (mechanotronics, architecture, design and business administration). Students developed ideas related to crop production, food processing, transportation and logistics. Linking REAL project courses to topics of global relevance (e.g. sustainability, mobility, energy and space) has proved successful for attracting external partners.

Source: OECD HEInnovate case studies, online available at *www.heinnovate.eu*.

Using European data, Hansson (2008) estimates that employer-provided training is the most important source of further education and training once individuals enter the labour market. A substantial portion of these human capital investments are financed by firms. Hansson found that individuals captured between 20% and 50% of the returns to such training, with the rest accruing to firms. Mismatches in the structure of firm-level and wider economic returns to training may lead to an undersupply of training.

Suggested policy avenues to support firm-level training include improving information about training opportunities, setting appropriate legal frameworks so that private parties can organise and finance their training (e.g. through contracts), and helping to support the portability of skills by improving information about the competencies and skills gained through various learning channels. Tax incentives to promote training and education might be a supplementary measure. Other policy suggestions include reinforcing public funding of vocational education and training (VET) to complement firms' training investments if these are insufficient, and helping small firms to provide training. Value might also be had in ensuring high levels of skills among the staffs of organisations that bridge the gap between young and small firms and the science base (such as science parks and business incubators). These staff members can help to identify skills needs in client firms.

Evidence from the OECD Survey of Adult Skills (PIAAC) suggests that a relatively large fraction of the adult working population may lack some of the skills that facilitate innovation. Levels of literacy and numeracy among workers vary across countries, and this variation is strongly correlated with the PISA results mentioned above (OECD, 2013c). Around half of all adults in the OECD area have low capabilities in "problem solving in a technology-rich environment" (Level 1 or below), an indicator that reflects both computer literacy and problem solving skills (Figure 3.2). Evidence from PIAAC also shows that adults with advanced skills in problem solving in a technology-rich environment are much more likely to receive job-related training than other workers.

Figure 3.2. **Proficiency in problem-solving in technology-rich environments among adults, 2008-13**

Percentage of 16-65 year-olds scoring at each proficiency level

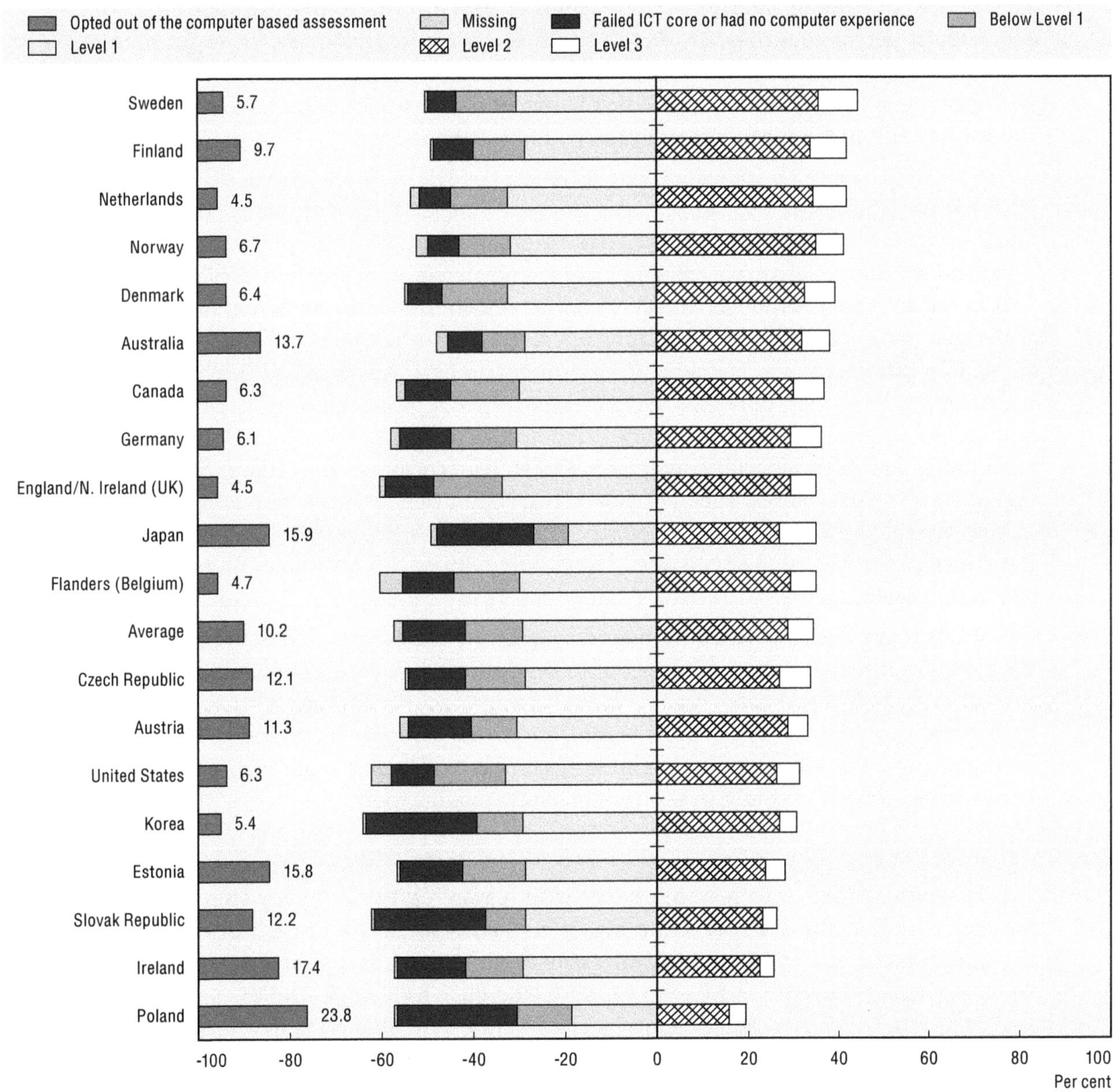

Notes: Countries are ranked in descending order of the combined percentage of adults scoring at Levels 2 and 3. Adults included in the missing category were not able to provide enough background information to impute proficiency scores.

Source: OECD (2013d), *OECD Skills Outlook 2013: First Results from the Survey of Adult Skills* http://dx.doi.org/10.1787/888932900612.

Policy should enable firms to adopt forms of work organisation that support innovation

Making the most of available skills for innovation depends in part on how the workplace is organised. For instance, Toyota's entrenched process of continuous incremental innovation reflects forms of workplace organisation that enable the collection and implementation of ideas from across its workforce (and this organisational asset has proven hard for competitors to copy). The OECD's PIAAC study shows a positive link between labour productivity and reading at work, even after controlling for average proficiency scores in literacy and numeracy. Indeed, once such adjustment is made, the average use of reading skills at work is found to explain 37% of the variation in labour productivity across countries (OECD, 2013c). In academia and industry, concepts such as employee engagement, high-performance working and the learning organisation are being widely studied. The evidence shows a link between management of human resources and innovation, although causality may run in both directions.

New OECD research indicates that different models of work organisation adopted by SMEs are associated with differences in their innovation performance. The effect is likely to operate through the impact of work organisation on the opportunities for the independent and creative use of employees' knowledge and problem solving abilities (OECD, forthcoming). Although causality cannot be confirmed, the evidence suggests that SMEs that adopt "learning organisation" or "discretionary learning" models – which are associated with teamwork, performance incentives and greater employee discretion in the planning and execution of tasks – have greater levels of product and process innovation and greater inter-organisational co-operation and knowledge exchange (relative to more traditional and hierarchically organised SMEs). Furthermore, the share of "learning organisation" SMEs varies substantially across countries. Micro-econometric analysis at the level of the firm, across 29 countries, shows a positive relationship between the national share of "learning organisation" SMEs and national innovation rates among SMEs (OECD, forthcoming).

While many decisions about human resources are the subject of practices internal to the firm, governments do have some scope to shape these decisions. Labour market policies that allow mobility and enable organisational change, while also supporting training, may help firms to adopt forms of work organisation that support innovation. Competitive markets – and the effective enforcement of competition policy – are also important in encouraging firms to innovate in terms of business organisation.

3.5. The participation of women in science and entrepreneurship

The participation of women in science may require particular policy attention. There are concerns that the skills of some highly trained women are underutilised and that the associated social and individual investments in education are at risk of being lost. Female scientists are concentrated in certain fields, such as biology. And the proportion of female scientists tends to fall as seniority rises. While participation is a result of personal choices, certain barriers to female participation may persist. These barriers include gender stereotypes, non-transparent nomination and appointment procedures, inadequate facilities for childcare, and insufficiently family-friendly workplace practices.

A variety of policies has been implemented to address gender issues in science. Countries have introduced equal opportunity legislation, units for women within science ministries, targets and quotas, networks and mentoring programmes, and policies on maternity and paternity leave. However, policies mainly aim at universities and public research institutions, not the private sector – although the Nordic and other countries have mandated quotas for female representation on the boards of publicly listed companies. Most of the relevant policies have also not been well evaluated.

OECD (2012c) shows that across the OECD area there are more male than female entrepreneurs, and the share of women who choose to run a business has not increased substantially in most countries. If women's intentions to engage in entrepreneurship are constrained by gender-specific conditions, society and the economy will fail to maximise entrepreneurial potential. Currently, more women than men become business owners out of necessity. On average, female-owned businesses register lower profits and labour productivity than male-owned businesses. These disparities can mostly be explained by differences in the size and capital intensity of female- and male-owned firms. Female entrepreneurs rely substantially less than men on external loans, but it is not clear if this is because women are less inclined to use external finance or because women experience discriminatory treatment in capital markets (or both). Female-owned firms do differ from male-owned firms in terms of innovation outcomes. But lower levels of product and process innovation in enterprises founded by women can be explained by the sectoral, investment and size characteristics of their firms, as well as by women's entrepreneurial experience prior to start-up.

Making the most of the available talent pool is also about ensuring that women have equal opportunities to contribute to innovation. Analysis of "gendered innovation" shows that removing gender biases can improve research and innovation and open up new market opportunities (European Commission, 2013). Examples included in the EU report note, "In engineering, for example, assuming a male default can produce errors in machine translation. In basic research, failing to use appropriate samples of male and female cells, tissues, and animals yields faulty results. In medicine, not recognising osteoporosis as a male disease delays diagnosis and treatment in men. In city planning, not collecting data on caregiving work leads to inefficient transportation systems." Taking better account of gender differences is therefore of great importance for science and innovation.

3.6. Competition and collaboration in the global market for internationally mobile talent

Increasingly, the international mobility of highly skilled individuals is a defining feature of the global innovation landscape. A range of innovation activities cannot be conceived without taking into account their global nature and the role played by mobile talent. This is particularly apparent in science, where progress relies on the circulation of knowledge, interaction between scientists, and the exchange of diverse views and evidence. Furthermore, businesses and academia often seek foreign staff for their specific knowledge and abilities. For talented individuals, mobility provides a means to exploit opportunities abroad, further develop their human capital, fulfil vocations and improve their livelihoods.

Global flows of highly qualified individuals, students, scientists and engineers have increased steadily over the past two decades (Docquier and Rapoport, 2012). Economic and cultural factors have contributed to making international mobility more affordable. For example, the cost of international flights is only a fraction of what it was in the early 1970s. And English as a working and teaching language is now widespread. Policies to attract talent and promote its circulation also appear to have been important.

Migration is increasingly skill-based. The number of highly skilled migrants increased by 70% during the past decade (Arslan et al., 2014).

Students – particularly tertiary-level students – are at the forefront of the increased international mobility of talent. During the period 2000-12 the number of foreign tertiary students enrolled worldwide more than doubled, with average annual growth of almost 7% (OECD, 2014a). Europe is the leading destination for tertiary-level students enrolled outside their country of origin, hosting 48% of these students, followed by North America and Asia.

Factors driving the increase in international student mobility range from the rapidly expanding demand for higher education worldwide and the perceived value of studying abroad, to government support for students in fields that are growing rapidly in the country of origin. In addition, some countries and institutions actively seek to attract foreign students (OECD, 2014a).

Internationalisation is even more marked in the upper tier of post-secondary education. International students account for nearly a quarter of all students in advanced research programmes such as doctorates (OECD, 2014e). Data from a recent OECD/UNESCO/Eurostat study on doctorate holders show that, on average, 14% of national citizens with a doctorate degree have had at least one experience of international mobility of three months or longer over the previous ten years. Individuals with doctorates who have already experienced international mobility are more likely to report an intention to move abroad, mainly for the purpose of knowledge acquisition. The data also show some significant differences across countries in the relationship between international mobility and earnings. International mobility can be associated with lower earnings in a number of countries, which suggests rigidities in the labour market for the highly skilled (Auriol, Misu and Freeman, 2013; OECD, 2013b).

Global databases on key actors in science and innovation systems, such as scientists and inventors, help to gauge the extent of brain circulation. For example, patents filed under the Patent Cooperation Treaty contain information on both the residence and the nationality of inventors. These data suggest that, in 2005, 10% of inventors worldwide had an experience of migration (Miguélez and Fink, 2013). An indicator developed for the *OECD Science, Technology and Industry Scoreboard 2013* tracks changes in the institutional affiliation of authors who published in scientific journals over the period 1996-2011. Large differences are seen to exist in scientist mobility: nearly 20% of authors based in Switzerland have had a previous affiliation abroad. However, in Japan, Brazil and the People's Republic of China (hereafter "China"), researcher mobility stands at less than 5%.

On average, the research impact of scientists who change university (or research centre) affiliation across national boundaries is 20% higher than those who never move abroad (Figure 3.3). If the performance of "stayers" could be raised to the level of internationally mobile researchers, many economies would catch up with leading research nations. Of course, causality in this relationship could go in both directions, as high-impact researchers may well have more opportunities to move internationally. But a lack of mobility and exposure to leading scientists and their institutions is likely to be a drag on the scientific performance of an institution or of a country as a whole, and could ultimately affect a country's innovation capacity.

There is new and compelling evidence that geographic, cultural, economic and scientific distance measures are good statistical predictors of mobility among scientists (Appelt et al., 2015) and among inventors (Miguélez and Fink, 2013). The analysis of bilateral flows of scientists also provides evidence of two mechanisms by which home countries can benefit from mobility. First, mobility is closely related to scientific collaboration. Economies that have higher rates of international collaboration tend to have higher average citation rates, and top-cited publications are more likely to involve scientific collaboration across institutions (especially internationally) than "average" publications (OECD, 2013b).

Secondly, the mobility of scientists is strongly related to student flows in the opposite direction. These findings lend support to a "knowledge circulation" perspective on scientist mobility, rather than a more traditional zero-sum view in which some countries win talent at the expense of others. Mobility among scientists appears to occur in the context of wider and more complex networks of mobile, highly educated and skilled individuals. The analysis also shows that mobility can be positively influenced by convergence in economic conditions and resources dedicated to R&D, as well as reduced visa-related restrictions.

Figure 3.3. **The impact of scientific authors, by category of mobility, 1996-2011**

Based on the median source-normalised impact per paper (SNIP)

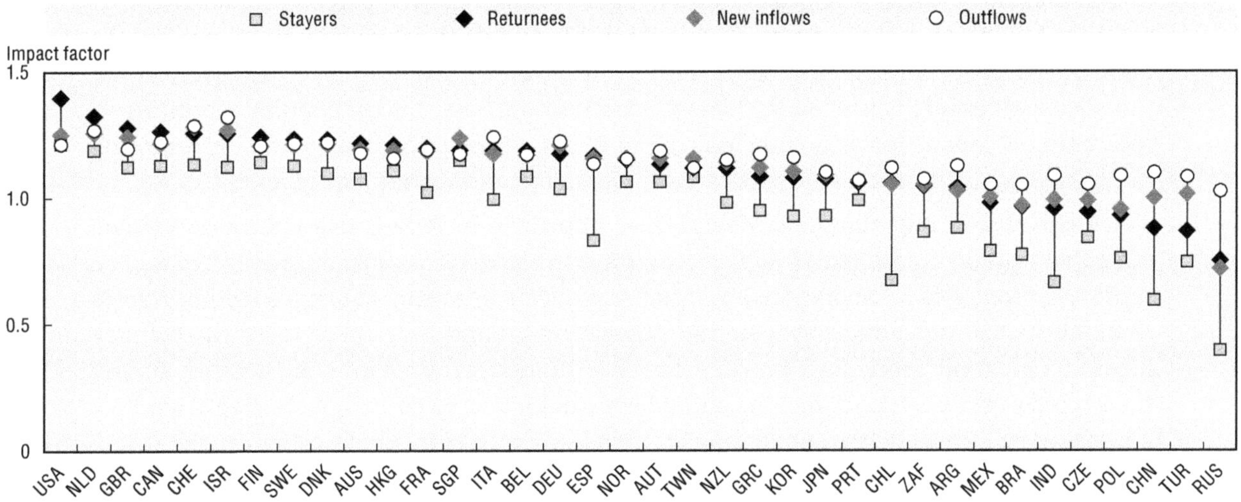

Note: This is an experimental indicator.

Source: OECD (2013b), *OECD Science, Technology and Industry Scoreboard 2013*, OECD Publishing, Paris, *http://dx.doi.org/10.1787/888932891549*. OECD calculations based on Scopus Custom Data, Elsevier, version 5.2012 and *SNIP2 Database*, *www.journalmetrics.com*.

The knowledge embodied in people is the object of strong global competition. But policy makers also need to be aware of the potential for different countries to simultaneously benefit from this knowledge. Policies should not be based on the idea that international mobility entails zero-sum competition. The recommendation in the 2010 OECD Innovation Strategy (OECD, 2010) that policy on mobility should support knowledge flows and enduring networks across countries thus appears to be validated by more recent evidence.

Educational accreditation standards and information are important in helping individuals demonstrate competences acquired elsewhere, thus removing major barriers for mobility and improving the efficiency of the global market for advanced skills. While countries sometimes enter into bilateral arrangements, international mobility is also promoted through multilateral programmes. For example, international co-operation in standardisation and grade recognition, and academic exchange among signatory countries, is part of the Bologna Process in the European Union, which also has initiatives to share information on funding opportunities and job vacancies for researchers in Europe (EURAXESS) (OECD, 2014e). UNESCO and the OECD have jointly developed guidelines for quality provision in cross-border higher education (OECD, 2005). These guidelines aim to facilitate the recognition of foreign degrees, ensure consumer protection for students and other stakeholders, and guarantee quality in the international mobility of students, researchers, educational programmes and institutions. Monitoring of the guidelines shows they have largely been implemented, but that some gaps remain in terms of information, transparency and consumer protection (Vincent-Lancrin and Pfotenhauer, 2012; Vincent-Lancrin, Fisher and Pfotenhauer, forthcoming).

Higher education policies focused on student mobility provide an opportunity to concentrate limited resources on educational programmes with potential economies of scale. And for host countries, *enrolling international students* can help raise revenues from higher education, and be part of a broader strategy to recruit highly skilled immigrants.

Some recruitment practices in the publicly controlled research system can have adverse effects on mobility. Evidence of an earnings penalty for international mobility paid by researchers in some countries may be a sign of dysfunctional personnel policies. If by moving abroad to acquire competences individuals find themselves in a worse position to take jobs in their home institutions, relative to those who stay, this may negatively affect mobility and research excellence. Some institutions address this problem by recruiting in international labour markets, precluding the hiring of incumbent students, or requiring mobility as a qualification for hiring, accompanied by relevant support incentives.

Restrictive immigration and visa policies appear to have negative effects on inflows of skilled workers. Even generic visa restrictions on short-term visits appear to hinder the most basic forms of collaboration. Immigration policies in several countries favour inflows of highly skilled individuals above other population groups. Barriers to the cross-country provision of some specialised services can also hinder certain forms of mobility among the skilled workforce. Policy makers should consider whether perceived shortages are best addressed by removing barriers to mobility or whether problems have other causes.

Financial assistance for mobility and support for the development of absorptive capacity are major policy approaches. Most OECD countries operate programmes to support the short-term outward mobility of students and researchers. These programmes differ with respect to the conditions and expectations placed on individuals upon their return. A major issue for policy makers is to develop coherent approaches for creating value from investments in acquiring skills abroad. This need not involve the creation of academic positions. Promoting the development of absorptive capacity in the business sector is a complementary option. Several countries offer schemes to attract the return of nationals working abroad or encourage the inward mobility of foreign-born individuals, even to a point where such measures become a central part of science and innovation strategies (OECD, 2014e).

Some countries provide tax relief for key foreign employees to help companies attract international expertise to their domestic operations. This is sometimes justified on the basis that short-term stayers do not get to fully benefit from local social welfare and pension systems. Such schemes have become increasingly popular in OECD countries. However, the schemes can become complex, imposing substantial compliance and administrative costs relative to the potential gains in employment or innovation (OECD, 2011b).

Main policy messages: Fostering talent and skills for innovation

Human capital policy for innovation must address a wide array of skills. A key principle should be the creation of an environment that enables individuals to choose and acquire appropriate skills and that supports the optimal use of these skills at work (the OECD's Skills Strategy sets out a comprehensive assessment of good practice in this area). An innovative economy and society requires the development, activation and use of skills in many disciplines and at many levels. There is no skills-related silver bullet for innovation.

Because international rankings often emphasise research, countries should give more incentives to institutions to improve the quality and relevance of their teaching. Broad curricula, updated pedagogical practices and the development of tools to assess innovation-related skills are all important in initial education. Beyond subject-specific expertise, tertiary education should also develop students' creativity, critical thinking and communication skills. Doing so ultimately depends on pedagogical approaches and the design of curricula.

Support for firm-level training requires a variety of steps. Possible policy avenues include improving information about training opportunities, setting legal frameworks so that private parties can organise and finance their training (e.g. through contracts), and increasing the portability of skills by improving information on the competencies and skills that are gained through various learning channels. Tax incentives to promote training might be a supplementary measure. Other policy suggestions include reinforcing public funding of VET to complement firms' training investments if these are insufficient, and helping small firms to provide training.

Individuals should have access to sufficient information – and be given incentives – to participate in work-related or other kinds of training over their life span.

Policy makers should assess the attractiveness of careers in academic research and improve these if necessary. Low starting pay, limited material rewards at senior levels, temporary contracts, and difficulties in moving institutionally and internationally because of tenure arrangements and pension rights lower interest in academic research careers.

At **a minimum, policy should ensure that barriers to women's participation in science and entrepreneurship are removed.** Gender stereotypes and non-transparent nomination and appointment procedures can all hinder female involvement in science. Shortages in regard to female experience of entrepreneurship should be addressed (for instance through innovations in the design and delivery of training programmes and support for networks of women entrepreneurs at multiple levels). Awareness programmes showcasing successful women in science and technology, and in high-growth firms, can provide useful role models for young women who may not otherwise consider such fields.

Policy should facilitate the development of enduring linkages and networks among researchers across countries. The knowledge embodied in people is the object of strong global competition. But policies should not be based on a view that international mobility entails zero-sum competition. Collaboration between countries often results in better outcomes. A key consideration is that migration regimes for the highly skilled should be efficient, transparent and simple, enabling movement on a short-term basis. Another consideration is the importance of facilitating the mutual recognition of skills, so as to allow efficient matching of mobile workers and jobs. Policy can also encourage inward and outward mobility. For example, for researchers, scientists and engineers, countries offer a range of economic incentives for inflows, including fellowships, grants and project funds, scholarships and tax benefits (although fewer options exist for those seeking to do research abroad). Individual institutions such as universities can also contribute. Their practices towards travel grants and support for mobile researchers can complement policies at national level. And recruitment practices in publicly controlled research systems should not create earnings penalties for internationally mobile researchers. Such practices may negatively affect both mobility and research excellence.

Notes

1. See for instance the speech given by US Federal Reserve Chairman Ben Bernanke to the conference on New Building Blocks for Jobs and Economic Growth. Available at: *www.federalreserve.gov/newsevents/speech/bernanke20110516a.htm*.
2. Intermediate technical skills are technical skills that are typically bounded at the lower limit by unskilled labourers and at the upper limit by university or polytechnic graduates engaged in management, research, design or production (Steedman, Mason and Wagner, 1991). Such intermediate technical skills are often developed through a mix of school- or workplace-based vocational education and training.

References

Adalet McGowan, M. and D. Andrews (2015a), "Labour market mismatch and labour productivity: Evidence from PIAAC data", *OECD Economics Department Working Papers*, No. 1209, OECD Publishing, Paris.

Adalet McGowan, M. and D. Andrews (2015b), "Skill mismatch and public policy in OECD countries", *OECD Economics Department Working Papers*, No. 1210, OECD Publishing, Paris.

Appelt, S. et al. (2015), "Which factors drive the international mobility of research scientists?", in Geuna, A., *Global Mobility of Research Scientists: The Economics of Who Goes Where and Why*, Elsevier.

Arslan, C. et al. (2014), "A new profile of migrants in the aftermath of the recent economic crisis", *OECD Social, Employment and Migration Working Papers*, No. 160, OECD Publishing, Paris, *http://dx.doi.org/10.1787/5jxt2t3nnjr5-en*.

Auriol, L., M. Misu and R.A. Freeman (2013), "Careers of doctorate holders: Analysis of labour market and mobility indicators", *OECD Science, Technology and Industry Working Papers*, 2013/04, OECD Publishing, Paris, *http://dx.doi.org/10.1787/5k43nxgs289w-en*.

Avvisati, F., G. Jacotin, and S. Vincent-Lancrin (2013), "Educating higher education students for innovative economies: What international data tell us", *Tuning Journal of Higher Education*, Issue 1, pp. 223-240.

Carlino, G. and R. Hunt (2009), "What explains the quantity and quality of local inventive activity?", *Federal Reserve Bank of Philadelphia Research Department Working Paper*, No. 09-12, Philadelphia.

Cressy, R. (1999), "Small business failure: Failure to fund or failure to learn?", in Acs, Z.J., B. Carlsson, and C. Karlsson, *Entrepreneurship, Small and Medium-Sized Enterprises and the Macroeconomy*, Cambridge University Press, Cambridge.

Docquier, F. and H. Rapoport (2012), "Globalization, brain drain and development", *Journal of Economic Literature*, Vol. 50, pp. 681-730.

European Commission (2013), *Gendered Innovations – How Gender Analysis Contributes to Research*, Publications Office of the European Union, Luxembourg, doi: *10.2777/11868*.

Hansson, B. (2008), "Job Related Training and benefits for Individuals: A Review of Evidence and Explanations", *OECD Education Working Paper* No. 19, OECD, Paris.

HLG (High Level Group on Increasing Human Resources for Science and Technology in Europe) (2004), *Europe Needs More Scientists*, European Commission, Brussels.

Hoidn, S. and K. Kärkkäinen (2014), "Promoting skills for innovation in higher education: A literature review on the effectiveness of problem-based learning and of teaching behaviours", *OECD Education Working Papers*, No. 100, OECD Publishing, Paris, *http://dx.doi.org/10.1787/5k3tsj67l226-en*.

Hytti, U. and C. O'Gorman (2004), "What is 'enterprise education'? An analysis of the objectives and methods of enterprise education programmes in four European countries", *Education + Training*, Vol. 46, pp. 11-23.

Jones, B. and D. Grimshaw (2012), "The effects of policies for training and skills on improving innovation capabilities in firms", *Nesta Working Paper Series*, No. 12/08, *www.nesta.org.uk/sites/default/files/the_effects_of_policies_for_training_and_skills_on_improving_innovation_capabilities_in_firms.pdf*.

Kärkkäinen, K. and S. Vincent-Lancrin (2013), "Sparking innovation in STEM education with technology and collaboration: A case study of the HP catalyst initiative", *OECD Education Working Papers*, No. 91, OECD Publishing, Paris, *http://dx.doi.org/10.1787/5k480sj9k442-en*.

Lin, J. (2009), "Technological adaptation, cities and new work", *Federal Reserve Bank of Philadelphia Research Department Working Paper*, No. 09-17, Philadelphia.

Lucas, B., G. Claxton and E. Spencer (2013), "Progression in student creativity in school: First steps towards new forms of formative assessments", *OECD Education Working Papers*, No. 86, OECD Publishing, Paris, *http://dx.doi.org/10.1787/5k4dp59msdwk-en*.

McKinsey and Company (2011), *Big Data: the Next Frontier for Innovation, Competition and Productivity*, McKinsey Global Institute, *www.mckinsey.com/insights/business_technology/big_data_the_next_frontier_for_innovation*.

Mevarech, Z. and B. Kramarski (2014), *Critical Maths for Innovative Societies: The Role of Metacognitive Pedagogies*, Education Research and Innovation, OECD Publishing, Paris, *http://dx.doi.org/10.1787/9789264223561-en*.

Miguélez, E. and C. Fink (2013), "Measuring the international mobility of inventors: A new dataset", *WIPO (World Intellectual Property Organization) Economic Research Working Paper*, No. 8.

OECD (forthcoming), *Skills and Learning Strategies for Innovation in SMEs*, OECD Publishing, Paris.

OECD (2014a), *Education at a Glance 2014: OECD Indicators*, OECD Publishing, Paris, *http://dx.doi.org/10.1787/eag-2014-en*.

OECD (2014b), *PISA 2012 Results: Creative Problem Solving (Volume V). Students' Skills in Tackling Real-Life Problems*, OECD Publishing, Paris, *http://dx.doi.org/10.1787/9789264208070-en*.

OECD (2014c), *Job Creation and Local Economic Development*, OECD Publishing, Paris, *http://dx.doi.org/10.1787/9789264215009-en*.

OECD (2014d), *Entrepreneurship at a Glance 2014*, OECD Publishing, Paris, *http://dx.doi.org/10.1787/888933064753*.

OECD (2014e), *OECD Science, Technology and Industry Outlook 2014*, OECD Publishing, Paris, *www.oecd.org/sti/outlook*.

OECD (2013a), *Supporting Investment in Knowledge Capital, Growth and Innovation*, OECD Publishing, Paris, *http://dx.doi.org/10.1787/9789264193307-en*.

OECD (2013b), *OECD Science, Technology and Industry Scoreboard 2013*, OECD Publishing, Paris, *http://dx.doi.org/10.1787/sti_scoreboard-2013-en*.

OECD (2013c), *Key Findings of the OECD-KNOWINNO Project on the Careers of Doctorate Holders*, OECD Publishing, Paris, *www.oecd.org/sti/inno/CDH%20FINAL%20REPORT-.pdf*.

OECD (2013d), *OECD Skills Outlook 2013: First Results from the Survey of Adult Skills*, OECD Publishing, Paris, *http://dx.doi.org/10.1787/9789264204256-en*.

OECD (2012a), *Better Skills, Better Jobs, Better Lives: A Strategic Approach to Skills Policies*, OECD Publishing, Paris, *http://dx.doi.org/10.1787/9789264177338-en*.

OECD (2012b), *Transferable Skills for Researchers: Supporting Career Development and Research*, OECD Publishing, Paris, *http://dx.doi.org/10.1787/9789264179721-en*.

OECD (2012c), *Closing the Gender Gap: Act Now*, OECD Publishing, Paris, *http://dx.doi.org10.1787/9789264179370-en*.

OECD (2011a), *Growing Unequal: Income Distribution and Poverty in OECD Countries*, OECD Publishing, Paris, *http://dx.doi.org/10.1787/9789264044197-en*.

OECD (2011b), "The taxation of mobile high-skilled workers", in *Taxation and Employment*, OECD Publishing, Paris, *http://dx.doi.org/10.1787/9789264120808-6-en*.

OECD (2010), *The OECD Innovation Strategy: Getting a Head Start on Tomorrow*, OECD Publishing, Paris, *http://dx.doi.org/10.1787/9789264083479-en*.

OECD (2007), "Summary report of the joint OECD-Spanish Ministry of Education and Science workshop: Research careers for the 21st century: Madrid, 26-27 April 2006", OECD Publishing, Paris.

OECD (2005), "*Guidelines for Quality Provision in Cross-Border Higher Education*", OECD Publishing, Paris, *www.oecd.org/education/innovation-education/35779480.pdf*.

Sagar, H. (2014), "Entrepreneurial Schools", prepared for the OECD in the framework of the Entrepreneurship360 project, a joint initiative with the European Commission DG Education and Culture.

Steedman, H., G. Mason and C. Wagner (1991), "Intermediate skills in the workplace: Deployment, standards and supply in Britain, France and Germany", *National Institute Economic Review*, Vol. 136/1, pp.60-76.

Vincent-Lancrin, S. and S. Pfotenhauer (2012), "Guidelines for quality assurance in cross-border higher education: Where do we stand?", *OECD Education Working Papers*, No. 70, OECD Publishing, *http://dx.doi.org/10.1787/5k9fd0kz0j6b-en*.

Vincent-Lancrin, S., D. Fisher and S. Pfotenhauer (forthcoming), *Guidelines for Quality Provision in Cross-Border Higher Education: 2014 Monitoring Report*, OECD Publishing, Paris.

Von Hippel, E., S. Ogawa and J.P.J. de Jong (2011), "The age of the consumer innovator", *MIT Sloan Management Review*, *http://sloanreview.mit.edu/the-magazine/2011-fall/53105/the-age-of-the-consumer-innovator/*.

The Innovation Imperative
Contributing to Productivity, Growth and Well-Being

Chapter 4

The business environment for innovation

Policies for innovation also rely on a sound business environment that encourages investment in technology and in knowledge-based capital; that enables innovative firms to experiment with new ideas, technologies and business models; and that helps them to grow, increase their market share and reach scale. New issues and policy learnings relevant to framework conditions that have emerged from the OECD's work on innovation relate to the appropriate framework conditions to benefit from investment in knowledge-based capital, tax policies related to innovation, the financing of innovation, as well as policies that enable experimentation and growth among young innovative firms. The work also emphasises the growing importance of global value chains (GVCs), and the implications this has for framework conditions that affect innovation, notably in regard to trade, investment and regulatory policies. Investment policies are important too, given the growing importance of investment in knowledge-based assets.

4.1. The role of framework conditions

The 2010 Innovation Strategy (OECD, 2010a) stressed the importance of sound framework conditions for innovation, including sound macroeconomic policy, competition, well-functioning product and labour markets, openness to international trade and investment, innovation-friendly tax systems, and financial systems that enable resources to flow to innovative activities. The main findings of that work were:

- Fiscal discipline and low and relatively stable inflation help to reduce uncertainty and enhance the efficiency of the price mechanism in allocating resources. Strong and stable rates of output growth also support firms seeking to introduce new products or undertake significant organisational changes.
- Strong competition encourages companies to innovate and develop new markets. Elimination of anti-competitive product market regulations is a powerful way to stimulate investment in innovation and supports the process of creative destruction. In addition, sound, proactive competition policies in line with international best practice can encourage innovation.
- More open trade in services, reduced barriers to merchandise trade across borders, modernised public procurement and a sound international framework for intellectual property rights (IPR), along with the effective enforcement of these rights, should foster innovation. The successful conclusion of the Doha Development Agenda by the World Trade Organization (WTO), and improved market access for goods and services, would be significant steps towards further market opening.
- Governments should consider the quality of their policy frameworks for investment. These frameworks are important in determining how much investment an economy receives. They also affect the extent to which this investment contributes to economic development and acts as a driver of innovation.

The importance of such framework conditions has increased in recent years as businesses and capital have become more mobile and seek the most favourable operating environments internationally. Moreover, the economic crisis and its aftermath have had wide-ranging impacts on many key framework conditions, notably the macroeconomic and financial environments. Reaping the benefits of innovation at national, regional and local levels requires governments and other stakeholders to undertake the investments and structural reforms that provide a good environment for engaging in innovation. The OECD's regular assessment of structural policies shows that much scope remains for productivity- and innovation-enhancing reforms, notably in product markets (OECD, 2014a).

Since the completion and release of the Innovation Strategy in 2010, a number of new issues and policy learnings relevant to framework conditions have emerged from the OECD's work on innovation. These particularly relate to the following themes:

1. The appropriate framework conditions to benefit from **investment in KBC**. In several OECD countries, investment in KBC – software, data, intellectual property, and

economic competencies such as brand equity, new organisational methods and firm-specific skills – is now larger as a proportion of gross domestic product (GDP) than investment in tangible capital. The growth of this form of investment raises several challenges for policy, including the need to improve a range of framework conditions.

2. **Tax policies**, including tax policies related to investment in knowledge-based assets.
3. The **financing of innovation**, which has grown in importance as a policy concern since the 2008 financial crisis, and which continues to be a challenge in many countries.
4. The framework conditions that enable **experimentation and growth among young innovative firms**. Such firms are a major source of employment growth and are also important for innovation, in particular for more radical innovation.
5. The growing importance of **global value chains** (GVCs), and the implications this has for framework conditions that affect innovation, notably in regard to trade, investment and regulatory policies.
6. **Investment**, notably ongoing revisions to the OECD's Policy Framework for Investment.

There are several other aspects of the environment for business that could be considered as framework conditions, e.g. trust, security and good governance. These are not discussed in this Chapter of the report, but will be addressed in Chapter 8.

4.2. Knowledge-based capital

KBC refers to a range of assets – typically intangible – that are based on investment in knowledge, including R&D, software and data, intellectual property, brand equity, firm-specific skills, and organisational know-how. Estimates for a wide range of OECD economies and a number of emerging economies show that business investment in KBC has risen rapidly during recent decades, often at a faster pace than investment in traditional physical capital (such as machines, equipment and buildings). During the global economic crisis, business investment in KBC declined less than investment in physical capital (Figure 4.1). Investment in KBC is also an essential ingredient for firms wishing to upgrade in GVCs, as it often facilitates differentiation from competitors. The growing importance of KBC raises a number of challenges for policy.

First, the impact of investment in KBC is greatly enhanced when resources can flow easily to KBC-intensive firms, so that these can grow and increase their market share (Andrews and Criscuolo, 2013). Given the inherent difficulties in allocating KBC efficiently, policies that facilitate the redeployment of tangible resources take on heightened importance. Specifically, well-functioning product, labour and capital markets and bankruptcy laws that do not overly penalise failure can raise the expected returns to investing in KBC. These benefits are realised partly through stronger competitive pressures and more efficient reallocation, which make it easier for successful firms to implement and commercialise new ideas. Benefits also arise from lowering the costs of failure, encouraging firms to experiment with uncertain growth opportunities.

Second, the rise of KBC implies an increasing importance – and in some cases changes in form or emphasis – for a number of types of policy (see OECD, 2013a). For example, ***competition policy*** faces new challenges in industries founded on KBC. This is particularly so in the digital economy, where never before have leading firms grown so large so quickly and where the nature of competition often differs in some respects from other sectors. Some experts have observed, for example, that unlike traditional manufacturing sectors, the digital economy's most meaningful competition takes place among platforms created by companies that use very different business models, rather than among companies that all follow more or less the same model. Apple, Google and Microsoft illustrate this point. They all compete in the market for mobile phone operating systems, but each uses a different business model. In such contexts, competition *among* the platform providers may be more important to innovation and consumer welfare than competition *within* the platforms (such as rivalry among companies that create apps for the iPhone). Competition policy should: 1) properly account for inter-platform competition; 2) promote the elimination of unnecessarily anti-competitive product market regulation; and 3) include the effective enforcement of competition law, which will protect and encourage innovation. Moreover, the previous work on KBC (OECD, 2013a) noted an important link between competition policy and IPR (see Chapter 5), noting that IPR systems must be coupled with pro-competition policies and efficient judicial systems.

Figure 4.1. **Business investment in knowledge assets weathered the crisis better and recovered earlier**

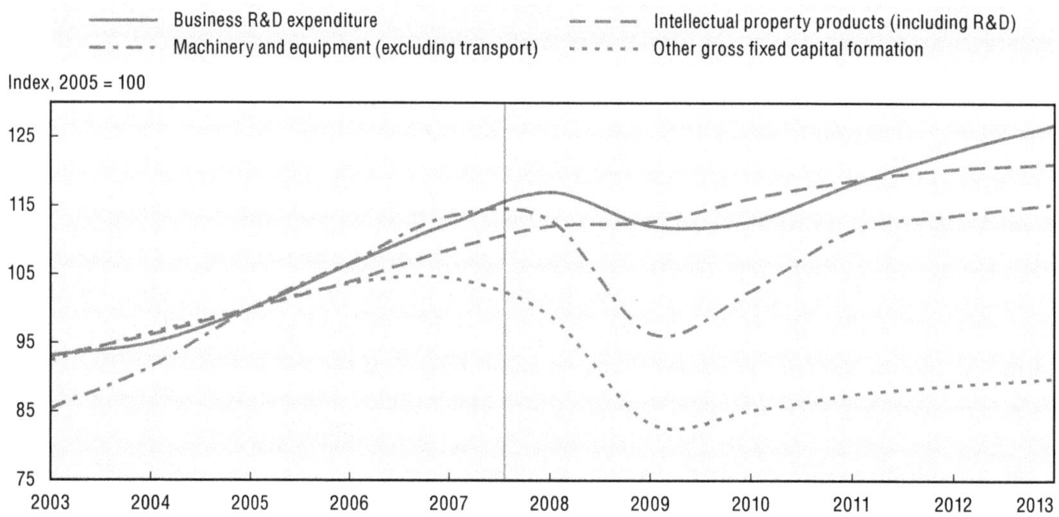

Note: In national accounts, spending on R&D activities is treated as expenditures and not as investment, and is therefore not capitalised. R&D capitalisation should be effective as from 2014. Intangible property products are non-financial fixed assets that mainly consist of mineral exploration, computer software, entertainment, and literary or artistic originals intended to be used for more than one year. Other gross fixed capital formation includes dwelling and transport investments.

Sources: OECD (2015a), *Main Science and Technology Indicators 2014-II*, OECD (2015b), *OECD National Accounts Database*, *http://stats.oecd.org/*, accessed on March 2015, based on OECD (2014b), *OECD Science, Technology and Industry Outlook 2014*, OECD Publishing, Paris, *http://dx.doi.org/10.1787/sti_outlook-2014-en*.

Another important framework condition that is relevant to KBC concerns **corporate reporting.** The value of many of the world's most successful companies resides almost entirely in their KBC. At the start of 2009, for example, physical assets accounted for only about 5% of Google's worth. Nevertheless, corporate financial reports provide only limited information on companies' investments in KBC. This may hinder corporate finance and governance. Governments could: 1) support better corporate disclosure by establishing voluntary recommendations and guidelines or backing existing private-sector reporting initiatives; 2) create mechanisms to facilitate companies' reporting of investments in KBC; 3) introduce frameworks for auditors; and 4) engage in global co-ordination, given that there is no global policy-related body addressing this issue.

The OECD's work on KBC also demonstrates there is a need to broaden the notion of framework conditions in an economy increasingly founded on investments in knowledge. For example:

- **Intellectual property rights:** As discussed in greater detail in Chapter 5, because intellectual property is becoming a more important asset for firms in many sectors of the economy, a well-functioning and efficient IPR system is essential to ensure that firms can benefit from their investments in intellectual assets, indicating that this has become a core framework condition.
- **Big data – creating economic value from large data sets:** As examined in Chapter 5, "big data" is part of the leading edge of business innovation and has become a core asset. Research shows that companies that base key decisions on data analytics tend to outperform other firms. While optimal policy is still unclear in this fast-evolving field, it is evident that to unlock major economic benefits, all OECD governments must do more to implement coherent policies in the fields of privacy protection, open data access, ICT infrastructure and ICT-related skills. Policy makers also need to make provision for new regulatory challenges that will come about because of data and machine intelligence.

Policy makers will also need to adopt an enlarged concept of innovation – beyond the conventional view in which R&D is pre-eminent. For example, most OECD governments try to help businesses gain easier access to advice and information on technology and research. They do this through various forms of programme, such as innovation vouchers and technical outreach initiatives. The OECD's work on KBC suggests that an exclusive focus in such schemes on STEM is too narrow. In the United Kingdom, for instance, nearly half of academics from the creative arts and media are engaged with business in some way. This reality, which reflects the changing nature of innovation, also needs to be reflected in government programmes.

A fuller understanding of innovation and growth also requires that governments do more to properly measure investments in KBC and agree on common measurement guidelines. For instance, OECD (2013a) reveals that business investment in organisational capital is perhaps twice as important as previously thought. Policy models reliant on forecasts of growth and productivity change are likely to be inaccurate to the extent that KBC is omitted or mismeasured. By omitting some important forms of KBC, conventional national accounts are mismeasuring the levels and rate of growth of aggregate savings and investment, of GDP and of labour's share of national income, and misallocating the sources of productivity growth. Arguments can be made for expanding the scope of national accounts to better capture additional elements of KBC, in addition to R&D that is just being capitalised in the national accounts (or for creating supplementary innovation accounts). Furthermore, if governments use innovation targets – such as the Lisbon Agenda's guideline for national R&D spending of 3% of GDP – then these should be based on the wider innovation indicators provided by KBC.

4.3. Tax policies

Tax policies are another framework condition that has important impacts on the decisions of firms and households to save or invest and on innovative activity. Important policy features include the level of taxes that are raised, the tax mix, the complexity of tax rules and a host of other factors (OECD, 2010b). The OECD analysis indicates that some forms of taxation, e.g. corporate taxes, are more harmful to growth and innovation than others, e.g. taxes on immovable property, with personal taxes and consumption taxes in between these two, resulting from their respective impacts on the economic decisions of firms and individuals. Growth- and innovation-enhancing policies aim at shifting the tax burden from incomes to consumption and/or residential property. Scope also exists in many countries to make the design of the various categories of tax more conducive to economic growth by levying them on a broader base, possibly at a lower rate, rather than providing targeted relief, except where such reliefs can be justified as externality-correcting.

Other tax policies that are important for the business environment for innovation are top marginal personal income tax rates, which should be set to avoid undue damage to human capital formation and entrepreneurship. Taxes that correct for specific externalities, e.g. environmental taxes, are also important for innovation, notably in supporting a shift in the direction of innovation towards greener growth (see Section 7.4). In general, a growth-oriented tax system may want to create as few obstacles as possible to the growth of economic activities. This implies also that tax systems should not discourage risk-taking or discourage the possible inflow of high-skilled and other foreign workers. Moreover, they should stimulate not only the creation but also the adoption of domestic and foreign created intellectual property. Tax systems can contribute to the creation of an attractive business climate, implying also that the restructuring of business activities for economic purposes should not be discouraged, although governments may want to ensure that they receive their fair share of tax revenues. Growth-oriented tax systems should also contribute to the creation of a favourable e-business and e-commerce environment.

Specific tax policies related to innovation matter too. Tax support to increase the incentives for businesses to undertake R&D is central to many governments' efforts to foster innovation. Most OECD countries offer significant tax relief for business spending on R&D, in recognition of the growth-enhancing spillover benefits of R&D for the general economy. The number of countries providing such relief, and its generosity, are rising. But overall tax relief for R&D by multinational enterprises (MNEs) could well be greater than governments foresaw when R&D tax incentives were initially designed. In particular, rates of tax incentive may not have taken into account tax relief on returns to R&D achieved by MNEs through cross-border tax planning strategies. Accordingly, many governments may be giving unintentionally high levels of total tax support for R&D.

Tax policy may also be encouraging the transfer of KBC to offshore holding companies, and the use of KBC in foreign rather than domestic production. Consequently, losses of tax revenue from R&D may be occurring, along with losses of the benefits from domestic knowledge spillovers. At the same time, relative to MNEs, pure domestic firms – including SMEs but also certain large companies that do not have foreign affiliates and thus do not have cross-border tax planning opportunities – may be placed at a competitive disadvantage in undertaking and exploiting R&D. These findings have significant implications for public finance and domestic employment, the cost-effectiveness and nature of government efforts to encourage business investment in R&D, and the efficiency of OECD economies in accumulating KBC (OECD, 2013a). The OECD's work on base erosion and profit shifting (BEPS) is seeking to address some of these important challenges (OECD, 2013b). Further discussion on R&D tax credits and other tax credits related to innovation is included in Chapter 6.

4.4. Financing of innovation

Access to finance is a key challenge for innovative enterprises. External financing is especially important when innovative firms, particularly young firms, begin to grow, at which point financing requirements become too large to be met by family and friends. Indeed, the financing gap that affects innovative firms is often a "growth capital gap". Traditional debt finance generates moderate returns for lenders and is therefore more appropriate for established businesses with a low-to-moderate risk profile. Furthermore, for innovative ventures with new technologies or untested business models, the problem of asymmetric information between entrepreneurs and financiers is especially severe. This is particularly so for seed and early-stage firms as well as for SMEs, which typically lack a track record and collateral and are often more opaque than large companies. Financing constraints can also be severe for firms reliant on intangible assets, which can be highly firm-specific and difficult to use as collateral in traditional debt relationships (OECD, 2010c, 2014c).

The availability of financing for innovation is influenced by a range of policies. For example, recent OECD work has examined the relationship between framework policies and the extent to which resources (labour and capital) are allocated to firms that have filed patents (an imperfect but useful proxy for innovation). The analysis indicates that cross-country differences in policy settings provide some explanation of the variation in expected returns from patenting (Andrews, Criscuolo and Menon, 2014). Specifically, Figure 4.2 illustrates how estimated resource flows to patenting firms vary with different policy settings in OECD countries. In this figure, the length of the bars indicates the percentage change in employment (Panel A) and capital (Panel B) associated with a 10% increase in the patent stock when the policy variable of interest is set equal to the minimum, mean and maximum values across the sampled countries. For example, capital and labour flow more readily to firms that patent in Sweden, where access to early-stage venture capital (VC) financing is conducive, than in Greece, where this is less so.

The data indicate significant benefits from increased access to seed and early-stage financing, as well as from increased efficiency in the judicial system. In addition, benefits arise from reforming product market regulations (PMRs) that inhibit competition, as well as reducing policy-induced barriers to exit (e.g. excessively strict bankruptcy codes). Reducing such barriers will accentuate competitive pressures, encouraging inefficient firms to exit, and channel resources to firms that are best able to make use of the resources. The latter effect can also be encouraged by less stringent employment protection legislation. While such measures are typically implemented with other policy objectives in mind, their unintended implications for innovation should be taken into account.

Figure 4.2. **Framework policies and resource flows to patenting firms, 2003-10**

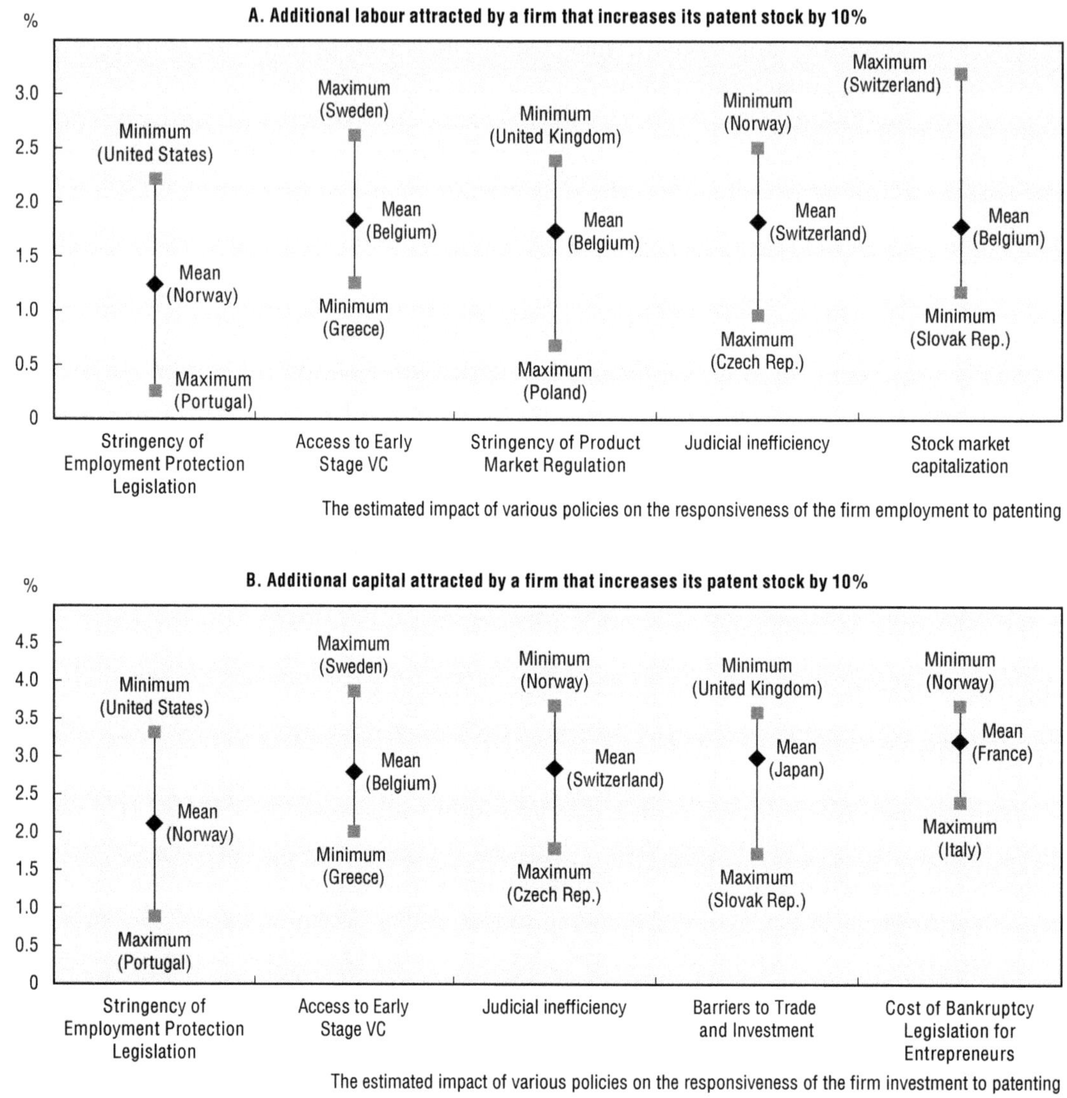

Note: This figure shows that the sensitivity of firm employment and capital to changes in the patent stock varies according to the policy and institutional environment. To calculate policy effects, coefficient estimates are combined with the average values of the policy indicators for each country over the sample period. The labels "minimum" ("maximum") denote the country with the lowest (highest) average value for the given policy indicator over the sample period.

Source: Andrews, D., C. Criscuolo and C. Menon (2014), "Do resources flow to patenting firms? Cross-country evidence from firm level data", *doi: 10.1787/18151973.*

The above examples suggest that framework policies could significantly affect the extent to which patenting firms attract the tangible resources required to implement and commercialise new ideas.[1] In general, the effects are far from trivial. For example, Panel A shows that the sensitivity of firm employment to changes in the stock of patents is more than three times larger when PMR is relatively low (e.g. the United Kingdom), compared with when PMR is very stringent (e.g. in Poland). Many OECD countries have improved their policies in recent years, and the values of the OECD indicator of PMR have decreased in all countries in the sample over the last decade, and in some cases by a considerable margin (e.g. Poland, Greece, Hungary, Slovak Republic).

Framework policies are one important factor that affect the financing of innovation. Another important factor concerns financial markets, including financing for SMEs and the provision of risk capital. In many OECD countries, the global financial crisis exacerbated the financial constraints experienced by SMEs. External financial resources dried up for many innovative enterprises. Following regulatory reforms in financial markets (e.g. Basel III), banks face more rigorous prudential rules, which risks restricting credit for business. The financial crisis also severely affected equity markets. In 2013, in most countries, the level of VC investment was still below the pre-crisis level, and in some cases even below the level reached in 2009 (Figure 4.3). Seed and early-stage capital have been impacted most, with a large number of VC funds shifting to later-stage investments. While assets under management in private equity funds experienced a dramatic surge in the pre-crisis period, the sector has stagnated since 2008, in spite of increasing investor interest in alternative investment instruments. In part, this stagnation reflects decreased exit opportunities for investors. In particular, stock markets declined as a destination for growth companies, as seen in the falling number of initial public offerings (IPOs) across the globe. Furthermore, over the period 2010-13, the number of VC-backed exits through mergers and acquisitions fell continuously.

Figure 4.3. **Venture capital trends, 2007 = 100**

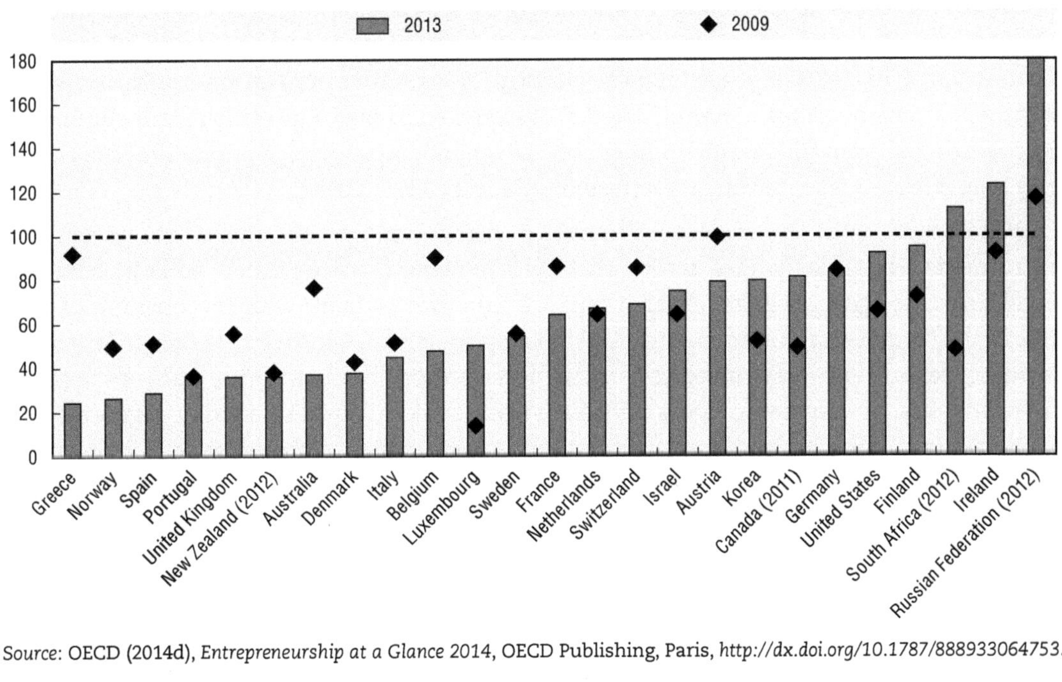

Source: OECD (2014d), *Entrepreneurship at a Glance 2014*, OECD Publishing, Paris, *http://dx.doi.org/10.1787/888933064753*.

The range of financing instruments available to the business sector needs to be broadened. Increasingly complex and interconnected financial markets offer opportunities to service the needs of innovative entrepreneurs and SMEs. Strengthening seed and early-stage equity finance, including VC and angel investment, can boost the creation and development of innovative ventures. And other mechanisms in the capital market, such as public listings for SMEs, can provide financial resources for established growth-oriented firms.

Over the last decade policy makers across OECD countries have intervened to boost the supply side of the equity market, providing *front-end* tax incentives (i.e. tax deductions on investments in seed and early-stage ventures), and *back-end* tax relief (which relate to capital gains and losses). Governments have also increased direct interventions to sustain the supply-side of the venture capital market by creating new government VC funds, and introducing fund-of-funds and public/private co-investment funds. Policies now also target training, mentoring and coaching for investors (Wilson and Silva, 2013).

However, the demand side of equity markets has received less policy attention. Investment readiness programmes help entrepreneurs to better understand the needs and expectations of potential investors and improve the quality and presentation of their business plans.

Other financing techniques exist in the broad risk/return spectrum. These include: asset-based finance, whereby firms obtain funding based on the value of specific assets, including intangible assets, rather than on the firm's overall credit standing; alternative forms of debt, such as corporate bonds, which can provide mid-sized to large firms with liquidity to undertake innovative investments; and hybrid instruments, which combine debt and equity features, and may serve both young and established companies that seek expansion capital but are not suitable for public listing or do not want the dilution of control that would accompany equity (OECD, 2014c). Since the late 2000s, crowdfunding, by which external finance is raised through web platforms, has been growing rapidly, although it still represents a very minor share of all business financing (and serves to finance specific projects rather than an enterprise). However, as crowdfunding becomes more regulated, it is expected to play a growing role, including for the financing of innovative ventures, as the online interaction with large numbers of customers may help entrepreneurs to validate untested products (OECD, 2014e).

Challenges exist to increase innovative firms' uptake of alternative investment instruments. These challenges can be met by improving finance-related skills in new and small firms, designing regulations that balance financial stability with the opening of new financing channels for entrepreneurs, and developing infrastructures to reduce information asymmetries and encourage investor participation. Furthermore, in the aftermath of the global financial crisis, as private investors withdrew from some higher-risk market segments, the public share of funding significantly increased. A key challenge therefore is to implement policies that lever private resources and develop mechanisms for sharing risk with the private sector. In recent years, programmes have developed in many OECD countries to favour co-investment, particularly in seed and early-stage ventures (see OECD, 2011a).

4.5. Fostering entrepreneurship and experimentation

In the past, a great deal of attention has been devoted to policy measures that seek to push out the technology and innovation frontier. The expectation is that the benefits of innovation will accrue not only to firms at the frontier, but will also spill over to other firms, increasing productivity in the economy more generally. However, it is becoming increasingly clear that the distribution of firm productivity is characterised by a large number of "average" firms, with just a minority of firms exhibiting high levels of productivity. For example, among four-digit manufacturing industries in the United States, firms at the 90th percentile of the total factor productivity (TFP) distribution are, *on average*, twice as productive as firms at the 10th percentile (Syverson, 2004). In China and India, the differential is fivefold (Hsieh and Klenow, 2010). Moreover, there is evidence that this productivity gap is increasing.

Owing to such productivity differentials, increased attention is being paid to the allocation of resources across firms in any given economy, and the role that policy can play in encouraging efficiency-enhancing reallocation. In particular, policy settings that affect firm entry and exit, and the growth of young firms, will affect the extent to which resources are allocated to more productive firms. Recent OECD work shows the extent to which, across countries, resources (in this case labour) flow to more productive firms (Andrews and Cingano, 2014). For example, in Sweden and the United States, the more productive manufacturing firms account for a higher share of employment than would be the case if labour were allocated randomly across firms. The efficiency of resource allocation is generally lower in market services than in manufacturing, reflecting less exposure to international competition and the fact that pro-competition product market reforms have generally been more extensive in the manufacturing sector than the services sector. Policy settings must provide incentives to ensure that dynamic reallocation occurs on a continuous basis. New and young firms are often the vehicles through which innovations enter the market. This occurs through the introduction of incremental innovations, disruptive technologies, new business models or other forms of KBC (e.g. new marketing strategies).

While patenting is an imperfect measure of innovation, a comparison of firms' patent filings and firm age can be used to document the importance of new and young firms for innovation (Squicciarini and Dernis, 2013). They find that most first patenting happens between the birth of a firm and its tenth year of existence. A notable proportion of firms also apply for patents *before* the firm is established. This may occur when start-ups are created to exploit intellectual property developed by founders, or when mergers and acquisitions involve firms with patents that pre-date the creation of the merging or acquiring firm. Moreover, in addition to their relatively higher patent counts, young firms also tend to have a stock of patents that reflect more radical inventions (Andrews, Criscuolo and Menon, 2014).

Given the importance of new and young firms as vehicles through which innovations enter the market, the role of entry is clearly significant. Furthermore, policy settings can play an important role in determining rates of entry. For example, PMRs can serve as barriers to entry, restricting competition in the market. In addition, capital market failures may particularly affect entrants and young firms, and affect productivity through a less efficient selection of firms at entry (Andrews and Cingano, 2014).

New micro-aggregated data collected by the OECD from 18 countries indicate that, across all countries in the sample, the share of start-ups has steadily decreased over the past decade (Figure 4.4). Indeed, evidence from the United States and other countries indicates that this trend has been in place for two decades or more. Furthermore, the trend has been exacerbated by the recent crisis (Criscuolo, Gal and Menon, 2014).

Subsidies to incumbents and other policy measures that delay the exit of less-productive firms might stifle competition and slow the reallocation of resources from less to more productive firms. Examples include regulations that are less stringent for incumbents (e.g. fiscal measures that favour well-established firms – such as R&D tax credits that do not have carry-forward provisions). Perhaps most importantly, bankruptcy legislation that excessively penalises failure is likely to reduce incentives for the efficient exit of less-productive firms, which would otherwise free up resources for more productive uses.[2] There are also potential complementarities among such policies; the efficiency of R&D tax credits might be enhanced by policies that facilitate the exit of poorly performing firms.

Figure 4.4. **Start-up rates have declined across countries**

Percentage of start-ups in all businesses

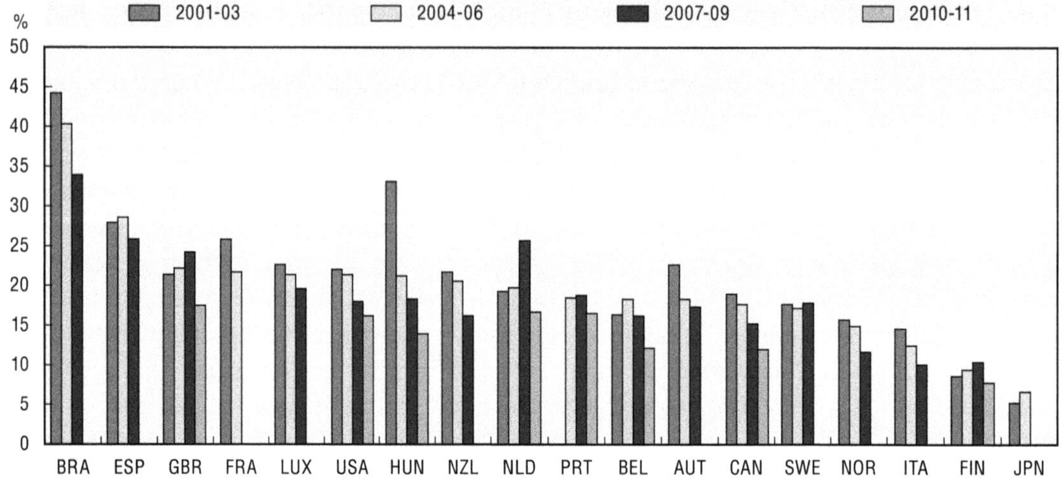

Note: The figure reports start-up rates (defined as the fraction of start-ups among all firms) by country, averaged across the indicated three-year periods. Start-up firms are firms aged between 0 and 2 years.

Source: Criscuolo, Gal and Menon (2014), "The dynamics of employment growth: New evidence from 18 countries", *http://dx.doi.org/10.1787/5jz417hj6hg6-en.*

Figure 4.5. **Average size of start-up and old firms across industries and across countries**

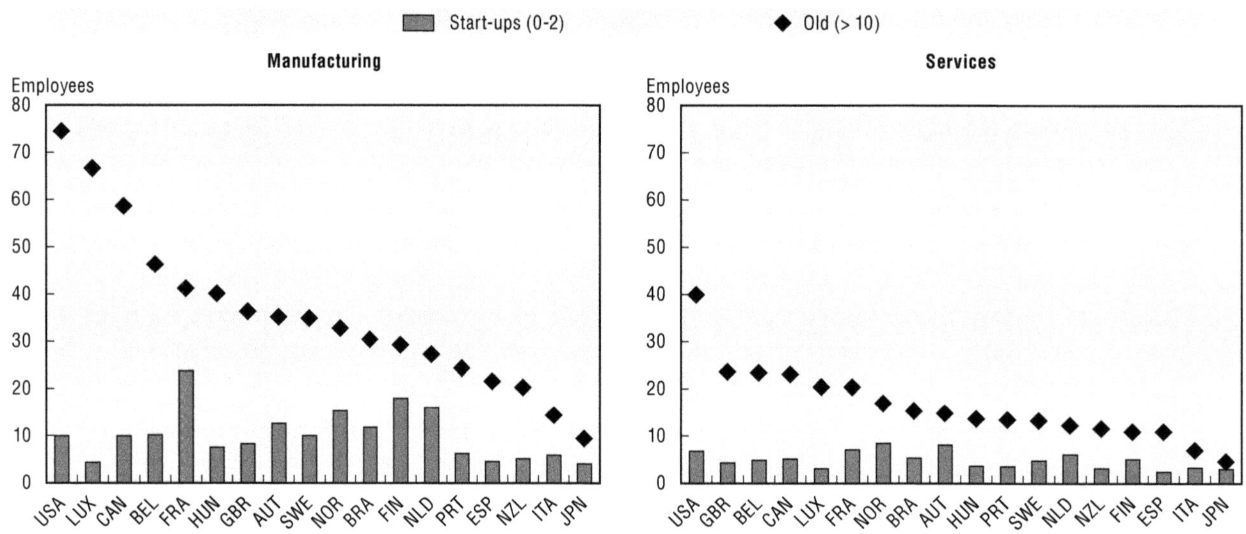

Note: The figure reports the average size of start-up firms (from 0 to 2 years old) and firms more than 10 years old, over the available years. The period covered is 2001-11 for Belgium, Canada, Finland, Hungary, the Netherlands, the United Kingdom and the United States; 2001-10 for Austria, Brazil, Spain, Italy, Luxembourg, Norway and Sweden; 2001-09 for Japan and New Zealand; 2001-07 for France; and 2006-11 for Portugal. The sectors covered are manufacturing and non-financial business services. Owing to methodological differences, figures may deviate from officially published national statistics. For Japan, the data are at the establishment level. For other countries, data are at the firm level. Data for Canada refer only to organic employment changes and abstract from merger and acquisition activity.

Source: Criscuolo, Gal and Menon (2014), "The dynamics of employment growth: New evidence from 18 countries", *http://dx.doi.org/10.1787/5jz417hj6hg6-en.*

While entry and exit are clearly important, post-entry dynamics are even more critical. Figure 4.5 shows differences across countries in the extent to which young firms grow. While there are some differences across countries in the size of start-ups at entry, these are not particularly striking. However, the situation is markedly different when considering older businesses. For instance, on average, an older manufacturing business in France is half the size of one in the United States, even though start-ups in France are larger than in the United States. In some countries, such as Italy, there is only a small difference between the size of start-ups and that of mature firms. In recent years, Italy has undertaken a range of reforms to improve the conditions for start-ups and help them grow (Box 4.1).

Box 4.1. **Recent policies in Italy to encourage start-ups**

In late 2012, Italy embarked upon reforms aimed at developing a fertile start-up ecosystem. The Italian Startup Act represents a package of tools affecting all stages of business life cycle, aimed at creating the enabling conditions needed for a quick go-to-market and scaling up. Innovative start-ups can profit from a vast array of benefits for five years, including:

- exemption from fees normally due to the Chamber of Commerce
- opportunity to remunerate workers and consultants through stock options and work for equity schemes that are tax-deductible
- opportunity to raise capital in exchange for shares through equity crowdfunding portals
- robust tax incentives by up to 27% on seed and early-stage investment amounting up to EUR 1.8 million (euros)
- streamlined, free-of-charge access to public guarantees by 80% on bank loans amounting up to EUR 2.5 million.

Recently, Italy has also launched the Italia Startup Visa programme, which enables citizens from outside the EU intending to establish a high-tech company in Italy to obtain an entrepreneurship visa within 30 days, following an online and streamlined procedure.

The Italian Startup Act is an ongoing process that also draws on the analysis and evaluation of its empirical impact through a structured monitoring system involving the National Statistics Institute. In recent years, the Italian innovation ecosystem has grown rapidly: to date it counts more than 3 600 highly innovative tech start-ups (with an average weekly increase of 40 start-ups) and involves more than 15 000 partners and employees (with 2 000 added in the last quarter of 2014).

Source: Information supplied by Government of Italy.

In a dynamic economy, the disproportionate contribution of young firms to job creation is a reflection of the "up or out" dynamics typical of this group of firms: they either go "up", resulting in higher-than-average rates of post-entry growth, or they go "out" (exit the market). The co-existence of high rates of success ("up") and failure ("out") in economies can be seen in emerging OECD work, which finds that countries where fast-growing firms expanded the most (in terms of employment) are also the countries that experienced the greatest shrinking at the bottom of the employment growth distribution.

Policies that (unwittingly) constrain the growth of firms should be assessed with particular care. Examples include both "sticks" (i.e. regulations that affect only firms above a certain size) and "carrots" (i.e. support mechanisms for which only smaller firms are eligible).

4.6. Global value chains

Openness to international flows of capital, goods, people and knowledge has always been essential for innovation. Globalisation increases the size of markets available to innovators and consumers. Conversely, the involvement of firms in globalisation often requires sufficient scale to overcome the fixed costs of entry in foreign markets. Globalisation also facilitates specialisation, increases competition, and facilitates the spread of knowledge, technologies and new business practices. These dynamics positively affect innovation and long-term economic productivity.

A range of recent OECD work (e.g. OECD, 2013c, 2014f) has called attention to the spread of value chains, i.e. the full range of activities that firms engage in to bring a product to the market, including design, production, marketing, logistics and distribution. Such value chains have become increasingly global, leading to a growing interconnectedness of the world economy, but also to a growing specialisation of firms and countries in specific tasks and business functions. Today, most goods and a rising share of services might be described colloquially as "made in the world".

The rise of GVCs underscores the importance of openness. Since imports are essential for exports, especially in complex value chains such as transport and electronics, tariffs and non-tariff barriers are effectively a tax on exports. Export restrictions, including in raw materials (OECD, 2014f), can also affect the efficient functioning of GVCs and raise costs. The negative effects of trade protection are compounded in GVCs when parts and components cross borders many times. By providing access to cheaper, more differentiated and better quality inputs, outsourcing and offshoring can also enhance export competitiveness in GVCs.

Strengthening the international competitiveness of firms in GVCs also requires strengthening factors of production that are "sticky" and unlikely to cross national borders. This implies investment in education, skills and high-quality infrastructure and the encouragement of strong industry-university linkages. The quality of institutions and government – and the overall business environment - are also important long-term factors in firms' decisions to engage in economic activity in a given country. To strengthen the benefits that countries obtain from participating in GVCs, governments need to support the upgrading process by enabling investment in knowledge assets such as R&D and design.

Trade-facilitating measures, such as fast and efficient port and customs procedures, permit the smooth operation of value chains that require goods to cross borders many times. The convergence of standards and certification requirements, and mutual recognition agreements, can also help alleviate burdens on exporting firms.

OECD analysis shows that services, e.g. business services, transport and logistics, account for over half of value creation in GVCs in many OECD countries, and over 30% in China (Figure 4.6). Global production networks crucially rely on well-functioning logistics, finance, insurance, communication and other business services. These services are necessary for the efficient transfer of goods, data, technology and know-how across borders, and the co-ordination of geographically dispersed activities.

Regulatory reforms and liberalisation of services trade are essential to enhance competition and increase the productivity and quality of services. Recent work on the OECD's Services Trade Restrictiveness Index (STRI) shows that domestic and international competition in some service sectors is severely hampered by regulatory frictions and state intervention. While tariffs are generally smaller or non-existent for services trade, the scope of potential barriers to trade in services is broader than trade in goods. Services are traded in a variety of modes, output is much more heterogeneous, and regulation (covering both domestic and cross-country transactions) is more complex. Some of the horizontal policies that restrict services trade include limitations on market entry (such as equity limits, licensing requirements or economic needs tests), restrictions to the movement of people, heterogeneous services standards, and competition regulation. Barriers may be discriminatory against foreign suppliers, but even when that is not the case, trade may be restricted through domestic regulation that prevents entry and competition. Much of this regulation is sector-specific. For instance, a wide range of rules in telecommunication services pertain to access and interconnection, number portability, local loop unbundling, and infrastructure sharing. These rules directly influence the decision of foreign providers to enter a given market.

Figure 4.6. **Services value added in manufacturing exports, OECD and BRIICS countries, 2009**

(as a percentage of total gross manufacturing exports)

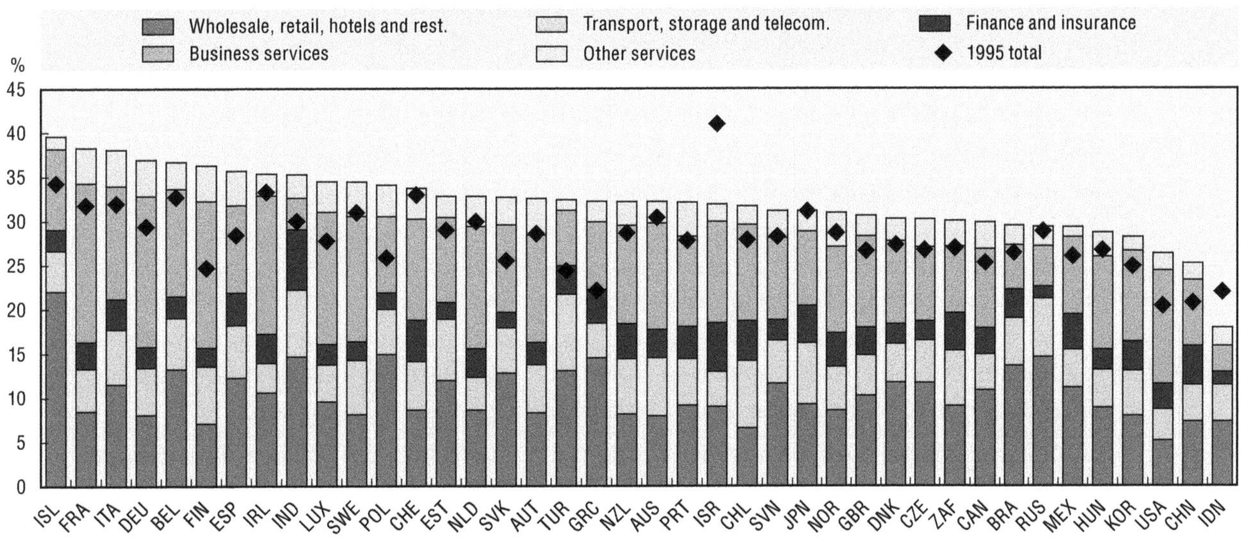

Note: BRIICS countries = Brazil, Russian Federation, India, Indonesia, China and South Africa.

Source: OECD/WTO (2013), *Trade in Value Added (TiVA)* (database), *http://stats.oecd.org/*.

Large MNEs control and co-ordinate activities in GVCs. Policy affects how international networks of buyers and suppliers are formed and where their activities are located. Given the important role of MNEs, lowering investment barriers is an efficient way for a country to become integrated in GVCs.

Finally, since GVCs involve activities contracted within and between MNEs and independent suppliers, the ability to enforce contracts is crucial. Countries with sound legal systems tend to export more in more complex industries. Tasks that require more complex contracts (e.g. R&D, design, branding) are also more easily carried out in countries with well-functioning contractual institutions.

4.7. Investment and innovation

A final key policy area that is an important framework condition for innovation concerns investment. As noted already in the section on KBC, firms in many OECD countries now invest as much in KBC as they invest in tangible forms of capital such as machinery, equipment and buildings. The framework conditions for investment therefore have an important bearing on innovation.

The OECD's Policy Framework for Investment (PFI) provides a broad framework for investment policy based on a set of questions to policy makers (OECD, 2006). Three principles apply throughout the framework. The first is **policy coherence**, based on an integrated approach to the interaction between various policy areas and the investment environment. For example, standards for investment protection and openness are of wide applicability to international as well as domestic investors including SMEs; effective competition and tax policies are important to ensure that investment, in particular in small businesses, is not deterred by unnecessary barriers to entry, dissuasive taxation, and poor legal compliance; and open trade policies contribute to realising the benefits of an open investment policy. Chapter 8 discusses the conditions for a sound regulatory framework within and across policy domains.

The second principle is the importance of a **transparent approach** to policy formulation and implementation, with government agencies accountable for their activities. Transparency reduces uncertainty and risk for investors and the transaction costs associated with making an investment, and facilitates public-private dialogue. Accountability reassures investors that government agencies are exercising their powers responsibly. How transparency and accountability in specific public policy domains foster an environment where investment flourishes is a theme taken up in detail in the PFI.

The third principle that applies across the framework is **regular evaluation** of the impact of existing and proposed policies on the investment environment. In this regard, the questions in the PFI seek to help evaluate how well government policies uphold established good practices in terms of fair treatment for all investors (foreign or domestic based, small or large in size) and opening opportunities to invest, taking into account the wider interests of the community in which investors operate. The questions attach a particular emphasis to the adaptability of the institutional framework and the role of periodic evaluations so as to identify early on new challenges and to be able to respond quickly to them.

While innovation is not explicitly included in the framework, many of the issues included have a bearing on innovation. IPRs protection is a key aspect of innovation on which the framework provides guidance for policy makers. IPRs give businesses an incentive to invest in R&D, fostering the creation of innovative products and processes (see also Chapter 5). They also give their holders the confidence to share new technologies, e.g. through joint ventures and licensing agreements. In this way, successful innovations are in time diffused within and across economies, contributing to higher productivity and growth.

The protection granted to IPRs needs to strike a balance between incentivising innovation and ensuring competitive markets, with new products priced affordably. The whole-of-government approach and policy coherence promoted by the framework can provide guidance on finding and maintaining this balance. The PFI is currently being updated to reflect new insights and policy needs as regards the enabling environment for investment.

4.8. Main policy learnings on the framework conditions for innovation

The main policy learnings from recent OECD work on the framework conditions for innovation are included below. One other element must be noted. The OECD provides many indicators that enable countries to compare and evaluate their performance on a wide set of framework conditions, including labour markets, product markets and financial markets, as well as in international trade and investment. Many of these indicators are published in the OECD's *Going for Growth* and other flagship reports. New indicators have emerged recently or are currently under development, including: indicators on specific aspects of KBC, including intellectual property; indicators on firm dynamics; and indicators on GVC, such as the *Trade in Value Added (TiVA)* database and the STRI. Further work on measurement and indicators can help to better compare and benchmark performance, and can also provide the basis for economic analysis and the impact assessment of specific policies.

Main policy messages related to the framework conditions for innovation

- **Knowledge-based capital:** As overall business investment in KBC increases – and because of KBC's particular economic features, especially its intangible nature – certain key policy settings need to be updated. Ensuring that policies conform to good practice is essential in the fields of taxation, entrepreneurship, competition, corporate reporting, intellectual property and policies that enable the exploitation of data as an economic asset.
- **Financing of innovation:** Access to external financing is especially important when innovative firms, particularly young firms, begin to grow. The range of financing instruments available to the business sector needs to be broadened. Strengthening seed and early-stage equity finance, including VC and angel investment, can boost the creation and development of innovative ventures. And other mechanisms in the capital market, such as public listings for SMEs, can provide financial resources for established growth-oriented firms. Along with efforts to boost the supply side of the equity market, demand-side initiatives – improving investment readiness – should also receive policy attention. Challenges exist to increase innovative firms' uptake of alternative investment instruments. These challenges can be met by improving finance-related skills in new and small firms, designing regulations that balance financial stability with the opening of new financing channels for entrepreneurs, and developing infrastructures to reduce information asymmetries and encourage investor participation. A further challenge is to implement policies that lever private resources and help to share risk with the private sector (such as through co-investment schemes for financing seed and early-stage ventures).
- **Global value chains:** An extremely broad array of policies and institutional conditions shape a country's role in any given GVC. Competitiveness in GVCs requires strengthening factors of production that are unlikely to cross national borders. This implies investment in education, skills and high-quality infrastructure and the encouragement of strong industry-university linkages. The negative effects of trade protection are compounded in GVCs, and trade-facilitating measures should be implemented where necessary. Convergence of standards and certification requirements, and mutual recognition agreements, can help alleviate burdens on exporting firms. Regulatory conditions that hinder domestic and international competition in services should be reviewed. Horizontal policies that restrict services trade should be reformed, as should unnecessary impediments to cross-border investment. The ability to enforce contracts is crucial. Consequently, sound legal systems and well-functioning contractual institutions must be developed and/or maintained.
- **Entrepreneurship and experimentation:** A number of policy recommendations can be drawn from this work:
  - *Policy must provide incentives to ensure that dynamic reallocation occurs on a continuous basis.* Policy settings that affect firm entry and exit, and the growth of young firms, will affect the extent to which resources are allocated to more productive firms.
  - *Policies that (unwittingly) constrain the growth of firms should be assessed with care.* Examples include both "sticks" (i.e. regulations that affect only firms above a certain size) and "carrots" (i.e. support mechanisms for which only smaller firms are eligible). Conversely, subsidies to incumbents and other policies that delay the exit of less-productive firms might stifle competition and hinder efficient resource allocation.
  - *Significant benefits can arise from increased access to seed and early-stage financing, as well as from increased efficiency in the judicial system.* Benefits also arise from reforming PMRs that inhibit competition, as well as reducing policy-induced barriers to exit (e.g. excessively strict bankruptcy codes).
  - *In framing employment protection legislation, the implications for innovation should be taken into account.*
- **Investment:** Policy coherence, transparency and regular evaluation are important characteristics of investment policies that help provide a sound framework for business investment in innovation and KBC.

Notes

1. It is also important to note that firms' initial investments in KBC will likely be shaped by their perceptions of the expected costs of implementing and commercialising new ideas and the ability to capitalise on the expected benefits or to exit at low cost (which will both depend on the ease of reallocation). In particular, firms' innovation strategies will be influenced by their perceptions regarding the extent of rigidities in the reallocation process. If the costs of reallocation are deemed to be high, entrepreneurs may focus on incremental innovations, rather than experiment with disruptive technologies, because it will be more difficult to realise the benefits of risky technologies when successful and contain losses when unsuccessful (Bartelsman, 2004; Andrews and Criscuolo, 2013).
2. See discussion and indicators of bankruptcy legislation included in the OECD's Innovation Policy Platform (*https://innovationpolicyplatform.org/content/bankruptcy-regulation*).

References

Andrews, D. and C. Criscuolo (2013), "Knowledge-based capital, innovation and resource allocation", *OECD Economics Department Working Papers*, No. 1046, OECD Publishing, Paris.

Andrews, D. and F. Cingano (2014), "Public policy and resource allocation: Evidence from firms in OECD countries", *Economic Policy*, No. 29(78), pp. 253-296.

Andrews, D., C. Criscuolo and C. Menon (2014), "Do resources flow to patenting firms? Cross-country evidence from firm level data", *OECD Economics Department Working Papers*, No. 1127, OECD Publishing, Paris.

Bartelsman, E.J. (2004), "Firm dynamics and innovation in the Netherlands: A comment on Baumol", *De Economist*, Vol. 152/3, pp. 353-363.

Bravo-Biosca, A., C. Criscuolo and C. Menon (2013), "What drives the dynamics of business growth?", *OECD Science, Technology and Industry Policy Papers*, No. 1, OECD Publishing, Paris, *http://dx.doi.org/10.1787/5k486qtttq46-en*.

Criscuolo, C., P.N. Gal and C. Menon (2014), "The dynamics of employment growth: New evidence from 18 countries", *OECD Science, Technology and Industry Policy Papers*, No. 14, OECD Publishing, Paris, *http://dx.doi.org/10.1787/5jz417hj6hg6-en*.

Hsieh, C.-T. and P.J. Klenow (2010), "Development accounting", *American Economic Journal: Macroeconomics*, Vol. 2/1, pp. 207-223.

OECD (2015a), *Main Science and Technology Indicators 2014-II*, OECD Publishing, Paris, *http://dx.doi.org/10.1787/msti-v2014-2-en*.

OECD (2014a), *Economic Policy Reforms: Going for Growth*, OECD Publishing, Paris, *http://dx.doi.org/10.1787/growth-2014-en*.

OECD (2014b), *OECD Science, Technology and Industry Outlook 2014*, OECD Publishing, Paris, *http://dx.doi.org/10.1787/sti_outlook-2014-en*.

OECD (2014c), *New Approaches to SME and Entrepreneurship Financing: Broadening the Range of Instruments*, OECD Publishing, Paris, *www.oecd.org/cfe/smes/New-Approaches-SME-full-report.pdf*.

OECD (2014d), *Entrepreneurship at a Glance 2014*, OECD Publishing, Paris, *http://dx.doi.org/10.1787/888933064753*.

OECD (2014e), *New Approaches to SME and Entrepreneurship Financing: Broadening the Range of Instruments. The Case of Crowdfunding*, OECD Publishing, Paris, *www.oecd.org/cfe/smes/New-Approaches-SME-full-report.pdf*.

OECD (2014f), *Global Value Chains: Challenges, Opportunities and Implications for Policy*, OECD, WTO and World Bank, report prepared for the G20 Trade Ministers Meeting, Sydney, Australia July, *www.oecd.org/tad/gvc_report_g20_july_2014.pdf*.

OECD (2014g), *Export Restrictions in Raw Materials Trade: Facts, Fallacies and Better Practices*, OECD Publishing, Paris, *www.oecd.org/trade/benefitlib/export-restrictions-raw-materials-2014.pdf*.

OECD (2013a), *Supporting Investment in Knowledge Capital, Growth and Innovation*, OECD Publishing, Paris, *http://dx.doi.org/10.1787/9789264193307-en*.

OECD (2013b), *Action Plan on Base Erosion and Profit Shifting*, OECD Publishing, Paris, *http://dx.doi.org/10.1787/9789264202719-en*.

OECD (2013c), *Interconnected Economies: Benefiting from Global Value Chains*, OECD Publishing, Paris, *http://dx.doi.org/10.1787/9789264189560-en*.

OECD (2011a), *Financing High-Growth Firms: The Role of Angel Investors*, OECD Publishing, Paris, *http://dx.doi.org/10.1787/9789264118782-en*.

OECD (2010a), *The OECD Innovation Strategy: Getting a Head Start on Tomorrow*, OECD Publishing, Paris, *http://dx.doi.org/10.1787/9789264083479-en*.

OECD (2010b), *Tax Policy Reform and Economic Growth*, OECD Publishing, Paris, *http://dx.doi.org/10.1787/9789264091085-en*.

OECD (2010c), *High Growth Enterprises: What Governments Can Do to Make a Difference*, OECD Studies on SMEs and Entrepreneurship, OECD Publishing, Paris, *http://dx.doi.org/10.1787/9789264048782-en*.

OECD (2006), *Policy Framework for Investment*, OECD Publishing, Paris, *http://dx.doi.org/10.1787/9789264018471-en*.

Squicciarini, M. and H. Dernis (2013) "A cross-country characterisation of the patenting behaviour of firms based on matched firm and patent data", *OECD Science, Technology and Industry Working Papers*, 2013/05, OECD Publishing, Paris, *http://dx.doi.org/10.1787/5k40gxd4vh41-en*.

Syverson, C. (2004), "Product substitutability and productivity dispersion", *Review of Economics and Statistics*, Vol. 86/2, pp. 534-550.

Wilson, K. and F. Silva (2013), "Policies for seed and early stage finance: Summary of the 2012 OECD Financing Questionnaire", *OECD Science, Technology and Industry Policy Papers*, No. 9, OECD Publishing, Paris, *http://dx.doi.org/10.1787/5k3xqsf00j33-en*.

The Innovation Imperative
Contributing to Productivity, Growth and Well-Being

Chapter 5

Knowledge creation, diffusion and commercialisation

Policies for innovation also *require a strong and efficient system for knowledge creation and diffusion that engages in the systematic pursuit of fundamental knowledge, and that diffuses this knowledge throughout society through a range of mechanisms. This chapter reviews policies on: the science system, including the promotion of research excellence and the role of open science in increasing the economic and social returns to public investments in scientific research, as well as the role of international co-operation in science and technology; emerging practices in commercialising publicly funded research trends; policy issues relating to the interconnected themes of ICT, "big data" and the open Internet; the evolving relationship between IPRs and innovation; and the development and functioning of knowledge networks and markets.*

The statistical data for Israel are supplied by and under the responsibility of the relevant Israeli authorities. The use of such data by the OECD is without prejudice to the status of the Golan Heights, East Jerusalem and Israeli settlements in the West Bank under the terms of international law.

5.1. Science and public research

As described in the 2010 Innovation Strategy (OECD, 2010a), while the relationship between science and innovation is complex, public investment in scientific research is widely recognised as an essential feature of effective national innovation systems. Public research plays a key role in innovation systems by providing new knowledge and pushing the knowledge frontier. Universities and public research institutions (PRIs) often undertake longer-term, higher-risk research and complement the research activities of the private sector. Although the volume of public R&D is less than 30% of total OECD R&D (OECD, 2014a), universities and PRIs perform more than three-quarters of total basic research (Figure 5.1).

Figure 5.1. **Basic research performed by the public sector, 2012 or latest available year**

As a percentage of total basic research

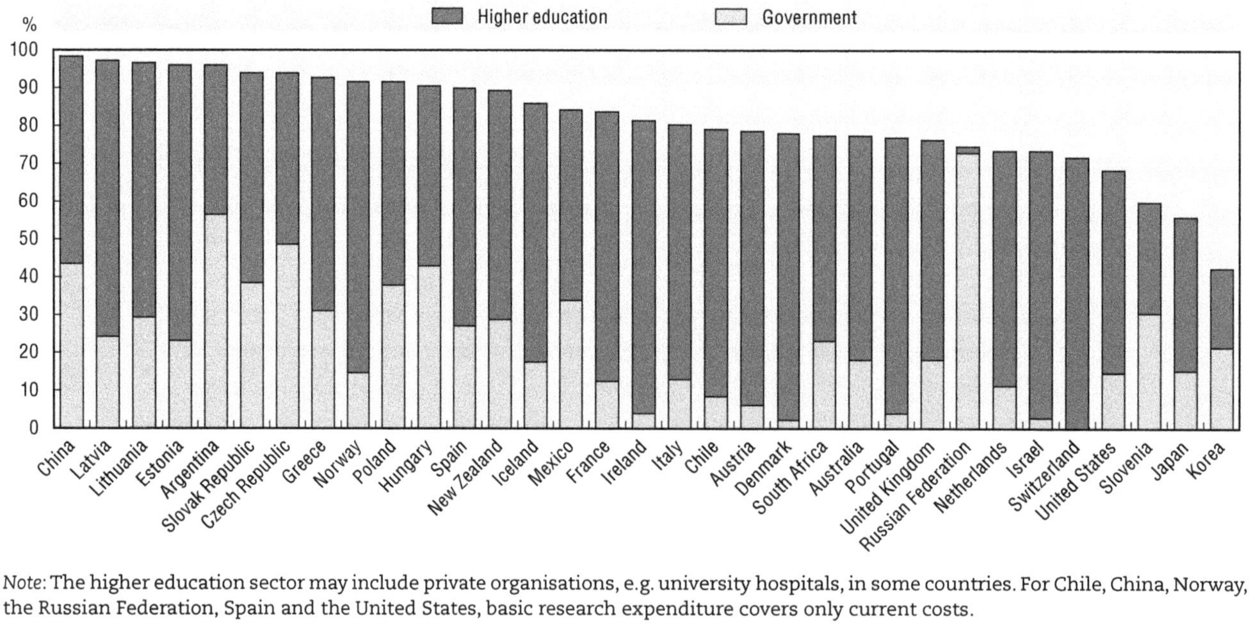

Note: The higher education sector may include private organisations, e.g. university hospitals, in some countries. For Chile, China, Norway, the Russian Federation, Spain and the United States, basic research expenditure covers only current costs.

Source: OECD (2014a), *Science, Technology and Industry Outlook 2014*, OECD Publishing, Paris, *http://dx.doi.org/10.1787/sti_outlook-2014-en*.

Basic research is particularly important, as it gives rise to significantly larger knowledge spillovers than applied research while making applied research much more productive (Akcigit, Hanley and Serrano-Velarde, 2014).[1] The history of science shows that many of the great breakthroughs resulting from scientific research were regarded as significant only in hindsight (Kirschner, 2013). They were not the result of a focused effort to achieve a specific impact, but instead reflected serendipity. Ensuring a balance between basic research, driven by excellence, and more focused, mission-oriented research is therefore an important challenge for public funding.

The immediate economic returns from investment in academic research have not always been easy to demonstrate,[2] although there is much evidence that many of the most important innovations over the past decade have their roots in public research, including the Internet and genomic technologies. In many OECD countries, there has been pressure on public funding for research over the past five years as a result of the economic crisis. Nevertheless, relative to GDP, public spending of R&D has held up quite well in most OECD countries since the crisis (Figure 5.2)

Figure 5.2. **Public R&D expenditure by type of research system**

HERD and GOVERD, as a percentage of GDP, 2012, and total HERD and GOVERD in 2007

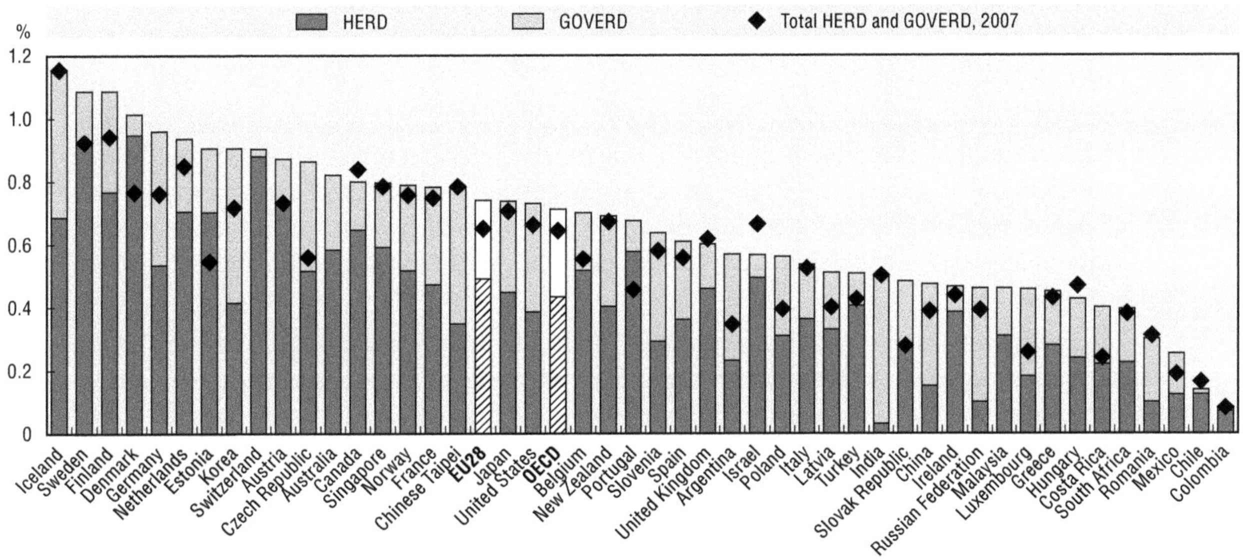

Note: HERD: higher education R&D; GOVERD = government intramural R&D.

Source: OECD (2014a), *Science, Technology and Industry Outlook 2014*, *http://dx.doi.org/10.1787/sti_outlook-2014-en* based on OECD (2014b), *Main Science and Technology Indicators 2014-I*, *http://dx.doi.org/10.1787/msti-v2014-2-en*.

Three key science-related trends and areas of policy interest were identified in the Innovation Strategy. These related to: 1) institutional financing mechanisms and promotion of multidisciplinary research; 2) the quality and relevance of research and research assessment; and 3) commercialisation, creation of spin-offs and support for centres of excellence. In addition, the shift towards open science was noted, and in this connection, access to research information and data resulting from public funding were highlighted as areas in which policy intervention would be useful.

The 2014 publication *Promoting Research Excellence: New Approaches to Funding* (OECD, 2014c) addressed all three of the above-cited areas of science policy interest. This work focuses on research excellence initiatives (REIs) and draws on the results of surveys completed by government research funding agencies, centres of excellence and their host institutions. The principal findings were that:

- National research systems face an increasingly competitive environment for ideas, talent and funds. The emergence of world rankings of university performance reflects such competition.[3] Governments have thus turned to more competitive forms of funding to promote efficiency and innovation. Among other steps, governments have shifted funds from institutional core funding to project funding, often on a competitive basis.

But research also requires a degree of funding stability, which is made difficult by an exclusive reliance on competition. REIs have emerged against this background. Today, over two-thirds of OECD countries operate REIs.

- REIs are designed to encourage outstanding research by providing large-scale, long-term funding to designated research units. They fund research, physical infrastructure, training, co-operation between research and industry, and the recruitment of outstanding researchers. The single most important goal is to increase national research and innovation capacities. Some countries operate a single REI while others have several. The average funding cycle is around six years. Such funding stability is especially important for new fields of research that might lead to significant scientific developments, but which are otherwise risky and difficult to develop through short-term project funding.
- Most REIs share the following traits: government funding of selected research units and institutions; exceptional quality in research and research-related activities; long-term funding (a minimum of four years); competitive funding distributed on the basis of peer-reviewed applications; funding applications made by institutions or research units (instead of individuals); and substantially larger funding than for project-based activities.

A first observation on the impact of REIs is that rigorous evaluations are lacking both of REI outputs and of how REIs affect broader dimensions of welfare. The available evidence mainly takes the form of expert opinion. However, multiple such accounts suggest that REIs have often been able to:

- promote excellence in research by providing researchers with better opportunities to work across disciplines than exist in many other research contexts
- bring exceptional researchers together in well-equipped working environments to open new lines of research, establish new interdisciplinary research and develop human capital
- raise the international reputation of domestic research institutions
- form long-term international linkages, in part by recruiting leading foreign researchers
- positively affect institutions not selected for funding by triggering intensified co-operation between traditional disciplinary departments and new interdisciplinary research initiatives.

A number of policy-relevant questions remain. For instance, what is the optimal balance among institutional, project and REI funding? The answer to this question may be case-specific and impossible to give unequivocally with existing data, but comparative qualitative research could provide important insights. Whether an REI is best used as a temporary tool to strengthen the research system or whether it should be institutionalised as part of the funding portfolio is not clear. If used as a temporary tool, the question arises of how to maintain excellence once REI funding ceases. But if REI funding is institutionalised, it is unsure whether constant competition for excellence status will improve system performance over the long term.

Looking to the future, many OECD countries are increasing their investment in research aimed at addressing global challenges. This raises issues of governance, including at the international level (see Chapter 8), but also highlights the need for new interdisciplinary and trans-disciplinary research environments that bring together diverse natural and social scientists. Such collaborations can be difficult to establish in traditional university settings, with their embedded disciplinary structures. It would be timely to analyse existing initiatives to assess which arrangements are most effective in bridging disciplinary barriers.

In order to evaluate the quality and relevance of publicly funded research, various forms of research assessment have been introduced in most OECD countries. In this regard, OECD (2010b) aimed to take stock of thinking and practice around performance-based funding for public research in tertiary education institutes (TEIs). Such funding involves *ex post* evaluation of research outputs and outcomes from universities and other tertiary institutions, and is generally based on peer review, bibliometric or other quantitative indicators. The results inform government decisions on which institutions to fund and how much funding they should receive.

In most countries some form of performance-based research funding (PBRF) has been introduced over the past two decades. Funding rounds are either annual or multi-year. The overall range of indicators of research outputs and outcomes used is similar across countries, although combinations and weightings differ.

In general, peer review is used for individual and departmental evaluations, while quantitative formulas (with or without additional peer review) are used for university-level evaluations. The direct and indirect costs of assessment can be large, but this is rarely discussed in the literature. While the amounts of money subject to PBRF may be small, the incentive effects could be strong, particularly if the results affect institutional prestige or access to other research funding.

The science community is still trying to define the meaning of "quality" in research outputs as well as the relationship between outputs and their impacts. All of the routinely used measures – from citation indices to patent numbers – are proxies for performance. In addition, the integration of innovation and research policy has given rise to new indicators of knowledge transfer and commercialisation. The rapid move towards open science is also likely to require new indicators for assessment, such as citations for databases.

Formal evaluations of PBRF are limited, but the available evidence suggests that it has positive effects on research output and management. However, there is intense debate on the intended and unintended consequences of PBRF on science systems. The effect of PBRF will depend on how institutions allocate funds internally, which in turn is affected by their degree of autonomy and internal governance practices. Institutional responses to assessment also differ. For instance, a poor assessment may lead one university to close a department, but lead another to make improvements. Negative effects have also been reported, such as a narrowing of research focus. There is a pressing need for structured studies to assess effects at national, institutional and departmental (and even individual) levels. Such research could be of great benefit to national authorities and universities in their efforts to increase the effectiveness and the efficiency of institutional funding.

Research integrity is an issue that has attracted growing attention over the past five years, and is linked to how research performance is incentivised and measured. Recent high-profile incidents of research misconduct threaten to undermine public trust in science. At the same time, the irreproducibility of some results purporting to be scientific breakthroughs raises questions about the rigour of scientific practice.

Substantial public and private investment in following up on some of this work has been wasted. Pressure to publish, extreme competition, short-term funding and uncertain career tenure are all variously blamed for creating distortive pressures in the scientific system. However, careful analysis of the effects of such factors on academic behaviour is sorely lacking. Some key policy messages related to the science system and its link to innovation are below.

Main policy messages: The science system and innovation

- Funding of basic research remains important for underpinning innovation and for tackling global challenges. Against a backdrop of fiscal consolidation, governments are under pressure to demonstrate social and economic returns from public funding, which calls for science policies that promote excellence, open access and impact.
- Long-term funding for curiosity-driven research must be preserved and project-based funding must be significant in order to allow more direct steering of public research by funding agencies and research ministries.
- Researchers are central to science systems and research careers must remain attractive, while training policies must respond to the increasingly data-driven, co-operative and multidisciplinary nature of science.
- As the role of large research infrastructures in scientific research and research budgets is increasing, sustainable funding and effective governance mechanisms are required.

5.2. Open science: Increasing the return on public investments in scientific research

"Open science" refers to a way of doing science based on unrestricted access to publicly funding research results, namely articles and data. Although associated with public research, open science can also be applied in the business sector and thus enable innovation. Open science also enables the increased engagement of citizens in scientific progress and innovation. Open science requires the interoperability of scientific infrastructure in order to share research results and data. This may involve the creation and long-term support of publication and data repositories, the creation and cleaning of metadata, open and shared research methodologies (such as open applications and informatics code), and machine-friendly tools (allowing, for example, text and data mining). The dissemination of government-funded research results has to date largely relied on scientific journals. However, this model is evolving. The Internet has greatly lowered the marginal cost of online publishing. The costs of data storage and archiving also continue to fall. And advances in computer science are creating opportunities to organise, share and reuse vast amounts of data generated by public research.

Governments and the scientific community have championed greater access to scientific data for a variety of reasons:

- To improve efficiency in science. Open science could increase research productivity by: 1) reducing research duplication and the re-creation of data; 2) allowing a more accurate verification of research results; 3) enabling more research to be done based on the same data; and 4) multiplying opportunities for domestic and global participation in research.

- To generate knowledge spillovers. Increased access to research results could spur knowledge spillovers, innovation and efficiencies across the economy and society.
- To open up new scientific research opportunities. Data-driven science – the exploration of data to generate new scientific hypotheses – has great potential in many areas. Being able to link data across different fields – for instance combining health records with genomic and biological data, or social sciences and environmental data – opens many exciting opportunities.
- To foster the use of public research among SMEs. While larger corporations have the resources to access scientific research results, many SMEs cannot afford potentially useful journals.
- To help address global challenges. Addressing global challenges requires access to and sharing of reliable data from many countries. The international Human Genome Project is an example of a large-scale research endeavour in which an openly accessible data repository has been used successfully by researchers all over the world, for different purposes in different contexts. Furthermore, for scientists in developing countries, greater access to international science and data can help meet social and economic goals.
- To strengthen the evidence base of policy. Public policies and decision making can benefit from scientific data. For example, administrative data from the institutions of OECD member countries, such as employment information, are now used extensively in social sciences and in policy making.

As key funders of public research, policy makers can take a variety of steps to promote access to and the use and reuse of scientific research results. In particular, they can remove barriers to open science by setting appropriate incentives, develop the infrastructure necessary to make open science happen and, in some cases, adopt mandatory rules for open disclosure of publicly funded research results. However, open access is not without costs. Currently, many governments and research institutions are bearing the costs of offering open access to articles and to data as well as the costs of storage and the preservation of data sets online. Given that the volumes of data being generated are increasing rapidly, public institutions will be challenged to find sustainable funding and business models. Public-private partnerships with private service providers may offer innovative solutions.

Universities and public research organisations also have a major role to play by adopting data management policies and ensuring researchers are aware of the IPRs related to scientific articles and data. Scientists often compete to advance science. They therefore have little incentive to share data and experimental material. Mechanisms that acknowledge the publication of data sets and other scientific material on researchers' curricula vitae (CVs) might promote the sharing of scientific information. Providing researchers with the skills to share and reuse data and scientific content in an open science environment is also important.

OECD member and non-member countries are increasingly developing legal and policy frameworks, guidelines and initiatives to encourage greater openness in science, with several countries implementing strategic approaches, e.g. Finland's Open Science and Research Initiative. However, there is heterogeneity in the approaches adopted in different countries and institutions. This is the case, for example, with scholarly publications – while the metadata of a published article are usually made available immediately, regulations on when the full text should be available differ across countries and institutions.

Examples of recent policy initiatives include the following:

- The creation of online repositories, databases, archives and digital libraries and platforms containing information on R&D projects and researchers' CVs.
- Mandatory access: Research funding agencies in many countries – including Australia, Costa Rica, Denmark, Estonia, Finland, Germany, Switzerland, the United Kingdom and the United States – have mandated public access to the results of the research they fund. Other OECD countries are also considering adopting rules for mandatory open access.
- Financial support. Funding agencies in Finland, Germany, Norway, the Netherlands, Switzerland and the United Kingdom have adopted funding mechanisms to cover some of the costs of open access publishing. Elsewhere, governments encourage universities or research organisations to allocate funding for open access initiatives directly.
- Open government data. Open science can also be promoted through the disclosure of government data. A number of OECD member and non-member countries have adopted policies in this respect.
- Modification of intellectual property rules for research or exemptions. Australia and Finland are currently discussing modifications to the existing legal framework for the publication of publicly funded research results to make the copyright legislation increasingly open science-friendly. Germany and the United Kingdom have amended their copyright legislation.

Several studies show that open access publishing improves the impact of scientific papers. Some studies have found a clear – if perhaps unsurprising – correlation between the number of times an article is cited and the accessibility of the same article free of cost online. The impact of open access on business innovation, science and the wider economy needs more extensive assessment. The research that is available suggests that the effects could be large. For instance, Houghton, Rasmussen and Sheehan (2010), estimated that expanding the National Institutes of Health (NIH) open access policy to all other science agencies in the United States could yield a net present value gain of around USD 51.5 billion.[4]

While there are strong arguments in favour of more open science, questions also arise of how the dissemination of lower-quality scientific results can be avoided, and how open access publishing and open data might be made more sustainable through market mechanisms. Indeed, the process of selecting, reviewing and publishing articles in open-access journals is not cost-free, even if it is less expensive to produce than conventionally published literature. In fact, most open access journals rely on subsidies or funding from universities, scientific societies and government agencies. Providing long-term access to quality-assured data is a challenge that is only beginning to be addressed. Some of main policy messages related to open science are incorporated below.

Main policy messages: Open science

The OECD *Principles and Guidelines for Access to Research Data from Public Funding* (OECD, 2007) provide an overarching framework for policy. In terms of scope, the principles and guidelines are meant to apply to research data supported by public funds for the purposes of developing publicly accessible scientific research and knowledge. The full text of the principles and guidelines is available online.[1]

Open science policies should be principle-based but adapted to local realities. For example, if a research project involves business sector partners and commercial interests are present, the requirements for sharing research results may be different from the case in which only public actors are involved. In other cases, privacy or confidentiality concerns may apply to the treatment of certain specific classes of data.

Consultative approaches that involve all relevant actors are key to successful open science strategies. Open science efforts involve different communities and actors. These include researchers, governmental institutions, universities and research centres, libraries and data centres, private non-profit organisations, business sector organisations (including private academic publishers), supra-national entities, and citizens. These actors do not necessarily have the same incentives, goals or expectations. A successful strategy needs to take this diversity into account and react accordingly.

Better incentive mechanisms are needed to promote data-sharing practices among researchers. While all public-sector researchers have an interest in sharing published research articles, the same is not true for research data sets, especially at the prepublication stage. In addition, data cleaning and curation (for example, by developing metadata) is a time-consuming activity that is rarely acknowledged in evaluations or grant allocation procedures. Most evaluations of universities and researchers are almost entirely based on teaching and bibliometric indicators, attributing little value to the sharing of prepublication inputs and post-publication outcomes. Extending citation mechanisms to data sets can partly address this issue.

Clear legal frameworks in relation to the sharing of publications and the reuse of data sets are needed at the national and international levels. A lack of clarity on the interpretation of national and international legal frameworks may prevent the sharing or reuse of research results. In addition, clear guidelines around text and data mining are needed as these tools will be used increasingly by researchers in future.

"Soft factors" such as the development of an open science culture are important. Recent surveys reveal that not all researchers are aware of the possibilities offered by open science. In some countries, different institutions regularly organise workshops and training to make researchers aware of these possibilities. Furthermore, tackling global challenges will require greater access to and the sharing of national public research datasets and hence, co-operation at a global level.

International collaboration is important for open science, especially to address global challenges. International collaboration is becoming more important as publications and data in electronic form cross national borders. Shared and inter-operable infrastructure is necessary to disseminate research results and promote scientific collaboration. Such collaboration can help to share investments and risk, and avoid duplication of effort. International co-ordination and co-operation will grow in importance as R&D and the global production of knowledge shift towards emerging economies. Furthermore, tackling global challenges will require greater access to and sharing of national public research data sets and hence co-operation at a global level.

The impact of open access on business innovation, science and the wider economy needs more extensive assessment, especially where the public-sector costs of openness are significant.

1. See *www.oecd.org/science/sci-tech/38500813.pdf*.

5.3. International co-operation in science and technology

As described in the OECD Innovation Strategy (2010a), science and innovation are global activities in which multiple actors from many countries simultaneously collaborate and compete. The international landscape continues to evolve, with the BRIICS countries and other emerging economies producing an increasing share of scientific knowledge.

While OECD countries no longer dominate, they are generally maintaining their historical strengths while seizing new opportunities for scientific and technological co-operation. At the same time, in many developing countries science is stagnating. This stagnation undermines efforts to address global challenges.

Over the past five years the OECD has focused on three areas where sound policy is necessary to promote effective international co-operation. These relate to: 1) research infrastructures and networks; 2) global challenges and governance; and 3) promoting co-operation with less-developed countries. The main findings in these three areas of work are summarised below.

International research infrastructures are a major catalyst for scientific co-operation between countries and an essential requirement for scientific progress in some fields, such as physics and astronomy. OECD has worked with science policy makers for more than two decades to improve processes for establishing, operating and assessing large-scale infrastructures. More recently, and spurred in part by moves towards open science and big data, issues around smaller distributed infrastructures have also moved up the policy agenda.

Large-scale science infrastructures can be extremely expensive. The complexities and potential pitfalls in setting up infrastructures are analysed in the 2010 OECD report *Establishing Large International Research Infrastructures: Issues and Options* (OECD, 2010c). While there is no single recipe for success, building on previous experiences and involving those who have this experience is clearly advisable. Empirical assessment of the socio-economic impact of large and long-term facilities is challenging. However, qualitative case studies can give important insights. This is the approach taken in OECD (2014d). This study focuses on the European Organisation for Nuclear Research (CERN) and its most prominent shared scientific facility, the Large Hadron Collider (LHC). CERN is here seen to have impacts far beyond its core scientific mission.

Special challenges affect geographically distributed infrastructures that are often decentralised administratively and financially. Issues relating to the continuity of personnel and funding, legal identity and heterogeneity of partners involved can create difficulties for such shared facilities (OECD, 2014e). One area in which significant progress has been made over recent years is the co-ordination of scientific collections. April 2013 saw the launch of Scientific Collections International (SciColl), a network of museums and other institutions possessing scientific collections. SciColl aims to promote access to these valuable, and often unique, research resources that can provide critical insights in areas as diverse as environmental change, societal development and disease epidemics.

Gauging the impact of large international research infrastructures will continue to preoccupy science policy makers for the foreseeable future. This may be a particularly acute issue for developing countries, where substantial investment in such facilities is a relatively new phenomenon. In this regard, the 2012 agreement to locate an important part of the Square Kilometre Array (SKA) radio telescope in South Africa, with Australia and several other African countries as co-hosts, is groundbreaking. Developing measures to ensure some return to Africa, in terms of scientific capacity and social and economic advances, is in this case an important policy challenge.

If they are to be effectively addressed, global challenges – such as climate change; food, energy and water security; and disease pandemics – require new knowledge and new technologies from science. Responsive and adaptable modes of governance combined with flexible funding and spending mechanisms are essential (OECD, 2012a). A tailored approach to knowledge sharing and intellectual property can be important, and participatory approaches and outreach efforts are indispensable for the successful uptake of innovations. The CGIAR Global Agriculture Research Partnership is one example of how this can be achieved. The inclusion of countries with weaker science, technology and innovation (STI) capacities as full partners is necessary and may require specific actions to build capacities. At the same time, the rapid evolution of open science should provide opportunities for more radical developments in the future governance of STI to help find solutions to global challenges. For example, the recently launched Future Earth initiative has a novel regionally distributed and multi-stakeholder governance and management structure that may provide a new model for the future.

The Ebola pandemic in Africa has highlighted not only the vulnerability to infectious diseases of the poorest countries but also, in a globally connected world, how difficult it can be to contain and effectively treat newly emerging diseases. The development, testing and deployment of new vaccines and therapeutic medicines are essential to the public health response to such outbreaks. In this regard, international clinical trials are critical. In 2012 the OECD Council issued a recommendation on the governance of clinical trials (OECD, 2012b). The recommendation focuses on three areas where policy can help: 1) reducing the administrative complexity of trial processes; 2) introducing a risk-based approach to the approval and management of clinical trials; and 3) improving training, infrastructure and patient involvement.

Natural hazards represent another important global challenge, particularly as more people migrate into urban areas, many of which are located in areas of risk. The Global Earthquake Model (GEM), established in 2009 with the OECD's assistance, is an example of an international science-based response to a natural hazard. GEM is a public-private partnership engaging a global community in the design, development and deployment of state-of-the-art models and tools for earthquake risk assessment. There is potential to expand this type of partnership between scientists and users in other areas of decision making.

In an era of rapidly developing open science, distributed data infrastructures will be critical for science and international collaboration (see the section on open science above). International bioinformatics databases have already played a key role in the development of molecular biology and biomedicine, and international data-sharing is critical for research that addresses global challenges. However, sustainable business models for funding many of these structures are urgently required. While some structures are supported by dedicated core funding, others are largely dependent on competitive short-term grants and/or public-private partnerships. Whatever the mechanism, demonstrating value for money and impact will be an important requirement as the size and number of data infrastructures grows.

The need to include developing countries in scientific initiatives to address global challenges was referred to in the previous section. Achieving this requires specific policy actions. One important area for consideration is the potential synergy between science funding and development assistance. Opportunities exist for science funding agencies and development co-operation agencies to work more closely to strengthen science in developing countries, particularly with regard to global challenges (OECD, 2011a).

Another area in which exchange of practices and experiences between developed and developing countries can be important is in relation to mechanisms and processes for the provision of science advice to governments. Most OECD countries have a variety of formal and informal science advisory structures and individuals that together make up a national advisory system. These national systems are complemented by a similarly diverse set of international structures. Many developing countries have relatively weak science advisory structures. This is an important area for capacity development. International co-operation and/or co-ordination between advisory structures are required, particularly in crisis situations. Some of the main policy messages related to the science system and international scientific co-operation are below.

Main policy messages: International co-operation in science and technology

- Developing effective science and technology initiatives to address global challenges requires responsive and adaptable modes of governance, combined with flexible funding and spending mechanisms.
- International collaboration to build science advisory mechanisms needs to be further developed. Doing so would help to provide reliable and coherent information in crisis situations as well as for meeting global challenges. There is also a need to build advisory capacity in developing countries and to better co-ordinate on common areas of interest across countries.

5.4. Commercialisation of publicly funded research

Publicly financed research in universities and PRIs has led to many landmark technological innovations, from recombinant DNA technology and the Global Positioning System (GPS) to MP3 technology and Siri voice recognition. Increased interest in the commercialisation of public research has various drivers. These include:

- The desire to improve national competitiveness.
- Concern with the fact that the number of patents, licences and companies created at universities and PRIs has slowed since the late 2000s. Indeed, universities face increasing pressure to combine excellence in teaching and research with commercialisation and fund-raising activities. Sweden has even amended its Higher Education Act so as to introduce the building of external partnerships into the mission of higher education institutions to encourage them to actively exploit research outcomes.
- The increasing cost of scientific research, leading many PRIs and universities to search for new funding sources (even if income from commercialisation at most PRIs accounts for a small part of all income).
- A trend towards greater business outsourcing of R&D, with firms increasingly looking to universities and PRIs for much of their basic research.
- An awareness of research that suggests that academic entrepreneurship does not detract from research productivity or lower the disposition to undertake basic research. Indeed, research from Sweden shows that there is a strong positive correlation between scientific excellence and the intensity of industry contacts of individual researchers (Bourelos, Magnusson and M. McKelvey, 2012).

While patents, licences and spin-offs remain important channels for commercialising public research, other channels appear to be increasing in importance. These include collaborative research, student and faculty mobility, contract research, faculty consulting, and student entrepreneurship. ICTs, and a push by science funding agencies for greater access to publicly funded research results and data, are also broadening the channels for commercialisation. OECD (2013a) – *Commercialising Public Research: New Trends and Strategies* – examines recent institutional and policy developments.

Nearly all OECD countries have now adopted specific legislative frameworks and policies to incentivise the commercialisation of public research. Best known among these is the Bayh-Dole Act in the United States, which allowed universities to own the patents arising from federal research funding, and provided incentives for their commercialisation. Bayh-Dole legislation has been widely emulated.

A policy convergence has also occurred whereby in most countries, IPRs are vested with universities. Universities can often overrule national university intellectual property (IP) regulations through university bylaws (for instance to negotiate different IP arrangements with third parties).

Despite the rise of new channels for knowledge transfer in most countries, institutions and policy makers are still focused on promoting commercialisation through patenting and licensing. However, with the exception of a few leading universities and government labs, patenting and licensing are, and will remain, a minor activity for most universities and PRIs. For instance, in the United Kingdom, higher education institutions generated external income of more than GBP 3 billion (British pounds) in 2011/12. However, only 2% to 4% of that amount was from licensing or the sale of shares in spin-off companies. Most of the revenue came from collaborative and contract research, consultancy and professional training (House of Commons, Science and Technology Committee, 2013). Across Europe, only 10% of universities account for around 85% of total licensing income.

There is increasing interest in improving access to scientific research findings in general and, in particular, the results of publicly funded research (see the previous section in this chapter, on open science).

An increasing share of public funding is being directed at co-operative research rather than individual organisations. While universities have long interacted with industry, this has intensified in recent years. Industry and universities (particularly technology transfer offices [TTOs]) can have different perceptions regarding the value of – and sharing of – income from IP. Diverging perceptions can also exist regarding how to share patented knowledge. Inexperience and lack of awareness of business' needs is a common industry complaint (Hertzfeld, Link and Vonortas, 2006). Intermediaries such as IP-based companies or government-backed patent funds are increasingly used to match supply and demand for IP between universities and SMEs.

A range of intermediary and bridging organisations has been created to facilitate knowledge transfer and commercialisation. These include TTOs; business incubators; business innovation centres; science parks; special agencies in chambers of commerce; industry liaison offices (ILO); proof-of-concept centres (which aim to close funding gaps when business angels and VC companies focus on larger or later-stage deals);[5] and libraries/institutional repositories. The missions of these entities can differ considerably, as illustrated in Box 5.1.

The most common goals of TTOs are enhanced licensing revenues, the maintenance or expansion of industrial research support, technology transfer, and, to a lesser extent, regional development. Licensing revenue is typically the most important criterion by which TTOs measure their success, although most TTOs do not generate positive net returns, or break even, from patenting and licensing (Bulut and Moschini, 2009). However, a small number of TTOs generate substantial licensing revenues.

Box 5.1. **Hungary's Innovation Marketplace and the Czech Republic's GAMMA programme**

Hungary – Innovation Marketplace: The National Research, Development and Innovation (NRDI) Office is Hungary's only innovation financing institution, and a center that aggregates all information that might be of interest to actors. It started its work on 1 January 2015 on the basis of the Hungarian Act 2014/XLLVI about scientific research, development and innovation. It is a national strategic and funding agency for scientific research, development and innovation, the primary source of advice on innovation policy for the Hungarian Government, and the primary funding agency.

NRDI is simultaneously contacted by start-ups looking for investors; by technology owners searching for foreign markets; by foreign agencies seeking technologies; by laboratories offering capabilities; and by enterprises needing such capabilities. To help link these different actors, NRDI has developed Innovation Marketplace, which channels all enquiries and turns them into business proposals that are accessible to both governmental and market players. The Innovation Marketplace is becoming a platform that is helping to turn project proposals into standardised business propositions, thus generating the deal flow indispensable for potential investors. Furthermore, the NRDI Office indirectly induces crowdsourcing through the platform, in acting as an intermediary from the state services provider.

Czech Republic – GAMMA programme: The GAMMA programme targets the lack of exploitation of public research results. It aims to support the transformation of R&D results achieved in research organisations into practical applications to enable their commercialisation and implementation. It also promotes co-operation between research organisations and enterprises through "learning by doing". The programme is divided into two sub-programmes:

Sub-programme 1 is aimed at research organisations and supports verification of the practical use of R&D results that were created in research organisations (mainly public) and that should have a high potential for commercial application. It also provides systemic support to identify, implement and develop commercialisation activities in research organisations.

Sub-programme 2 is aimed at supporting applied research and experimental development by enterprises exploiting the research results from public research organisations. Support is provided for projects involving the completion of functional prototypes, verification of their properties, verification of test series, and evaluation of all technological, economic, social, health and other impacts of new products or services. The programme is aimed at enterprises, but research organisations can be project participants.

Many TTOs have expanded their missions from simply administering technology transfer (such as managing invention disclosures, filing patents, etc.) to a range of IP management and support activities (e.g. patent scouts, consulting), marketing non-patent services, administering seed funds, and creating a culture of innovation.

Many universities have sought to reform or replace their TTOs, given their mixed record of success. Alternative models and proposals include:

- **Technology transfer alliances (TTAs):** Given the limited ability of many universities to generate enough income to cover the costs of their TTOs, it has been proposed to share services through TTAs. In theory, TTAs could bundle inventions across universities, lower unit operational costs and increase access to expertise. However, TTAs might raise some co-ordination/communication costs. An example of a TTA is the Innovation Transfer Network (ITN) in the United States. Established in 2006 with public support, the ITN serves as the TTO for 13 smaller colleges, each of which is represented on the ITN's board.

- **For-profit models:** For both cost and efficiency reasons, some institutions have created privately funded TTOs. These take the form of limited liability corporations. Some universities have operated such private TTOs since the late 1980s. (An example is Isis Innovation, a wholly owned subsidiary of the University of Oxford, created in 1988.) In Israel, the majority of TTOs operate under a limited liability model, partly or wholly owned by universities.
- **Internet-based models:** Internet-based platforms can complement existing TTO structures. These platforms respond to the needs of technology transfer professionals and application-oriented researchers for easier access to information. And they help to showcase university technologies to the corporate sector. An example is Flintbox, at the University of British Columbia.
- **A free agency model:** TTOs are sometimes viewed as revenue maximisers reluctant to explore alternative commercialisation paths. This has led to proposals that researchers be allowed to choose between their university TTO or an agent elsewhere (i.e. a free agency model). In theory, this could improve efficiency in TTOs by creating competition. However, reservations exist on the extent to which TTO performance can be enhanced through competition, among other considerations.

While prominent in policy discourse, spin-offs are less prevalent than is often thought. Recent data from the United States' Association of University Technology Managers show that the number of spin-offs per university per year among the top 100 United States research universities is just 2, with a maximum of 22 (at Massachusetts Institute of Technology [MIT]). Spin-off creation varies markedly across OECD countries. On average, Europe has a higher rate of spin-off formation (2.4 spin-offs per USD 100 million of research expenditure during 2004-10) than the United States (1.1 for 2004-11), Canada (1.1 for 2004-11) and Australia (0.7 for 2004-11) (OECD, 2013i).

Research has highlighted the importance of student entrepreneurship, which is being encouraged by universities and governments in many OECD countries. Åstebro, Bazzazian and Braguinsky (2012) show that recent graduates are twice as likely as faculty to create a business and that such firms are often of high quality.

New approaches to financing commercialisation are also emerging. Many universities and PRIs are complementing government funding for university start-ups by setting up their own proof-of-concept and seed funds. But studies point to constraints arising from the lack of business expertise, skills and networks among many academics (Wright, Clarysse and Mosey, 2012). Accordingly, approaches to nurture academic entrepreneurship that focus only on funding gaps may be insufficient.

Case studies show that universities can achieve high levels of entrepreneurial activity through good programme design. This can be the case even when universities have low R&D expenditures, low research/financing capacities and low VC availability (Åstebro, Bazzazian and Braguinsky, 2012).

When universities can override national regulations by developing internal patent regulations and processes, some have experimented with alternative settings. For example, some have provided preferential treatment to researchers wishing to license technologies they have developed. Other universities allow professors to establish new ventures, granting leaves of absence or allowing tenure clock stoppage while faculty pursues commercialisation activities. Of 64 universities surveyed in the United States and Canada, 16 were found to consider patenting and commercialisation achievements in tenure and promotion decisions (Stevens, Johnson and Sanberg, 2012).

Universities in OECD countries also increasingly face the issue of ownership of IP by graduate students and other non-faculty/employees engaged in research. This may give rise to tensions between universities and students. Owing to these changes, and to avoid IP disputes between students and universities, the University of Missouri in the United States established a policy in 2011 that under certain conditions allows students to own inventions created during their enrolment.

National funding agencies and individual institutions have developed standard licensing agreements for academic inventions (e.g. the United Kingdom's Lambert Toolkit, Germany's model R&D co-operative agreements, and Denmark's Schlüter model agreements). Standardised agreements can help address industry concerns of difficulties negotiating licences with PRIs. And some OECD countries have started to sponsor the creation of patent funds specifically for PRIs. Some key policy messages related to the commercialisation of publicly funded research are included below.

Main policy messages: Commercialisation of publicly funded research

A central role for government is to set the basic rules and institutional frameworks that reflect the public interest and provide the right incentives to firms, public researchers and PRIs. In so doing, governments, research ministries and business must work together to develop coherent policy frameworks.

Policy for commercialisation should go beyond the TTOs of universities and PRIs. In most countries, institutions and policy makers are still focused on promoting commercialisation through patenting and licensing. But patenting and licensing are a minor activity for most universities and PRIs. Other channels play important roles. These include public-private collaborative research, student and faculty mobility, contract research, faculty consulting, and student entrepreneurship.

Because many intermediary organisations have arisen as channels for commercialisation, government can help in identifying and disseminating information on best practice.

Especially because of digital technology, governments play an increasingly important role in developing the legal frameworks to increase access to and use of scientific research and data.

5.5. ICT, "big data" and the open Internet

Almost no business today is run without the help of ICTs. In 2014, 95% of enterprises in OECD had a broadband connection (Figure 3.13), although with considerable variation among small enterprises (Figure 5.3). More than 75% of all OECD enterprises had a website or a homepage in 2014, up from about 70% in 2009. As for broadband access, web presence is lower among small firms. The speed of adoption depends in some cases on prior uptake. It took 15 to 20 years for slightly more than three-quarters of enterprises to develop a website, but only a few years for around 30% of businesses to become active on social networks.

Figure 5.3. **Broadband connectivity, by size, 2010 and 2013**

Percentage of enterprises in each employment size class

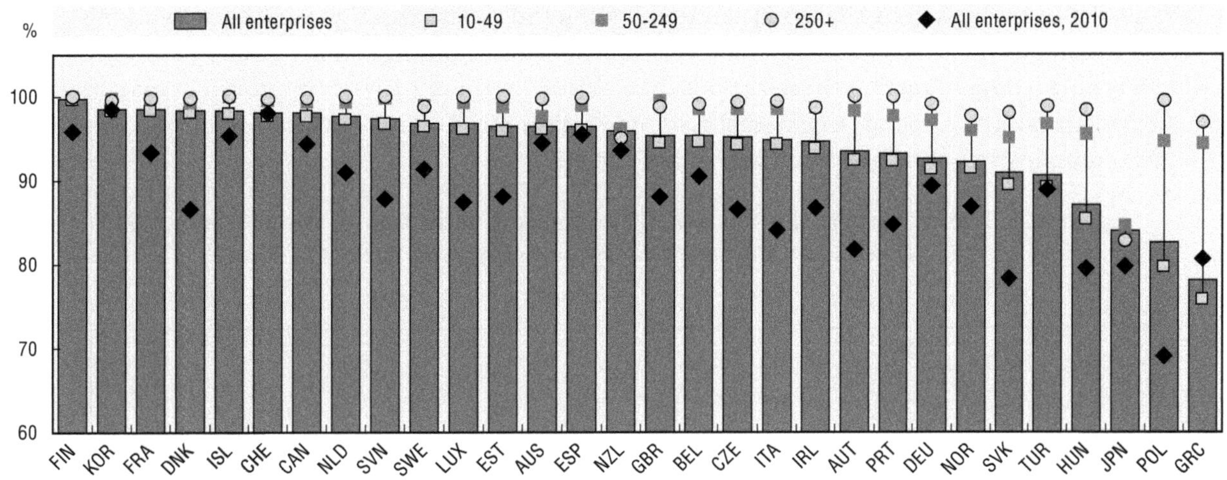

Source: OECD (2014f), *Measuring the Digital Economy*, OECD Publishing, Paris, *http://dx.doi.org/10.1787/888933148520*.

In most cases, a web presence is still used as a window to provide information on the enterprise. Indeed, figures on participation in e-commerce are much lower. On average, 21% of firms with at least ten people employed in reporting OECD countries received electronic orders, 4 percentage points more than in 2009 (Figure 5.4). Differences among countries remain considerable, however, which closely follows the differences in shares of smaller firms among countries. For enterprises with 250 or more employees, participation in e-commerce is about 40%, and the share is above 30% even in some lagging countries. The use of more sophisticated ICT is also less frequent. These include ICT applications used to manage information flows, where implementation requires changes in business organisation, and radio-frequency identification (RFID), where uptake is limited to certain types of businesses.

The high level of uptake of ICT across the economy shows how important the technology has become, increasingly, also for innovation. Improvements in the transmission of information have often underpinned innovation. For example, the introduction of telegraphy in the 19th century made railway systems more efficient as speedy communication about the condition of rail tracks allowed many trains to use the same track. The development of information theory in the first half of the 20th century brought the possibility of coherently manipulating data. And the advent of silicon-based integrated circuits, containing billions of transistors, made it possible to apply these theories to information on a previously unimaginable scale. New communications technologies were essential to realising the innovation potential of information technology. Today, the two cannot be seen separately, and "ICT" has become a nearly ubiquitous abbreviation. Around 55% of VC invested in the United States is invested in firms producing ICT-related goods and services. In the past two decades this share has averaged about 60%, with much of the remaining 40% being linked to ICT in indirect ways.

In most OECD countries, information industries[6] account for 20% to 25% of total business expenditure on R&D (BERD) (OECD, 2014f). In Finland, Israel, Japan, Korea and the United States, the sector accounts for 30% to 50% of BERD. It is estimated that across the

OECD, 70% of firms in ICT industries introduce innovations, against an average of 50% for businesses overall (OECD, 2014f). Much innovation in ICTs today is aimed at making data available, through better sensors; communicating those data through various networking technologies; storing data in the cloud; big-data analysis; defining actions through machine learning; communicating those actions; and acting on them through improved actuators (actuators allow a change in physical state, and take many forms, from lasers to ink nozzles to complicated valves operating with magnetic fluid).

Figure 5.4. **The diffusion of selected ICT tools and activities in enterprises, 2014**

Percentage of enterprises with ten or more people employed

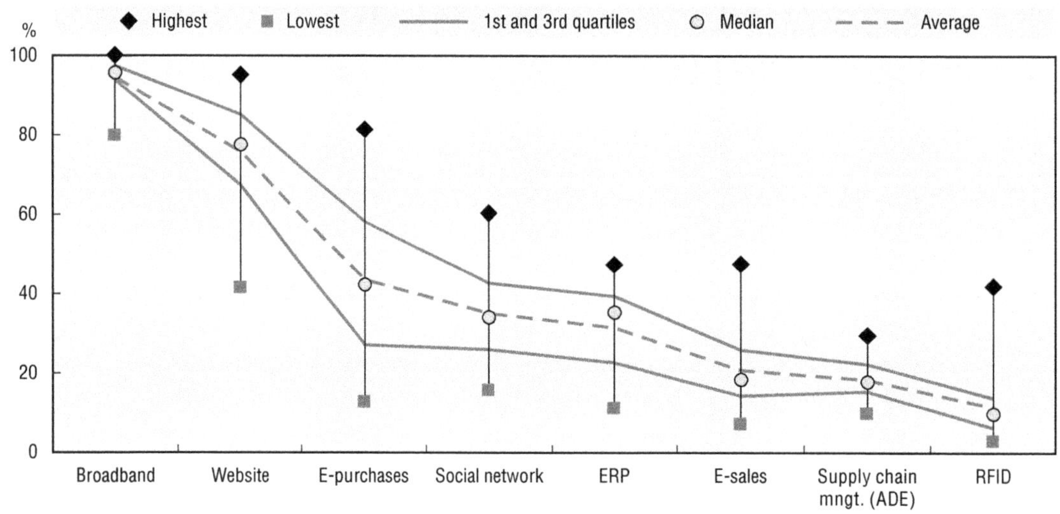

Note: ERP = enterprise resource planning; ADE = automated data exchange.

Source: OECD (2015), *Digital Economy Outlook 2015*, OECD Publishing, Paris, *http://dx.doi.org/10.1787/888933224847*.

One way in which ICT underpins innovation is by facilitating the dissemination of knowledge. In some academic fields, such as physics, mathematics and biology, almost all science is first disseminated as a pre-print via arXiv.org. As many as 7 000 preprints are submitted each day. arXiv.org was one of the first examples of the open access movement in science (see the section on open science earlier in this chapter). ICT has also enabled a rapid internationalisation of research. For example, in 1998 only a few countries experienced more than 10 000 international collaborations, but by 2011 this level of international collaboration was common in developed countries.[7] In a further illustration of the knowledge-dissemination effect of ICTs, many universities are putting their courses online. The effect of massive open online courses (MOOCs) is as yet hard to determine but they have the potential to completely revolutionise higher education. Some examples of the impact of ICT on society are included in Box 5.2.

Business models have also changed in response to the possibilities offered by ICTs. For example, today micro-multinationals are common: these are medium-sized enterprises that, despite their size, operate on a global scale. Such firms sometimes have employees and freelancers working around the globe on the same projects, thanks to ICTs. ICT-enabled platforms such as Kickstarter, Indiegogo and Quirky have also given innovators a way of connecting customers and potential investors. By receiving instant feedback through donations and through comments from potential customers, innovators glean an idea of the potential success of their products. Seventy-seven per cent of companies in the OECD have a website, and around 21% sell their products electronically (OECD, 2014f). Over 80% of enterprises use e-government services, with some countries, such as the Netherlands, requiring online filing of taxes for all businesses. Consumers are also pushing companies to change their business models, with as many as 77% of consumers in Denmark, the Netherlands and the United Kingdom purchasing online (OECD, 2014f).

Box 5.2. **Examples of innovation in ICTs affecting society**

In agriculture, ICTs are becoming essential. Dairy farmers increasingly operate automated farms. Robots that feed, milk and clean cows can change the operation of a farm, with these activities being performed to the schedule of individual cows, rather than the farmer's timetable. Geolocation data allow optimal use of fertiliser and pesticide across fields. And greenhouses monitor crops through sensors, but also, because of ICT, can become an integral part of the energy production sector. In the Netherlands, since 1994, greenhouses have entered the energy production market with combined heat power exchange systems that produce warmth and carbon dioxide (CO_2) to grow crops. Such systems reduce the costs of crop production by as much as 20% (Koolwijk and Peeters, 2011).

In transport, vehicles today are a combination of 80 to 200 sensors and processors. Functions such as motor management, anti-lock braking and traction control all rely on information processing. Without ICT's, vehicles would not be as efficient and safe as they are today. And self-driving vehicles promise greater efficiency and safety, as well as increased autonomy for disabled and elderly people. Nissan and Audi expect to sell autonomous vehicles by 2017.

GPS-enabled ICT applications in cars and mobile devices will boost efficiency in transportation systems, yielding significant savings in time and CO_2 emissions. For example, TomTom, a leading provider of navigation hardware and software, has collected more than 9 trillion data points from its navigation devices and other sources. These describe the time, location, direction and speed of travel of individual anonymised users. TomTom now adds 6 billion data points a day. TomTom's data analyses are fed back to its navigation devices to inform drivers of current and predicted traffic conditions. Significant time savings and reduced congestion have resulted, notably in cities. Indeed, the global pool of personal geolocation data is estimated to have grown by 20% a year since 2009. By 2020, this data pool could provide USD 500 billion in value worldwide in the form of time and fuel savings, or 380 megatonnes (million tonnes) of CO_2 emissions saved (TomTom, 2014). The data are also used by governments to understand the effect of proposed and realised infrastructural changes on traffic flows.

5.6. "Big data" analytics

The growing number of computer-mediated transactions and the accelerating migration of social and economic activities to the Internet have led to the generation of a huge volume of (digital) data, commonly referred to as "big data". Big data are now leveraged by organisations, often in highly creative ways, to generate innovations in products, processes, organisational methods and markets. Data and data analytics have become a driving force in innovation. The exploitation of data and data analytics has already created significant economic value for many businesses, and more is expected to follow. A range of studies suggests that the use of data analytics can boost firms' productivity by 3% to 13% (Brynjolfsson, Hitt and Kim, 2011; Bakhshi, Bravo-Biosca and Mateos-Garcia, 2014; Tambe, forthcoming). Some estimates put the global market for big data technology and services at USD 17 billion in 2015, with a growth rate of 40% on average every year since 2010 (IDC, 2012).

Fifty per cent to 85% of all data stored by businesses may be unstructured (Shilakes and Tylman, 1998; Russom, 2007). Data analytics can allow the cost-effective extraction of information from unstructured data sources such as text documents and e-mails, videos, images and audio streams. Furthermore, data analytics has empowered organisations to base their decisions on (near to) real-time data. For businesses, this means reduction of time to market, as well as benefits due to first- or early-mover advantage. And for governments, data analytics can permit real-time evidence-based policy making (Reimsbach-Kounatze, 2014).

As organisations increasingly use data and analytics, a shift can be seen in the way decisions are made, with greater reliance on correlation instead of causation. For example, a company such as Wal-Mart may change the product placement in its stores based on correlations with purchases, without the need to know *why* the change affects consumer behaviour. As Anderson (2008) explains: "Who knows why people do what they do? The point is they do it, and we can track and measure it with unprecedented fidelity." Anderson (2008) has even challenged the usefulness of model-building in an age of massive data sets, when machines can detect complex patterns in vast databases that are otherwise invisible to researchers.

Data analytics also empower autonomous systems, which use machine learning algorithms to improve performance with every data set analysed. These systems are now becoming mainstream thanks to the widespread availability of large volumes of data. Such systems can perform a growing range of tasks that previously required human intervention. Google's driverless car is an example of this potential. The car takes data from multiple on-board sensors, including video cameras and radar systems, and combines them with data from Google Maps and Google Street View (for landmarks, traffic signs and lights). Another example is automated or algorithmic trading systems (ATS), where stocks are bought and resold within fractions of a second. In the United States, algorithmic trading is estimated to account for more than half of all trades (Figure 5.5).

Figure 5.5. **Algorithmic trading as a share of total trading**

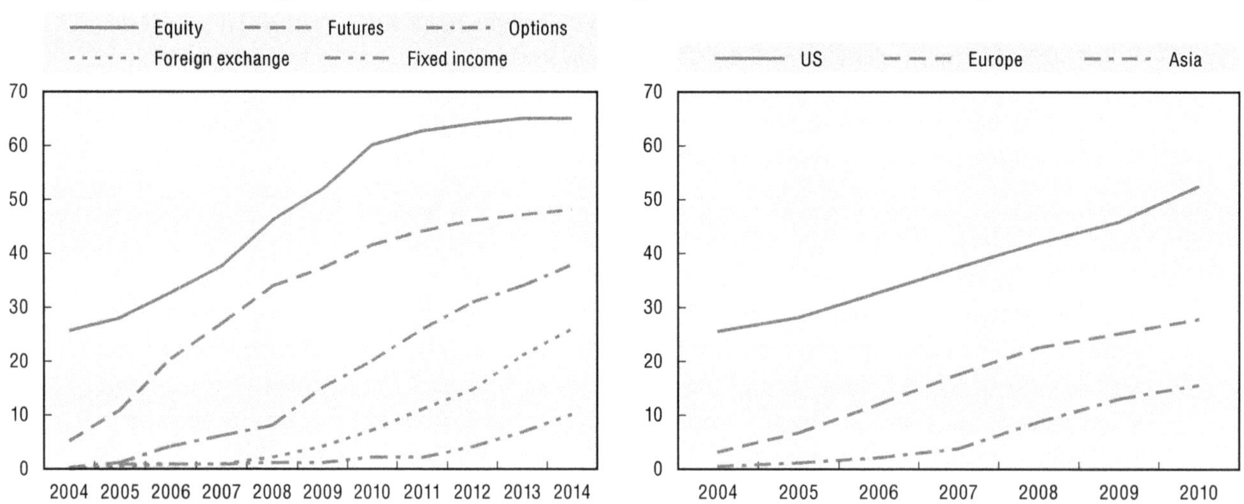

Note: 2013-14 based on estimates.

Source: OECD calculations based on *The Economist* (2012), "High-frequency trading: The fast and the furious", *www.economist.com/node/21547988*, and Aite Group (2012), *The Next Generation of Execution Consulting Services: Leveraging Technology to Build Relationships.*

5.7. The Internet of Things

In 2012, the OECD estimated that an average family with two teenagers had ten Internet-connected devices at home (OECD, 2013b). By 2022 there could be as many as 50 devices per home, totalling 14 billion such devices across the OECD. Estimates by Ericsson, Cisco and Intel show that the number of Internet-connected devices might grow to 50 billion over the next two decades. This phenomenon is known as the Internet of Things (IoT).

The IoT is also being linked to other advances in ICT, particularly in the areas of big data, the cloud, machine-to-machine (M2M) communication, and advanced sensors and actuators. The combination of cloud computing and big data analytics leads to improved machine learning applications. The combination of remote-controlled machines and systems with machine learning will lead to increasingly autonomous machines and systems (Figure 5.6). The economic benefits of the IoT could take the form of consumer surplus, new revenues and higher GDP growth as technologies are commercialised as well as increased firm productivity (for instance through new process optimisation and preventive maintenance strategies) (McKinsey & Company, 2013).

The development of autonomous machines is a goal of many innovations in ICT. Some examples of current developments include London's traffic lights, which operate with machine learning algorithms; the NEST learning thermostat; self-driving cars; and fully automated self-organising warehouses. Some services on smartphones, such as Google Now and Apple's Siri, demonstrate the possibilities of machine learning and autonomous systems (for example when they notify the user that it is time to leave for an appointment based on current traffic conditions). Such systems will increase in sophistication and their impacts will be felt across increasingly wide areas of the economy.

Figure 5.6. **Main elements of the Internet of Things**

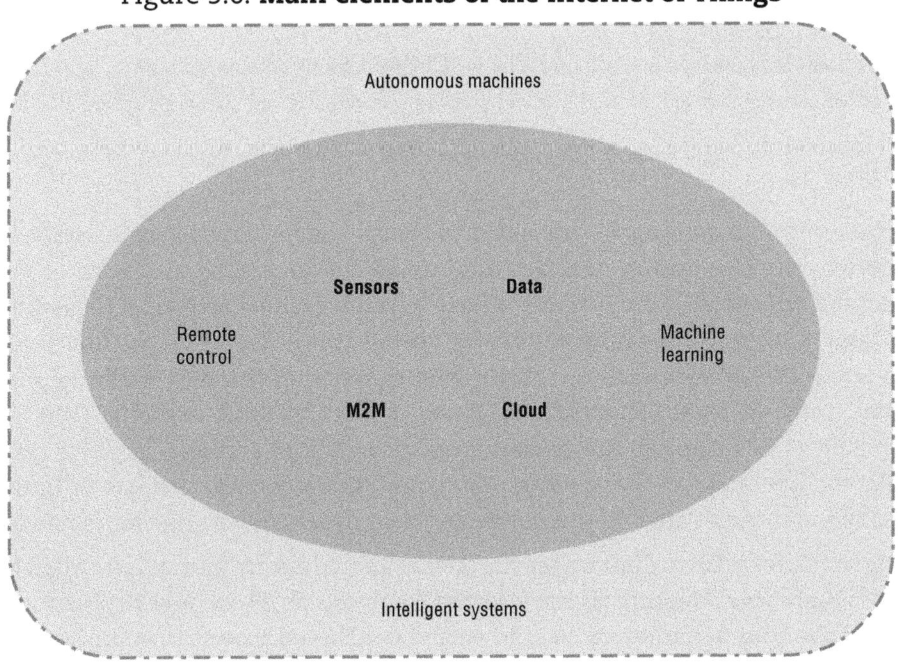

5.8. The need for an open Internet

An open and accessible Internet with high fixed and mobile bandwidth is essential. The Internet's permission-free innovation, end-to-end connectivity and lack of gatekeepers gave companies such as Google, Skype, eBay, Hotmail and Alibaba an environment where they could experiment and refine their ideas. The open Internet doesn't just enable the exchange of data, information and knowledge; it also fosters global competition by allowing users to choose between service providers. For example, Internet users can make choices as to their Internet service provider, their browser and many other criteria (Clark, 2012). Furthermore, an open Internet enables the formation and management of GVCs in which companies increasingly spread production and supply-chain systems internationally. In addition, the economies of many countries are heavily rooted in cross-border ICT-related services. Countries such as Canada, France, Germany, Ireland, Japan, the Netherlands, the United Kingdom and the United States, among others, are particularly important in terms of their share of hosted web services and data centres as well as exported ICT-related services (Figure 5.7).

Figure 5.7. **OECD and other major exporters of ICT services, 2000 and 2013**

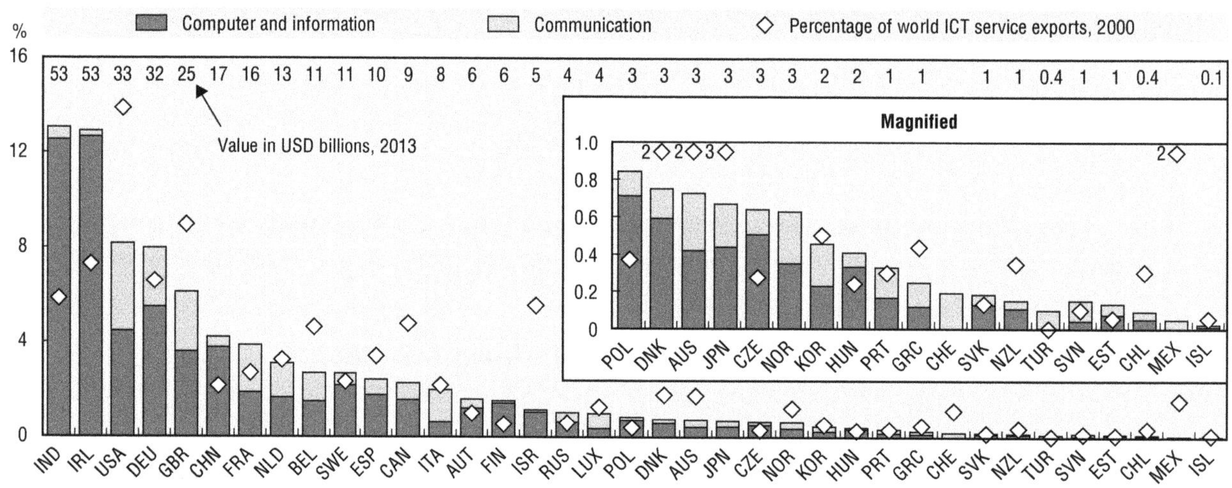

Source: OECD (2014f), *Measuring the Digital Economy*, *http://dx.doi.org/10.1787/888933148882*, based on UNCTADstat, *http://unctad.org/en/Pages/Statistics.aspx*, June 2013.

Barriers to the open Internet are likely to create significant economic costs. Some of the barriers to the free flow of data are the intended or unintended results of measures that affect the openness of the Internet. These barriers include technical measures (some of which aim to optimise the flow of data for specific purposes) such as Internet Protocol packet filtering, or "data localisation" requirements, whether through territorial restriction of Internet traffic or legal obligations to locate servers in local markets. Others aim to protect public values through the regulation of privacy and security, but as a result may also affect the openness of the Internet. The social and economic effects of limiting the openness of the internet are still unknown, and a comprehensive analysis is needed.

Many countries wish to find consensus on how to maintain a vibrant and open Internet. The OECD's High-Level Meeting on the Internet Economy, on 28-29 June 2011, discussed the openness of the Internet and how best to ensure continued growth and innovation in the Internet economy. The resulting draft communiqué led to the *OECD Council Recommendation on Principles for Internet Policy Making* (2011b). That communiqué contains basic principles for Internet policy making which, if followed, would help ensure that the Internet remains an open and dynamic platform for innovation and growth. Some of the key policy messages related to ICT, big data and the Internet are included below.

Main policy messages: ICT, big data and the Internet of Things

While a focus of important innovations, the use of data and data analytics brings major economic and societal challenges that governments need to address. These include enabling data-driven innovation by:

- Stimulating investments in broadband, smart infrastructure and the IoT as well as in data and analytics with a strong focus on SMEs and high value-added services (i.e. data analytic and data-driven services). This also includes investment in R&D.
- Fostering data-driven innovation in the public sector, including healthcare, science and education. However, domain-specific policy issues deserve consideration as there may be significant differences in the policy issues faced in these different fields of policy.
- As described elsewhere in this report, framework conditions for business must encourage organisational change and entrepreneurship in the private and public sectors.
- Promoting skills and competences in data analytics. Data specialists account for around 0.5% of total employment in most OECD countries. A lack of skills is a frequent barrier to the adoption of data-driven innovation. Also needed are domain-specific competencies to make informed data-based decisions, and to identify opportunities for data-based innovations.
- Removing unnecessary barriers to the development of the IoT. Because the IoT will affect so many aspects of society, many rules and regulations will need to be revisited. Rules regarding telecommunications networks and services (numbering), e-health (certification, compensation for doctors), transportation (self-driving vehicles, remote-controlled aircraft, regulations on taxi services), construction (building codes, energy savings) and many others may need to be updated.

Governments also need to work with other stakeholders to preserve the open Internet, and take measures to understand the economic and social costs of barriers to an open Internet. Governments therefore need to address the following challenges:

- **Preserving the open Internet and promoting the free flow of data across the global data ecosystem so as to facilitate data-driven innovation.** This includes encouraging data sharing and promoting open access to data, as well as the interoperability of data-driven services, through open standards and application programming interfaces. It also includes promoting data portability across applications.
- **Encouraging multi-stakeholder co-operation.** Multi-stakeholder processes have been shown to provide the flexibility and global scalability required to address Internet policy challenges.
- **Addressing individuals' concerns about harms caused by privacy violations.** Data-driven innovation can infringe core societal values around autonomy, equality and free speech. Key means of addressing such concerns include enhancing the transparency of data processing, promoting the responsible use of personal data and the effectiveness of privacy enforcement, and encouraging privacy risk management. The most difficult policy challenge is the operational and legal definition of the boundaries determining when responsible data use must apply and when decision automation is permissible.

Main policy messages: ICT, big data and the Internet of Things *(cont.)*

- **Addressing concerns related to the appropriation of returns on investments in data-driven innovation.** IPRs may need to be recalibrated to enable sharing of data in some cases, particularly because data ownership remains a challenging concept. Alternative incentive mechanisms including a copyright regime such as Creative Commons, and open source software licences, in combination with mandatory data citations in publications, should be considered further.
- **Assessing market concentration and barriers to competition.** Data-driven innovation may challenge traditional approaches to ensuring competition. How to define the relevant market and assessing the degree of market power and potential consumer detriments due to privacy violation deserve further consideration. Achieving policy coherence will also require promoting a dialogue among regulatory authorities, in particular competition, privacy and consumer protection authorities, so that: 1) potential consumer detriments due to data-driven innovation are taken into account; 2) synergies in the enforcement of rules controlling privacy violations, anti-competitive practices and mergers are realised; and 3) firms are incentivised to compete on and invest in privacy-enhancing goods and services as well as privacy-enhanced goods and services.
- **Improving measurement to prevent erosion of the tax base.** The global data ecosystem challenges the ability of tax authorities to determine the location of tax-relevant economic activities. Improved measurement will help in assessing the economic value of data assets and identifying where such activities occur.
- **Promoting a culture of digital risk management across society.** The openness and interconnectedness of the data ecosystem challenge the applicability of a traditional approach to digital security, which favours a closed digital environment. A modern risk-based approach to digital security is needed.

5.9. Intellectual property rights and innovation

Why is IP important for innovation? The economic rationale for IP rights is that it is in everyone's long-term interest for people and businesses that create knowledge to have well-defined, enforceable rights to exclude third parties from appropriating their inventions and creative works, or the expression of such works, without permission. Failing to put restrictions on appropriating others' inventions and creations dilutes the rewards to investment in innovation, thereby reducing the incentives for making such investments.

IP is pervasive today. Until recently, IP policy mainly affected just a few specific sectors such as pharmaceuticals and artistic content. IP's influence is now economy-wide, affecting a wide swathe of sectors and demand. Today, a mobile phone may have as many as 3 000 different patents. The development of technologies such as digitisation and the Internet has brought consumers into more direct and frequent contact with copyright laws by making it easier, faster and cheaper to create, duplicate and disseminate content. Consequently, IPRs have become a mainstream framework condition that has a broad effect on innovation.

In the past, firms that used IP tended to rely more frequently on one particular type, and to the extent that they owned multiple kinds of IP, these may have been used in very distinct parts of the firm's business operations. For example, media companies could rely almost exclusively on copyright while maintaining trademarks on their brand names. Today, more companies use a bundle of IP rights. For instance, in-house software used in product design and manufacturing is common at larger firms and is typically protected by copyright, while the products themselves may be protected by patents, trademarks and, again, copyrights. Indeed, evidence suggests that firms worldwide increasingly rely on the joint use of patents, trademarks and industrial designs.

Looking at evidence on IP's aggregate economic role, three main points emerge: IP's economic importance has grown over time, investment in IP-protected assets was resilient during the economic crisis, and that investment is growing much faster than investment in physical assets (see also OECD, 2014a). Figure 5.8, for example, illustrates all three points:

Figure 5.8. **Total UK investment in tangible and intangible assets, including intellectual property, 1990-2011**

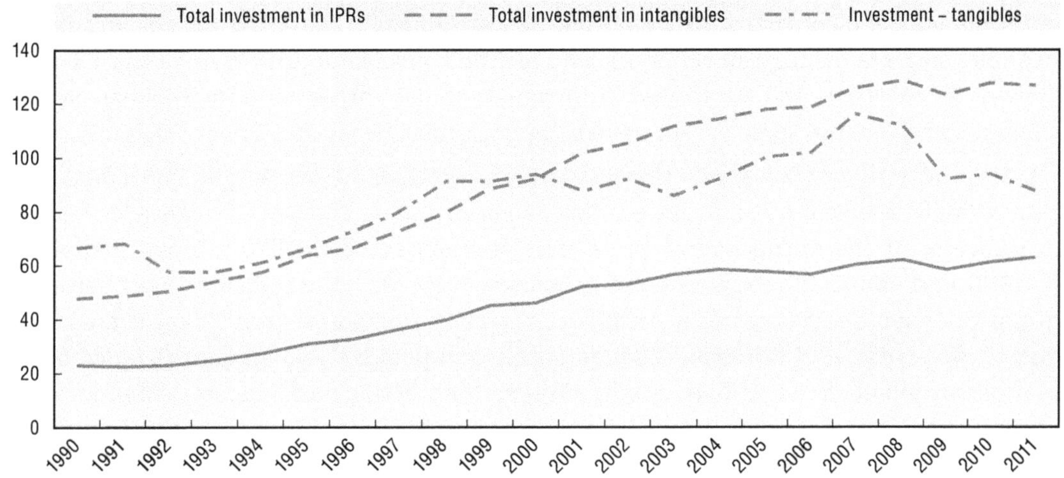

Source: Derived from data in Goodridge, Haskel and Wallis (2014), "Estimating UK investment in intangible assets and intellectual property rights", and from supplementary data provided by Professor Goodridge.

Copyright and trade secrets have a bigger role than some might have thought. Indeed they may in some respects be the most economically significant forms of IP. Investment in copyrighted works in the United Kingdom has grown more than investment in any other form of IP-protected asset (with the possible exception of trade secrets), more than tripling on a nominal basis between 1990 and 2011. As of 2011, by far the largest component of IP-protected investment in the United Kingdom involved copyrighted works, which drew more than twice the amounts invested in unregistered designs and trademarks, and nearly five times the amount invested in patents (Goodridge, Haskel and Wallis, 2014). Furthermore, copyright has also had a relatively strong impact on jobs in the United States in comparison with the impact of other forms of IP. Employment data show that job growth in copyright-intensive industries far outpaced that in trademark- and patent-intensive industries from 1990 to 2011. In fact, during that period, employment actually contracted in the latter category of industries, and markedly so in the patent-intensive group (US Department of Commerce, 2012).

Yet the economics of copyright and trade secrets has been less researched, mainly because data are less readily available.[8] While it is difficult to obtain data on trade secrets due to their nature, more could be done to improve data availability with respect to copyright. For example, governments could fund research and surveys to estimate the benefits of copyright registration. Then they could encourage voluntary copyright registration by enhancing the protections available for registered copyrights. They could also change the accounting rules that apply to creative industries to enable better data collection.

IP frameworks and stakeholders have been and continue to be affected by a number of broad developments, including the rise of cloud computing, the growth of the Internet, digitisation and globalisation. These developments have created new challenges for IP, for instance by facilitating piracy and industrial espionage. But they have also created new opportunities for IP to stimulate and diffuse invention and creativity. For example, new business models and research tools (based for instance on text and data mining, open access and e-content) hold the promise of stimulating diffusion.

New OECD indicators, including several composite indices, have been tested with data from the European Patent Office. The composite indices are consistent in that they show: 1) The average technological and economic value of inventions protected by patents has eroded over time, at least up to and including 2004, possibly reflecting application backlogs and strategic behaviour such as defensive patent filings; 2) patented micro- and nano-technologies have the highest economic and technological value; and 3) Australia, Canada, Norway, South Africa, and the United Kingdom are the countries with the highest average economic and technological patent values (Squicciarini, Dernis and Criscuolo, 2013).

To provide a way to study the relationship between the strength of trade secret protection in an economy and that economy's performance, the OECD recently developed an indicator of the stringency of protection of trade secrets. With a broad sample of 37 OECD and non-OECD countries and data from 1985-2010, the indicator has been used to test the hypothesis that more stringent protection of trade secrets is associated with greater innovation and diffusion. The results show a positive and statistically significant relationship between the stringency of trade secret protection and indicators of innovation inputs (Lippoldt and Schultz, 2014). While these results do not mean that ever-stronger protection will yield similar results, the positive and statistically significant relationships identified do indicate that adequately protecting trade secrets may be an appropriate policy for supporting certain key aspects of economic performance.

While experiencing growth in terms of the number of industrial designs contained in applications, some studies suggest that design rights do not seem to be terribly important to innovation. While, according to one study, design is significant for 85% of businesses in the United Kingdom, a mere 4% of these businesses use registered designs (and only another 4% use unregistered designs). Nevertheless, earlier work indicated that shares of companies that were "effective users of design" (but not necessarily design rights) outperformed the stock market in the United Kingdom by 200% between 1994 and 2004 (Design Council, 2005). That raises questions about the effectiveness of design rights for motivating investment in design; further empirical work is clearly needed in this area.

The patent arrangement – granting exclusive rights in exchange for more inventions and better dissemination – could be more fully achieved if certain steps are taken to improve both the disclosure and the diffusion of information. Some surveys suggest that the information contained in patents is not in fact very useful to disclose information on the innovation. Opinions tend to vary by sector, though, with respondents in biotech, medical devices and computer hardware having relatively favourable views and those in the software and nanotech sectors being less favourable. Views also vary by firm size, with SMEs being four times less likely to consider patent disclosures important than large manufacturing firms. Experts at a recent OECD workshop suggested several ideas for improving the effectiveness of disclosure requirements, including more enforcement, ensuring that patent information is more up to date by reducing the time between filing and publication, and continuing to improve access by moving patent databases online and making them freely available.

IP can facilitate business finance, especially for SMEs, in two ways. First, IP can serve as a signal of a firm's quality (both managerial and technological), which helps to compensate for information asymmetries. Second, IP can boost profitability because it confers exclusive rights to use inventions or creations, which can lead to competitive advantages. If there is a well-functioning secondary market for IP, the IP can also be sold if the firm that owns it has trouble repaying its loans. In other words, IP can serve as collateral in debt financing. Indeed, a substantial body of empirical work has found that young, high-growth firms with IP assets receive more funding than firms without IP. Nevertheless, IP-based finance is significantly underused, especially by SMEs, which are most in need of it. One reason for this is a lack of opportunities to sell IP in secondary markets. Policy makers in several countries are striving to support IP markets. Generally, their efforts fall into two categories: 1) supporting greater transparency of IP ownership and transfer information via disclosure requirements or measures to foster greater clarity in patent claims; for example, to enhance clarity in patent claims, the US Patent and Trademark Office (USPTO) has strengthened the technical training of patent examiners and made it easier for external experts to contribute to that training; and 2) creating new IP market infrastructures. Another approach that governments can take is to help manage the risks associated with collateralising IP. Government agencies and development banks can do that through risk-sharing mechanisms. Some key findings from the OECD's work on IPR and innovation are included below.

Main policy messages from the OECD's work on IPR and innovation

- IP's overall role in economies has evolved from a niche policy area that was relevant to only a handful of industries to a force that influences a wide swathe of demand and sectors. Consequently, IP policy has become a mainstream framework condition that has a broad effect on innovation, among other areas such as trade, competition, taxes and consumer protection.
- Copyright appears to be the type of IP with the most impressive economic performance, and it is undergoing more statutory change than other IP types, yet there are fewer empirical studies about copyright than about patents. Encouraging and enabling the collection and availability of more data on copyright would facilitate data-driven copyright policy.
- SMEs are better at creating jobs than large companies, but it has become harder for SMEs to find financing. Making it easier to use IP as a basis for obtaining financing would help SMEs to drive job growth and spur innovation.
- To lever the economic benefits of patents, steps should be taken to improve the diffusion of patent information.
- Recent, exploratory OECD work on trade secrets suggests a link between trade secret protection and innovation. Further work in this area would be worthwhile.

5.10. Knowledge networks and markets

The OECD introduced the concept of knowledge networks and markets (KNMs) in the 2010 Innovation Strategy (OECD, 2010a). KNMs are the set of systems, institutions, social relations, networks and infrastructures that enable the exchange of knowledge and associated IPRs. The term "KNM" has grown in popularity and has been applied to a very diverse set of agreements, institutions, organisations and intermediaries in the innovation system. Broad categories of KNMs can be defined on the basis of whether their focus is on:

- facilitating the transfer of disembodied knowledge, as in the case of searchable registers and repositories of existing data and information
- providing platforms for sourcing solutions to *ad hoc* problems and challenges (including platforms for innovation prizes or identifying consultants to assist with new R&D projects)
- resolving ownership of, and the transfer of rights to, disembodied knowledge (IP brokers, patent pools and funds primarily deal with the allocation of IP rights and the management of financial assets and liabilities attached to these rights)
- the transfer of knowledge embodied in people
- transforming the nature of the knowledge embodied in goods or people (for example, standard-setting organisations codify existing know-how and best practices embodied in a community of practice).

KNMs are many and varied. (OECD [2013c] discusses schema for classifying KNMs.) This section briefly outlines developments and challenges associated with online knowledge marketplaces and KNMs in the field of synthetic biology.

Online knowledge marketplaces (OKMs) manage platforms that communicate, match and transact innovative knowledge (Dushnitsky and Klueter, 2010). In general, OKMs are independent entities, unaffiliated with either knowledge owners or seekers. Many operate as for-profit companies, but some are not-for-profit ventures that rely to different degrees on member subscriptions, fees or other support. OKMs share similarities with more widely known online marketplaces for goods and services, such as the ambition to exploit economies of scale and scope.

Dushnitsky and Klueter (2010) studied 30 prominent websites that act as marketplaces in which owners of knowledge (e.g. a patent owner or an entrepreneur with an innovative business idea) interact with knowledge seekers (e.g. potential licensees or prospective investors). They find that IP-related OKMs systematically require entrepreneurs and inventors to disclose their inventions and/or pay upfront fees as a prerequisite for participation. Both mechanisms appear to alleviate the problem of adverse selection (but their effectiveness as an inducement to widespread market participation may be limited).

While OKMs may attract owners of high-quality inventions, their anonymity and speed might result in domination by low-quality ideas. Anonymity has benefits, such as lower risks when information is disclosed, but it also helps to dilute the reputational ties that bind parties together and contribute to building trust. To be efficient, OKMs typically require standardised procedures to collect and convey information about the knowledge procured and the knowledge offered. Otherwise, they would not deliver services at a significantly lower cost than the sum of individual, uncoordinated search efforts. In this connection, emerging semantic technologies can be important for organising and communicating information about knowledge.

It is important for policy makers to consider how knowledge markets support the growth of promising new general-purpose technologies, drawing lessons from experience in other technology domains (OECD, 2012c). For example, the emerging field of synthetic biology relies heavily on engineering and computer science (OECD, 2014g). As noted by

Torrance and Kahl (2012), among others, disciplinary features of synthetic biotechnology require more consideration of standards setting, interoperability and interchangeability than is usual in other areas of biology. By operating at the intersection of biotechnology and information technology, synthetic biotechnology has the potential to be affected by IP problems that exist in both fields. For example, synthetic biologists have argued that strings of DNA bases are comparable to source code and that DNA strings could therefore be covered by copyright. However, Rai and Boyle (2007) have questioned the appropriateness of invoking copyright protection in this domain, owing for example to the wide scope for expressive choice when constructing DNA sequences with base pairs that do not exist in nature.

Probably the best-known KNM in the field of synthetic biology is the BioBricks Foundation (BBF), which has created a registry and repository of standard biological parts (the building blocks of synthetic biology). Scientists can browse the BioBricks catalogue and contribute new parts that conform to the foundation's specifications. BioBricks has created a technical standard, an open technology platform and a repository open to anyone interested in building new biological parts.

Among the standards-setting groups that have formed in the synthetic biology community, most express a preference for standards that remain open and accessible to the community as a whole. In this early development stage, academics play an important role, and the public ethos is quite visible. Synthetic biology also illustrates a potentially symbiotic relationship between open and proprietary innovation models. For example, the dissemination of synthetic biology "parts" on a free and open basis would likely increase demand for various proprietary DNA-synthesis platforms. Some of key policy messages related to knowledge networks and markets are included below.

Main policy messages: Knowledge networks and markets

Because KNMs are so diverse, only highly generic policy messages will have broad relevance. In this connection, key points are that:

- Promoting specific knowledge markets should not be considered as a policy objective in its own right. Rather, KNMs represent a set of potential instruments for achieving a wide range of policy goals.
- Policy should not continue to support networks once they are established and their benefits are clear to participants. At this stage, all participants should have found and put in place mechanisms for contributing fairly to costs, while sharing benefits. The government's role should shift to addressing problems that may be due to established networks (such as detrimental effects on competition in product markets).
- Improved measurement is needed. Through the OECD's work on KNM indicators, four broad areas for measuring knowledge flows have been identified: 1) **skills mobility and knowledge flows**. The knowledge embodied in people and the very different types of data required for tracing such flows warrant special measurement efforts; 2) **disclosing and accessing knowledge**. Analysing access to and the use of knowledge sources, including repositories of disclosed information on science and technology, is of key importance; 3) **transactions on knowledge and knowledge rights**. Traditional and new evidence sources should be used to shed light on how different actors transact with other parties to procure knowledge; and 4) **co-creating knowledge**. Beyond transactions, good indicators are needed of collaboration in the creation of knowledge.

Notes

1. Returns to investments in basic science vary across countries. This point, which has not been subject to extensive study, could have significant policy implications. For instance, it is relevant to the issue of whether small economies might benefit from pooling research resources, rather than employing those resources individually.
2. Forthcoming OECD work does point to an important contribution of investment in basic research to long-term productivity growth, however.
3. In the first half of the 2000s, two important and widely referenced university rankings were established: *The Academic Ranking of World Universities* (first published in 2003), known as the "Shanghai Ranking", and the *Times Higher Education World University Ranking* (first published in 2004). A host of similar rankings followed.
4. The NIH has made its public access policy mandatory: all funded researchers must submit an electronic version of their final peer-reviewed manuscripts to PublicMed Central.
5. A recent survey of European universities identified 59 proof-of-concept/seed funds, of which half are in the United Kingdom (19) and Belgium (11) (Toschi, 2013).
6. In 2007 the OECD defined the information economy as the aggregate of the ICT and digital media and content industries. This aggregate includes ISIC Rev. 4 Division 26 (manufacture of computer, electronic and optical products) and Section J (information and communication services) consisting of Divisions 58-60 (publishing and broadcasting industries), 61 (telecommunications) and 62-63 (computer programming and information services). ICT trade and repair activities (in Groups 465 and 951) are also included, but are not considered in this chapter due to data availability.
7. Furthermore, OECD data indicate that international co-operation increases the quality of research (OECD, 2013).
8. Some countries have undertaken special statistical surveys on the management of intellectual property, e.g. Canada's Survey of Intellectual Property Management: *www23.statcan.gc.ca/imdb/p2SV.pl?Function=getSurvey&SDDS=5183&lang=en&db=imdb&adm=8&dis=2.*

References

Aite Group (2012), *The Next Generation of Execution Consulting Services: Leveraging Technology to Build Relationships*, *www.streambase.com/wp-content/uploads/downloads/AITE_Group_Execution_Consulting_Services_report_Early_Access.pdf.*

Akcigit, U., D. Hanley and N. Serrano-Velarde (2014), "Back to basics: Basic research spillovers, innovation and growth", *NBER Working Paper Series*, No. 19473.

Anderson, C. (2008), "The end of theory: The data deluge makes the scientific method obsolete", *Wired Magazine*, 23 June, *www.wired.com/science/discoveries/magazine/16-07/pb_theory/.*

Åstebro, T., N. Bazzazian and S. Braguinsky (2012), "Startups by recent university graduates and their faculty: Implications for university entrepreneurship policy", *Research Policy*, Vol. 41, pp. 663–677.

Bakhshi, H., A. Bravo-Biosca and J. Mateos-Garcia (2014), *Inside the Datavores: Estimating the Effect of Data and Online Analytics on Firm Performance*, Nesta, London, *www.nesta.org.uk/sites/default/files/inside_the_datavores_technical_report.pdf.*

Bourelos, E., M. Magnusson and M. McKelvey (2012), "Investigating the complexity facing academic entrepreneurs in science and engineering: The complementarities of research performance, networks and support structures in commercialisation", *Cambridge Journal of Economics*, Vol. 36/3, pp. 751-780.

Brynjolfsson, E., L.M. Hitt and H.H. Kim (2011), *Strength in Numbers: How Does Data-Driven Decisionmaking Affect Firm Performance?*, *http://papers.ssrn.com/sol3/papers.cfm?abstract_id=1819486.*

Bulut, H. and G. Moschini (2009), "US universities' net returns from patenting and licensing: A quantile regression analysis", *Economics of Innovation and New Technology*, Vol. 18, pp. 123-137.

Design Council (2005), *Design Index: The Impact of Design on Stock Market Performance*, The Design Council, London.

Dushnitsky, G. and T. Klueter (2010), "Is there an eBay for ideas? Insights from online knowledge marketplaces," *European Management Review*, Vol. 8(1), pp. 17-32.

The Economist (2012), "High-frequency trading: The fast and the furious", *The Economist*, 25 February, *www.economist.com/node/21547988*.

Goodridge, P., J. Haskel and G. Wallis (2014), "Estimating UK investment in intangible assets and intellectual property rights", Report 2014/36 commissioned by the UK Intellectual Property Office, Intellectual Property Office, *https://www.gov.uk/government/uploads/system/uploads/attachment_data/file/355140/ipresearch-intangible.pdf*.

Hertzfeld, H.R., A.N. Link and N.S. Vonortas (2006), "Intellectual property protection mechanisms in research partnerships", *Research Policy*, Vol. 35, pp. 825-838.

Houghton, J., B. Rasmussen and P. Sheehan (2010), *Economic and Social Returns on Investment in Open Archiving Publicly Funded Research Outputs*, Report to SPARC (Scholarly Publishing and Academic Resources Coalition), Centre for Strategic Economic Studies, Victoria University, Victoria, British Columbi, *www.sparc.arl.org/sites/default/files/vufrpaa.pdf*.

House of Commons, Science and Technology Committee (2013), *Bridging the Valley of Death: Improving the Commercialisation of Research*, The Stationery Office Limited, London.

IDC (International Data Corporation) (2012), *Worldwide Big Data Technology and Services 2012-2015 Forecast*, IDC, Framingham, Massachusetts.

Kirchner, M. (2013), "A perverted view of impact", *Science*, Vol. 340, 14 June.

Koolwijk, E. and S. Peeters (2011), "WKK en BioWKK in de Glastuinbouw", *TVVL Magazine*, June.

Lippoldt, D. and M.F. Schultz (2014), "Uncovering trade secrets – An empirical assessment of economic implications of protection for undisclosed data", *OECD Trade Policy Papers*, No. 167, OECD Publishing, Paris, *http://dx.doi.org/10.1787/5jxzl5w3j3s6-en*.

McKinsey & Company (2013), *Disruptive Technologies: Advances that Will Transform Life, Business, and the Global Economy*, McKinsey Global Institute, *www.mckinsey.com/insights/business_technology/disruptive_technologies*.

OECD (2015), *Digital Economy Outlook*, OECD Publishing, Paris, *http://dx.doi.org/10.1787/9789264232440-en*.

OECD (2014a), *OECD Science, Technology and Industry Outlook 2014*, OECD Publishing, Paris, *http://dx.doi.org/10.1787/sti_outlook-2014-en*.

OECD (2014b), *Main Science and Technology Indicators 2014-I*, OECD Publishing, Paris, *http://dx.doi.org/10.1787/msti-v2014-2-en*.

OECD (2014c), *Promoting Research Excellence: New Approaches to Funding*, OECD Publishing, Paris, *http://dx.doi.org/10.1787/9789264207462-en*.

OECD (2014d), *The Impacts of Large Scale Research Infrastructures on Economic Innovation and on Society*, OECD Publishing, Paris, *www.oecd.org/sti/sci-tech/CERN-case-studies.pdf*

OECD (2014e), *International Distributed Research Infrastructures: Issues and Options*, OECD Publishing, Paris, *www.oecd.org/sti/sci-tech/international distributed research infrastructures.pdf*.

OECD (2014f), *Measuring the Digital Economy: A New Perspective*, OECD Publishing, Paris, *http://dx.doi.org/10.1787/9789264221796-en*.

OECD (2014g), *Emerging Policy Issues in Synthetic Biology*, OECD Publishing, Paris, *http://dx.doi.org/10.1787/9789264208421-en*.

OECD (2013a), *Commercialising Public Research: New Trends and Strategies*, OECD Publishing, Paris, *http://dx.doi.org/10.1787/9789264193321-en*.

OECD (2013b), "Building blocks for smart networks", *OECD Digital Economy Papers*, No. 215, OECD Publishing, Paris, *http://dx.doi.org/10.1787/5k4dkhvnzv35-en*.

OECD (2013c), "Knowledge networks and markets", *OECD Science, Technology and Industry Policy Papers*, No. 8, *http://dx.doi.org/10.1787/5k44wzw9q5zv-en*.

OECD (2012a), *Meeting Global Challenges through Better Governance*, OECD Publishing, Paris, *http://dx.doi.org/10.1787/9789264178700-en*.

OECD (2012b), "Recommendation on the governance of clinical trials", OECD Publishing, Paris, *www.oecd.org/sti/sci-tech/oecdrecommendationonthegovernanceofclinicaltrials.htm*.

OECD (2012c), *Knowledge Networks and Markets in the Life Sciences*, OECD Publishing, Paris, *http://dx.doi.org/10.1787/9789264168596-en*.

OECD (2011a), *Opportunities, Challenges and Good Practices in International Research Cooperation between Developed and Developing Countries*, OECD Publishing, Paris, *www.oecd.org/sti/sci-tech/47737209.pdf.*

OECD (2011b), *OECD Council Recommendation on Principles for Internet Policy Making*, OECD Publishing, Paris, *www.oecd.org/sti/ieconomy/49258588.pdf.*

OECD (2010a), *The OECD Innovation Strategy: Getting a Head Start on Tomorrow*, OECD Publishing, Paris, *http://dx.doi.org/10.1787/9789264083479-en.*

OECD (2010b), *Performance-Based Funding for Public Research in Tertiary Education Institutions*, OECD Publishing, Paris, *http://dx.doi.org/10.1787/9789264094611-en.*

OECD (2010c), *Establishing Large International Research Infrastructures: Issues and Options*, OECD Publishing, Paris, *www.oecd.org/science/sci-tech/47027330.pdf.*

OECD (2007), *OECD Principles and Guidelines for Access to Research Data for Public Funding*, OECD Publishing, Paris, *www.oecd.org/science/sci-tech/38500813.pdf.*

Rai A. and J. Boyle (2007), "Synthetic biology: Caught between property rights, the public domain, and the commons", *PLoS Biol*, Vol. 5/3, p. e58, *http://dx.doi.org/10.1371/journal.pbio.0050058.*

Reimsbach-Kounatze, C. (2014), "The proliferation of data and implications for official statistics and statistical agencies: A preliminary analysis", *OECD Digital Economy Papers 245, OECD Publishing, Paris.*

Russom, P. (2007), "BI search and text analytics: New additions to the BI technology stack", *TDWI Best Practices Report*, Second Quarter, TDWI (The Data Warehousing Institute).

Shilakes, C. and J. Tylman (1998), *Enterprise Information Portals: Move Over Yahoo! The Enterprise Information Portal Is on Its Way*, Merrill Lynch.

Squicciarini, M., H. Dernis, and C. Criscuolo (2013), "Measuring patent quality: Indicators of technological and economic value," *OECD Science, Technology and Industry Working Papers*, 2013/03, OECD Publishing, Paris, pp. 59-62, *http://dx.doi.org/10.1787/5k4522wkw1r8-en.*

Stevens, A.J., G.A. Johnson and P.R. Sanberg (2012), "The role of patents and commercialization in the tenure and promotion process", *Technology & Innovation*, Vol. 13, pp. 241-248.

Tambe, P. (forthcoming), "Big data investment, skills, and firm value", *Management Science*, *http://ssrn.com/abstract=2294077.*

Torrance, S. and L. Kahl (2012), "Synthetic biology standards and intellectual property", the National Academies. Draft presented to the Symposium on Management of Intellectual Property in Standards-Setting Processes, *www.synberc.org/content/articles/standards-and-intellectual-property.*

Toschi, L. (2013), "Mapping university seed funds and proof of concepts funds in Europe: Initial evidence from the FinKT project", workshop presentation on *Financing Knowledge Transfer in Europe*, Bologna, 6 February 2013.

US Department of Commerce (2012), *Intellectual Property and the U.S. Economy: Industries in Focus*, Washington, DC.

Wright, M., B. Clarysse and S. Mosey (2012), "Strategic entrepreneurship, resource orchestration and growing spin-offs from universities", *Technology Analysis & Strategic Management*, Vol. 24, pp. 911-927.

The Innovation Imperative
Contributing to Productivity, Growth and Well-Being

Chapter 6

Effective innovation policies

The fourth set of policies for innovation relate to the specific actions that encourage firms to engage in innovation and entrepreneurial activity. This includes R&D tax incentives, place-based policies, policies for high-growth firms, new approaches to industrial policy, including smart specialisation strategies, demand-side innovation policies, as well as consumer policy. In designing such policies, the appropriate mix of policies is often an important consideration for government.

The statistical data for Israel are supplied by and under the responsibility of the relevant Israeli authorities. The use of such data by the OECD is without prejudice to the status of the Golan Heights, East Jerusalem and Israeli settlements in the West Bank under the terms of international law.

This chapter sets out new empirical findings and policy messages from a broad range of recent OECD work covering: the policy settings that shape business innovation, including the role of R&D tax incentives; the role of place-based policies for innovation; policies for high-growth firms; evolving approaches to what has been termed "new industrial policy", including smart specialisation strategies; demand-side innovation policies, many of which are oriented to meeting global challenges in areas such as health and the environment; and the influence of consumer policy on innovation, particularly in a set of fast-changing, complex and data-driven industries shaped by ICTs.

All of these policies are aimed at overcoming specific weaknesses and barriers to innovation that may be present in the economy, as discussed in Chapter 1, that contribute to low economic returns, or a low appropriability of such returns. Most countries have a wide range of policies in place to address such weaknesses and support innovation. However, with such a wide range of policies comes questions about the balance and the appropriate mix of specific policies. For example, many countries have a tendency to add further policies to the existing mix, rather than to replace existing policies with new ones. This can lead to duplication of policies, lack of critical mass and also to a complex environment for businesses that are seeking to understand the policy support that they might be eligible for. The final part of this chapter briefly discusses some questions on the overall policy mix, which is further discussed in Chapter 8.

6.1. Direct and tax support for business innovation

Chapter 4 already discussed the role of framework policies in channelling resources to innovative firms and found that several framework policies matter. However, in addition to framework conditions, the role of more targeted policies that support innovation also needs to be assessed. Governments offer direct support through public procurement for R&D and a variety of grants, subsidies, loans or equity funding. They also provide indirect support through fiscal incentives, such as R&D tax incentives. Direct funding allows governments to target specific R&D activities and steer business efforts towards new R&D areas or areas that offer high social returns relative to private returns, e.g. green technology, social innovation or other novel areas. Direct funding instruments typically depend on discretionary decisions by government agencies, which can help reduce deadweight loss, but creates opportunities for rent-seeking and can lead to lock-in. Tax incentives reduce the marginal cost of R&D and innovation spending; they are usually more neutral than direct support in terms of industry, region and firm characteristics, although this does not exclude some differentiation, most often by firm size (OECD, 2010a). While direct subsidies are more targeted towards long-term research, R&D tax schemes are more likely to encourage short-term applied research and boost incremental innovation rather than contribute to radical breakthroughs.[1]

A recent *OECD Science, Technology and Innovation Outlook* survey shows that direct financial support is typically offered through competitive grants and debt financing, such as loans for R&D projects (OECD, 2014a; Table 3.1). Risk-sharing mechanisms are widely used to provide lenders with insurance against the risk of default and improve firms' access to credit. Some direct support is also linked to public procurement (see discussion on stimulating demand for innovation later in this section). For example, in France and the United States, a large share of public support for R&D is provided to firms in the defence industry to develop military equipment and potentially civil applications. Many OECD countries also have schemes and funds to access early-stage finance, particularly for equity. Support is provided to the VC industry, with some governments actively providing equity funding (OECD, 2011a; Wilson and Silva, 2013). A common approach is to facilitate the growth of venture funding through public VC funds, co-investment funds with private investments and "funds of funds" (see discussion on financing in Chapter 4).

Direct support for innovation, other than R&D-related schemes, also includes measures to facilitate the commercialisation of innovation, support the development of networks, promote regional innovation hubs, and ease access to information, expertise and advice (OECD, 2011a). Innovation vouchers or technology consulting services and extension programmes are major policy instruments in this respect. Tax incentives, including corporate and personal income taxes, are also widely used to encourage private investments in R&D and the exploitation of IP assets, to attract business angels and leverage early-stage finance, and to attract foreign talent or foreign multinationals.

Over the past decade, OECD member countries have increasingly turned to tax incentives (rather than grants or other direct forms of support) as a means to support investment in R&D (OECD, 2014a). The majority of OECD countries use such tax incentives, as do many of the BRIICS economies (Figure 6.1). Moreover, in many countries the level of generosity of these tax incentives has increased in recent years (OECD, 2014a). The following paragraphs discuss the use of these incentives as part of innovation policy; however, it should be noted that in many countries they are also considered as part of the broader framework for corporate taxation.

However, the benefits of R&D tax support may be skewed. In particular, large, incumbent and multinational firms may be best placed to reap the benefits from such measures. This is due in part to their capacity to exploit international tax-shifting opportunities (see Chapter 4). It may also be due to the design of the tax incentive itself. For example, if there are no carry-forward provisions, new firms may not be able to benefit. Bravo-Biosca, Criscuolo and Menon (2013) provide evidence of the impact of R&D tax subsidies on the distribution of employment growth in R&D-intensive sectors. This work shows that support for R&D has a positive impact on employment growth only in incumbent firms with relatively low growth rates, while it has a negative effect on firm entry and on the employment of firms in the top of the growth distribution. These results suggest that R&D tax incentives might favour incumbent firms and slow down the reallocation process. The effect of the design of incentives on overall firm dynamism is, therefore, of great importance.

It is therefore important that R&D tax incentives are refundable or contain carry-over provisions so as to avoid overly favouring less dynamic incumbents at the expense

Table 6.1. **Major policy instruments for financing business R&D and innovation and some country examples**

Financing instruments			Key features	Some country examples
Direct public funding	Grants, subsidies		Most common funding instruments. Used as seed funding for start-ups and innovative SMEs. Granted on a competitive basis and in some cases, on the basis of private co-funding. No repayment is usually required. Supply-side, discretionary instruments.	ANR subsidies (Argentina), Central Innovation Programme for SMEs (Germany), R&D Fund (Israel), Small Business Innovation Research (SBIR) Program (US), TUBITAK 1512 (Turkey)
	Debt financing	Credit loans	Government subsidised loans. Require sorts of collateral or guarantee. Obligation of repayment as debt. The investor/ lender does not receive an equity stake.	Novallia (Belgium), High-Tech Gründerfonds (Germany), Public Investment Bank (France), Microfinance Ireland, Slovene Enterprise Fund, British Business Bank (UK)
		Repayable grants/ advances	Repayment required, partial or total, sometimes in the form of royalties. Could be granted on the basis of private co-funding.	Repayable Grants for Start-Ups (New Zealand)
		Loans guarantees and risk-sharing mechanisms	Used widely as important tools to ease financial constraints for SMEs and start-ups. In the case of individual assessment of loans, can signal ex ante the creditworthiness of the firm to the bank. Often combined with the provision of complementary services (e.g. information, assistance, training).	Small Business Financing Program (Canada), Mutual guarantee schemes (Confidi) (Italy), 7(a) Loan Program (US), R&I Loans Services (European Commission)
	Debt/Equity financing	Non-bank debt/equity funding	New funding channels. Innovative lending platforms and non-bank debt or equity funds.	Business Finance Partnership (UK)
		Mezzanine funding	Combination of several financing instruments of varying degrees of risk and return that incorporate elements of debt and equity in a single investment vehicle. Used at later stage of firms' development. More suitable for SMEs with a strong cash position and a moderate growth profile.	Guarantees for Mezzanine Investments (Austria), PROGRESS Programme (Czech Rep.), Industrifonden and Fouriertransform (Sweden), Small Business Investment Company (US)
	Equity financing	Venture capital funds and funds of funds	Funds provided by institutional investors (banks, pensions funds, etc.) to be invested in firms at early to expansion stages. Tends to increasingly invest at later -less risky-stage. Referred as patient capital, due to lengthy time span for exiting (10-12 years). The investor receives an equity stake.	Innpulsa (Colombia), Seed Fund Vera (Finland), France Investment 2020, Yozma Fund (Israel), Scottish Co-investment Fund (UK), Venture Capital Funding Program – TUBITAK 1514 (Turkey).
		Business angels	Provide financing, expertise, mentoring and network facilities. Tends to invest in the form of groups and networks. Financing at start-up and early stage.	Seraphim Fund (UK), Tech Coast Angels and Common ANGELS (US)
	Public procurement for R&D and innovation		Create a demand for technologies or services that do not exist, or, target the purchase of R&D services (pre-commercial procurement of R&D). Provide early-stage financial support to high-risk innovative technology-based small firms with commercial promise.	Small Business Innovation Research (SBIR) Program (US) and SBIR-type of programmes (UK)
	Technology consulting services, extension programmes		Expand the diffusion and adoption of already existing technology, and contribute to increase the absorptive capacity of targeted firms (especially SMEs). Provide information, technical assistance, consulting and training etc. Of particular importance in low income countries.	Manufacturing Extension Partnerships (US)
	Innovation vouchers		Small lines of credit provided to SMEs to purchase services from public knowledge providers with a view to introducing innovations in their business operations.	Innovation Voucher (Austria, Chile, China, Denmark, etc.)
Indirect public funding	Tax incentives	Tax incentives on corporate income tax	Used in most countries. Broad range of tax arrangements on corporate income tax, including tax incentives on R&D expenditure and, less frequently, tax incentives on IP-related gains. Indirect, non-discriminatory.	SR&ED tax credit (Canada), R&D Tax Credit (France), exemption on payroll withholding tax (Netherlands), patent box (UK), R&D Law 5746 (Turkey)
		Tax incentives on personal income tax and other taxes	Available in many countries. Broad range of tax incentives on R&D and entrepreneurial investments and revenues that apply to personal income tax, value added tax or other taxes (consumption, land, property ec.). Indirect, non-discriminatory.	Personal wage tax reduction for foreign researchers and key staff (Denmark), wealth tax exemption for business angels (France), Business Expansion and Seed Capital Schemes (Ireland)

Source: OECD (2014a), *OECD Science, Technology and Industry Outlook 2014*, OECD Publishing, Paris, *http://dx.doi.org/10.1787/sti_outlook-2014-en*.

Figure 6.1. **Government funding of BERD, direct funding and R&D tax incentives, 2012**

As a per cent of GDP

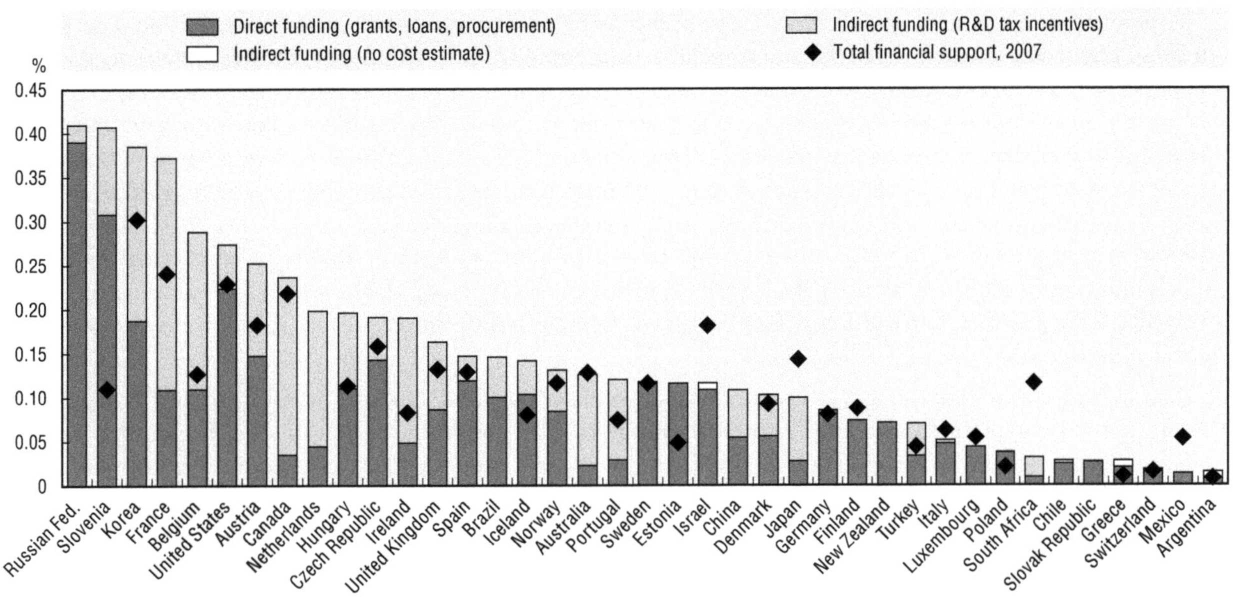

Source: OECD (2014a), *OECD Science, Technology and Industry Outlook 2014*, OECD Publishing, Paris, *http://dx.doi.org/10.1787/sti_outlook-2014-en*. For more details, see *www.oecd.org/sti/rd-tax-stats.htm*.

of dynamic young firms. The implicit subsidy rate of R&D tax incentives increases with the profitability of the firm, and many young innovative firms are typically in a loss position in the early years of an R&D project. Thus, these firms will not benefit from the programme unless it contains provisions for immediate cash refunds for R&D expenditure or allows such firms to carry associated losses forward to deduct against future tax burdens.

Complementing R&D (expenditure-based) tax incentives are tax incentives on intellectual property (or so-called patent boxes) that are being used by a number of OECD countries (Box 6.1).[2]

Box 6.1. **Tax incentives on intellectual property**

IP assets, such as patents, copyrights, trademarks or brands, are highly mobile and therefore can be located away from the activity that generated them. In particular, multinationals often locate their intangible assets in low-tax jurisdictions to reduce their corporate tax liabilities, thus eroding tax revenues in countries with high tax regimes –BEPS. Some countries have decided to introduce special IP tax regimes, generally called "patent boxes". For example, France introduced such a scheme in 2001, Hungary in 2003, and the Netherlands and Belgium in 2007, followed by Spain and Luxembourg in 2008. The United Kingdom introduced a scheme in 2013 and Italy and Ireland have recently proposed such schemes.

A question that can be raised is whether such patent boxes are likely to be effective and represent "value for money". The debate on the impact of such R&D income support schemes has raised some concerns as to whether 1) they tackle the fundamental market failures of investment in innovation; and 2) they bring long-term increases in tax revenue or are prone to spur tax competition that ultimately results in a fall in tax revenues for all countries concerned (Griffith et al., 2010).

Box 6.1. **Tax incentives on intellectual property** *(cont.)*

A patent box by its very nature gives an *ex post* reward only to successful innovators that already hold a monopoly right on their inventions and receive an income from it. The policy, therefore, will not foster experimentation per se. Experimentation is a risky activity, which naturally entails high rates of failure, and is an important feature of policies designed to foster entrepreneurship, ensure reallocation, and promote learning and growth at the innovation frontier (Andrews and Criscuolo, 2013). Secondly, a patent box might push firms to focus on innovations that lead to outcomes susceptible to protection by IPRs, and therefore may lead firms to focus more on applied research (Akcigit, Hanley and Serrano-Velarde, 2014) or towards products that are closer to market that in the long run might not be a productivity-enhancing strategy. The policy could also push firms to seek patent protection for innovations for which they would not have sought patent protection in the absence of the policy. In fact, as innovation surveys reveal, many innovative firms choose not to seek any IP protection.

In addition, credit-constrained, innovative firms need funds to conduct their research as early as possible. Policies that provide funds with a lag, relative to research effort, might not be suitable for this group of firms and might even make the playing field uneven (Andrews and Criscuolo, 2013). The patent box does seem to suffer from this limitation and thus is unlikely to support research activities by credit-constrained firms. Given the long lags that characterise the patenting process, firms are likely to benefit from the support only years after having carried out the R&D investment. The patenting lag may raise a further issue of additionality, because the scheme might provide discounted tax rates for income deriving from R&D activity already conducted before the introduction of the scheme.

Finally, while IP boxes may in some instances cover several IP assets, concern has been expressed that a large share of patents tend to be held by a small number of large multinational corporations, particularly when focusing on high-revenue patents. This would imply that the benefits of the patent box would accrue mainly to multinational firms, which are precisely the firms that might find ways of using the patent box to shift profits across jurisdictions. For example, calculating the income eligible for the tax breaks when firms directly use the IP is extremely difficult, both because of the difficulty of clearly identifying the stream of profits generated by a single patent when multiple patents – often granted at different points in time – are used to produce a complex product (such as semiconductors or microchips), and because the income flow will have to be imputed in the absence of an explicit price for the use of the IP. This difficulty for the tax authority of knowing the share of profits concerned might allow firms to abuse the system in order to shift profits across jurisdictions.

For the reasons outlined above, grants and other forms of direct support may be valuable as a complementary form of support for R&D, perhaps targeting the firms that are unlikely to benefit as much from tax incentives (e.g. young firms). In other cases it may be necessary to provide direct forms of support for more mission-oriented innovation that has strong public good elements (e.g. public health, climate change, national security). However, in such cases the award selection process must be designed so as to ensure efficiency, avoid rent-seeking activities and avoid problems of adverse selection. A recent OECD survey shows that direct funding instruments, especially competitive grants, remain major levers of innovation policy (Figure 6.2, Panel A). Direct support is provided through an increasing variety of tools for an increasing variety of purposes (e.g. to encourage knowledge transfer, growth of high-technology start-ups, VC activity, green innovation; see OECD [2014a]). Moreover, several of these instruments, e.g. innovation vouchers and credit loans, have increased in importance in recent years (Figure 6.2, Panel B).

Figure 6.2. **Relevance of major funding instruments in the policy mix for business R&D and innovation, 2014**

As a percentage of total country self-reported responses

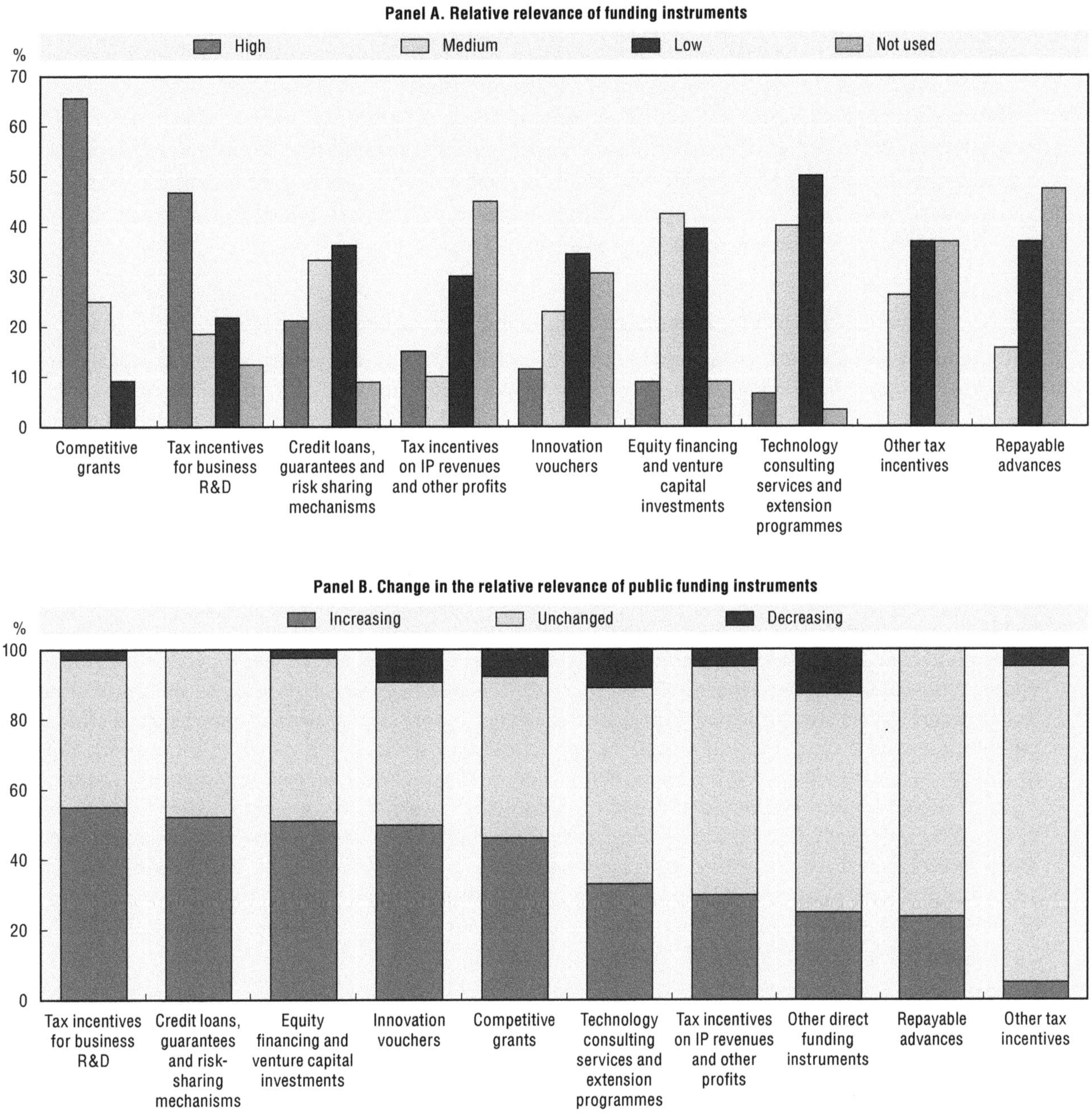

Note: Simple counts of country responses to the question: "Which of the following are the principal instruments of public funding of business R&D and innovation in your country? How has the relative balance between these instruments changed recently, if at all? Please rate the relative relevance of the following financial instruments in your country's policy mix and indicate whether their share in the total has increased/decreased or has remained unchanged." Responses are provided by delegates to the OECD Committee for Scientific and Technological Policy.

Source: OECD (2014a), *OECD Science, Technology and Industry Outlook 2014*, OECD Publishing, Paris, *http://dx.doi.org/10.1787/sti_outlook-2014-en*.

In many countries, direct support mechanisms involve a wide range of instruments, tailored at specific firms, innovation barriers and stages of the innovation process. Partly linked to the wide range of existing instruments, evaluations are typically more limited than for R&D tax credits, and the impact of individual measures on overall innovation performance is often harder to assess. Further work on the evaluation of direct support instruments would help establish a better understanding of good practices in this area. Good design can help ensure that individual instruments are as efficient as possible, but governments also need to evaluate the overall mix to reduce duplication, ensure sufficient scale for each instrument and ensure that the combination of policy instruments addresses the key innovation barriers. Careful consideration of programme logic is also required when establishing such schemes (not least because a clear articulation of programme logic is essential for evaluation). Policy makers must also ensure – among other things – that these programmes do not distort market mechanisms. Some key policy messages of the OECD's work on business innovation are presented below.

Main policy messages: Fostering business innovation

- **R&D tax incentives should be designed to meet the needs of "stand-alone" firms without cross-border tax planning opportunities and young, innovative firms.** Domestic stand-alone firms that perform R&D may be put at a competitive disadvantage in relation to MNEs unless other measures, such as ceilings and differentiated rates, are put in place to ensure a level playing field. Young firms may also benefit less if they have not yet generated taxable income to make immediate use of (non-refundable) R&D tax incentives. This may inhibit innovation and growth as such firms have particular strengths as R&D performers (e.g. in creating radical innovations) and job creators unless measures such as cash refunds, carry-forwards, or the use of payroll withholding tax credits for R&D-related wages are used. But care must be taken that tax relief is not so high that it hampers the process of creative destruction that is essential to a dynamic innovative ecosystem.
- **Policy makers should balance indirect support for business R&D (tax incentives) with the use of direct support measures to foster innovation.** Recent OECD analysis suggests that direct support measures – e.g. contracts, grants, awards for mission-oriented R&D or support for networks – may be more effective in stimulating R&D than previously thought, particularly for young firms that lack the upfront funds to start an innovative project (Westmore, 2013). It is important, however, that any allocation of direct support should be non-automatic and based on competitive, objective and transparent criteria (e.g., by involving independent international experts in the selection process). Moreover, selection processes must be designed to ensure efficiency (including minimal bureaucracy), avoid rent-seeking activities and avoid problems of adverse selection. More broadly, a well-designed and transparent system of direct support measures can be complementary to the use of R&D tax incentives as it may help direct public funding to high-quality projects with high social returns.
- **Governments should ensure that R&D tax incentive policies provide value for money.** In many countries, overall tax relief for business R&D may be greater than governments intended when they designed support of R&D business expenditure. This may be compounded by the rising generosity of tax relief for R&D observed over the past decade, the full cost of which is not always transparent because R&D tax incentives are "off budget" as a tax expenditure. As a result, governments should undertake systematic evaluation of tax relief measures to assess the continuing validity of their rationale and objectives and whether their targeting and design remain appropriate. Important aspects of R&D tax schemes that require review include the scope of eligible R&D, the firms that qualify, the treatment of large R&D performers, and carry-back and carry-forward provisions. Governments should also focus on the policy package – including the interactions and complementarities – as well as related fiscal measures concerning R&D workers to ensure that R&D tax incentives provide value for money.
- **The effectiveness of R&D tax incentives depends upon the broader regulatory environment and its stability over time.** OECD evidence shows that well-functioning product, labour and risk capital markets, and bankruptcy laws that do not overly penalise business failure, can raise the returns to investing in

Main policy messages: Fostering business innovation *(cont.)*

knowledge-based assets. OECD analysis also suggests that in countries that have experienced a large number of R&D tax policy reversals, the impact of R&D tax credits on private R&D expenditure is greatly diminished (Westmore, 2013). It is therefore important that governments do not repeatedly tinker with such policies to minimise policy uncertainty for firms.

- **Incentives to encourage business R&D must be properly designed.** OECD governments increasingly use tax incentives to support business investment in R&D. But R&D tax incentives might favour incumbent firms (for instance when they do not include carry-forward provisions) and slow the reallocation process.

6.2. The role of place-based dynamics for innovation

Another area of policies to foster innovation that is of growing importance in many countries concerns place-based policies. The potential to benefit from a country's innovation policy and contribute to its innovation performance varies considerably across locations. Striking local variation can exist for some important inputs to the innovation process, such as R&D and skilled workers (Figure 6.3). Variation also exists for intermediate outputs, such as patenting. For example, in 2010, the top 10% of small regions in the OECD accounted for just under 40% of both population and GDP, but contributed to 58% of total patenting (OECD, 2013a).

Despite the dramatic changes that ICT has brought in terms of connecting people and firms, geographic proximity continues to matter in the innovation process. The academic literature is replete with studies that show the persisting, and in some cases increasing, importance of geographic proximity for innovation. As examined in research dating back to Alfred Marshall, the agglomeration of firms and their suppliers can confer a competitive advantage on the enterprises involved. Agglomeration, or clustering, can permit locally concentrated labour markets, specialisation in production and the attraction of specialised buyers and sellers. Such concentration and specialisation can bring a range of (self-reinforcing) benefits, including: more efficient sharing of infrastructures and facilities; a greater division of labour between firms (offering greater scale economies for individual firms); and more efficient matching between economic agents (employees and employers, financiers and firms looking for funding, firms looking for partners, and buyers and sellers).

By operating in close proximity, firms can also more easily subcontract to competitors those orders that exceed their own capacities – because proximity allows greater knowledge of the capabilities of potential contractors – which may allow firms to retain valued customers. The clustering of firms can likewise facilitate the flow of ideas and information, and may also facilitate the access of firms to risk capital. Such flows occur formally and informally, for example when employees change employer, through contacts with common suppliers, and through social exchanges. Carlino and Kerr (2014) provide a recent review of how agglomeration relates to innovation.[3]

Policies to support "clusters" are frequently used by subnational authorities. The exact definition of a "cluster" is a subject of debate among academics and practitioners. However, the term is here used to refer to geographic concentrations of inter-related firms, higher education and research institutions, and other public and private entities. In recent years, policies to foster enterprise clusters have been common in both OECD and developing economies, in wealthy and lagging subnational regions, and in jurisdictions with *laissez-faire* and *dirigiste* approaches to economic development. Policy makers have devoted significant public resources to cluster policies. Most OECD countries have some form of cluster or sector-based approach that supports innovation (OECD, 2014a). Cluster approaches are often used in ways that overlap with industrial policy, science and technology policy, and regional development policy (Table 6.2) (OECD, 2007).

Figure 6.3. **Significant inter-regional variations in innovation inputs contribute to disparities in performance**

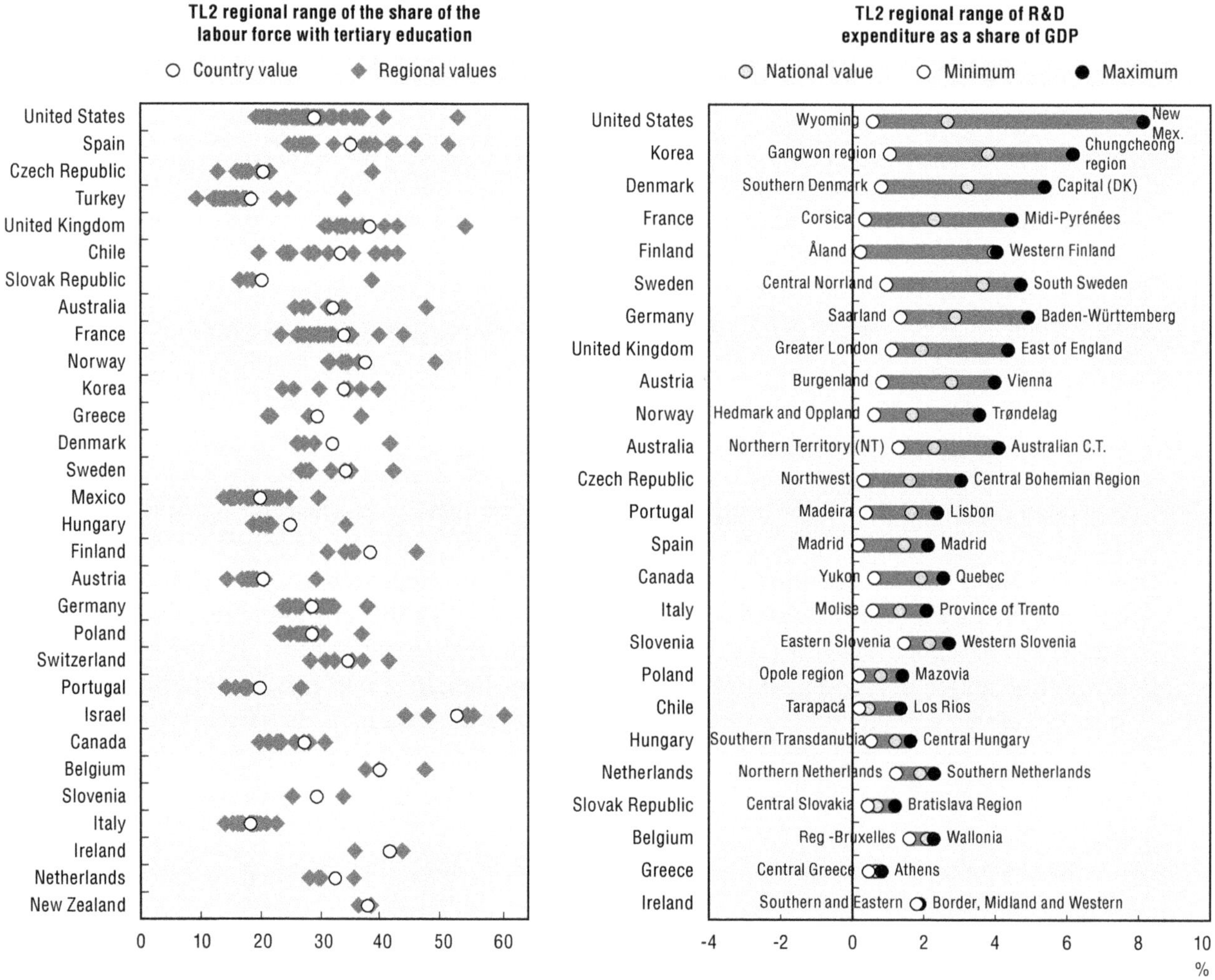

Note: OECD large regions (TL2 = Territorial level 2) represent the first administrative tier of subnational government. In the left panel, a diamond represents the share of the labour force with a tertiary education in each large region within the country. The dot represents the country average. Data are for 2012 except for: Australia (2011); Greece (2006); Israel (2005); Italy (2011); Portugal (2011) and Slovenia (2008). In the right panel, the line represents R&D expenditure as a percentage of GDP for large regions. The white dot represents the region with the lowest value. The grey dot represents the country average. The black dot represents the region with the highest value. Data are for 2010 except for: Australia (2009); Austria (2009); Belgium (2009); Czech Republic (2011); Denmark (2009); France (2009); Germany (2009); Greece (2005); Israel (2008); Netherlands (2009); Slovak Republic (2011); Sweden (2009) and the United Kingdom (2009).

Source: OECD (2013a), *Regions at a Glance*, using data from the *OECD Regional Database*. Left graphic: *http://dx.doi.org/10.1787/888932913722*. Right graphic: *http://dx.doi.org/10.1787/888932913779*.

Table 6.2. **Cluster development support policies in selected OECD countries**

Policy goal	Actions	Countries reporting policy
Creating and consolidating clusters	Creation of new clusters through co-ordinated action for R&D activities (e.g. public funding programmes)	Argentina, Chile, Norway
	Promotion of network structures, service support for entrepreneurs, cluster co-ordination	Argentina, Austria, Australia, Belgium, Canada, China, Colombia, Denmark, France, Germany, Greece, Ireland, Japan, New Zealand, Norway, Sweden
Internationalisation	Cluster competition and cluster excellence programmes	Austria, Belgium, Germany, France, Ireland, Japan, Netherlands, European Commission
Networking platforms	Science-science (e.g. promotion of collective research centres, centres of excellence)	Belgium, Canada, Denmark, France, Norway, South Africa, Spain, Switzerland, Turkey
	Industry-science (e.g. promotion of public-private networks, science parks)	Argentina, Australia, Belgium, Canada, Colombia, Denmark, Finland, France, Germany, Italy, Norway, Poland, Portugal, United Kingdom
	Industry-industry (e.g. promotion of sectoral networks)	Belgium, Colombia, Denmark, Germany, Poland, Portugal, Spain, United Kingdom

Note: The networking platforms are not always explicitly place-based. Regional development policy approaches may not be included.

Source: Extracted from OECD (2014a), *OECD Science, Technology and Industry Outlook 2014*,OECD Publishing, Paris, *http://dx.doi.org/10.1787/sti_outlook-2014-en*. Based on country responses to the OECD STI Outlook Policy questionnaire 2014.

In policy discussion, it is important to distinguish between the economic benefits to firms of cluster membership and the separate question of the rationale for (and effects of) government action. A number of observations are relevant in this connection:

- The available evidence suggests that while clustering can bring benefits to firms, the magnitude of these benefits is modest, especially when viewed in relation to changes in cluster size (Rosenthal and Strange, 2004).
- Micro-level evidence from France shows that firms are able to appropriate some of the productivity gains from belonging to a cluster (Martin, Mayer and Mayneris, 2011). To the extent that firms capture these gains, the need for policy to alter firms' location behaviour is reduced.
- Some of the productivity or other benefits of cluster membership might not be caused by the cluster, but might reflect a process whereby higher-productivity firms choose to locate in clusters (in part because their higher productivity allows them to locate in attractive places where factor costs, such as land values, would deter lower-productivity firms).
- The mere fact that firms in a cluster might be more productive than firms elsewhere is not an economic justification for policy support (as it is a state of affairs that could be consistent with efficient markets). Policy needs to start from the identification of market failure(s) that could merit correction.
- When the stated rationales for cluster policy do refer to market failures, they usually concern market failures that equally affect firms not located in clusters. This may suggest there are perhaps no frequently offered forms of support for clustered firms from which there is a basis for excluding firms outside clusters. On the other hand, firms that are already in a cluster may have a greater chance to deliver economic benefits, which suggests concentrating resources on such firms.[4]
- The goals of cluster policy can appear sound from a local standpoint, but might be economically unsound from a national perspective. For instance, wasteful competition is evident when – as surveys show sometimes occurs – different subnational bodies pursue

policies to develop clusters in the same sector or industry (typically, "sunrise" industries such as biotechnology, new materials and information technology).

Cluster policies often have marked similarities. Frequently, business networking or promotion of public-private partnerships is at the centre of a cluster programme. Many programmes concentrate on SMEs. Others provide generic information on business and economic trends as well as cluster-specific information on parameters such as markets, technologies and competitors. The specific infrastructure and training requirements of a cluster are a common focus. Governments sometimes also provide business services ranging in sophistication from basic research to advice on bookkeeping. And schemes sometimes seek to attract direct investment to a cluster.

While there are recurrent features to cluster programmes, important differences can also exist. For example, differences exist in the level of government involved: the public sponsors of cluster strategies have included local, regional, national and even supranational governments. Programmes can likewise differ as to whether they focus on developing the existing economic base, attracting firms into the cluster or a combination of the two.

Terminology in this area of policy is sometimes loose, and various interventions are often subsumed under the same generic category ("cluster policy") with little differentiation. Policies that support competence/excellence centres or business networks are cases in point. Business networks operate with varied forms and objectives. Some aim at general sharing of information, while others tackle more specific goals. Distinctions between clusters and business networks are important. The two types of initiative can entail different resource requirements, objectives and, consequently, different evaluation metrics. Business network programmes might be easier to design and implement among firms that are located near one another, although they can also operate beyond the geographic boundaries of a cluster.

Industrial parks, or science and technology parks, are other place-based instruments that can affect innovation. These instruments are a form of physical infrastructure. As such, their effectiveness can depend on complementary activities or policies that affect innovation-oriented collaboration among co-located firms (e.g. funding for joint research projects). While the first generation of science and technology parks focused on universities and the links between universities and large firms, a second generation shifted to smaller parks with incubators aiming at new firm creation. The most recent wave focuses on city strategies and the provision of facilities for multinational and domestic firms. In all cases, it is important to distinguish between infrastructure provision that is primarily driven by political aims (e.g. local job creation) and policies that address genuine market failures or barriers to innovation.

Despite their popularity, rigorous evaluations of cluster-related policies are few and show mixed results. The shortage of useful evidence may have a number of causes. One may be that policy towards enterprise clusters often encompasses many different types of intervention, some with multiple and inter-related objectives. Uyarra and Ramlogan (2012) and others note that many policies that are likely to have a major impact on clusters are almost never assessed. These include transport, land-use planning and labour market policies. Instead, evaluations focus on activities implemented under enterprise-oriented and time-bound initiatives. There also appears to be little evidence on how specific design and implementation features of cluster policy have contributed to policy outcomes.

Various evaluations using statistical controls are cited in a recent OECD study (Warwick and Nolan, 2014). A salient finding is that policy effects for the clusters programmes are often modest, and that assessments of long-term impacts are almost entirely lacking. Network-oriented cluster support schemes appear to give more positive outcomes. The main policy conclusions of recent OECD's work on clusters and place-based policies are presented below.

Main policy messages: Place-based approaches and enterprise clusters

- **National and regional innovation policies should seek opportunities to foster innovation dynamics in specific places.** Despite global connections, proximity still matters for certain forms of knowledge-sharing and innovation. Explicit place-based policies can help adapt innovation policy instruments to the specificities of particular places. Policy makers should be aware, furthermore, that some policies may not target specific places, but can end up having place-specific consequences.
- **While cluster-type policies are popular at all levels of government, it remains to be proven which, if any, cluster policy measures are effective, and to what extent they might increase innovation or productivity.** Caution in policy development is therefore prudent. Policies that focus on removing barriers to innovation rather than "creating" innovation are least likely to be wasteful.
- **Policy should explicitly target market failures.** Several forms of market failure may be relevant. These include undersupply of public goods and co-ordination failures. Also important can be failures affecting small and medium-sized firms more generally, such as in the supply of industrial real estate and, in some cases, the provision of certain financial services. Nevertheless, the fact that markets can fail in some of the above fields does not imply that they will be failing everywhere. Indeed, many clusters have thrived in the absence of policy.
- **Government should work with existing and emerging clusters rather than trying to create clusters from scratch.** A policy aimed at developing entirely new groups of firms in selected sectors can entail high costs and high risks, and give rise to destructive competition should many regions follow the same policies in pursuit of identical industries. Furthermore, there are few if any examples of a significant cluster of firms having emerged as an intentional outcome of policy.
- **A policy on clusters should provide a framework for dialogue and co-operation among firms, the public sector (particularly at local and regional levels of government) and non-governmental organisations.** This dialogue could identify and lead to the development of inter-firm networks, an improved quality of policy and government action (such as in the provision of information, infrastructure supply and better decisions in co-locating complementary public investments with related concentrations of private investment).
- **The general business environment, rather than location subsidies, should be the focus for attracting and embedding firms in specific places.** Direct financial subsidy of firms' location decisions appears to have only modest effects but could create multiple inefficiencies. Wider determinants of cluster success include transport, land-use planning, housing, the quality of public amenities and labour market policies.
- **Local adaptability in university-industry relationships is desirable.** Incentive structures should encourage local linkages to industry. The conditions in universities that shape incentives for collaboration are many (from rules on IP to civil servant mobility constraints and evaluation criteria that give little weight to outreach or economic development). Local firms, universities and training institutions can also interact in many different ways. For example, university-industry partnership mechanisms can range from grants and fellowships to targeted research contracts, collaborative research and consortia agreements, training, mobility, and networking programmes. A brokerage function between universities and SMEs will often need some degree of third-party financial support.

Main policy messages: Place-based approaches and enterprise clusters *(cont.)*

- **Initiatives should be matched to the most suitable level of government.** The ideal level of government will correspond to the scale of the cluster, and have substantial influence over relevant programmes and expenditures.
- **If firms benefit from business networks then it is natural to expect that they will invest in such networks of their own accord. However, governments might justify a facilitating and/or co-ordinating role in network development owing to the fact that in some places and industries there may have been no, or limited, prior familiarity with the opportunities that networks afford, or limited numbers of firms to take the initial lead in network formation.** Public action, at least in a catalytic role, may be needed. However, funding should be modest, and should be phased out as participants start to engage more formally and obtain benefits. Policy must operate with realistic time frames. And precise market-oriented objectives should be set by, or in conjunction with, private actors.

6.3. The role of business accelerators

The incidence of high-growth firms varies at national and local levels with regional disparities often as wide as national disparities (Figure 6.4).[5] Recent OECD analysis shows that large urban areas host proportionally more high-growth firms because they possess core assets such as infrastructure, skilled human resources and high-quality business services. This prevalence in large urban areas is particularly marked in services (Hart and Temouri, 2013).

In sectoral terms, high-growth firms are more common in services than in manufacturing. However, they are not disproportionally present in technology-based or knowledge-intensive sectors. In Scotland, for example, high-growth firms are found primarily in business-to-business services (Brown and Mawson, forthcoming). In the Swedish region of Scania, high-growth firms are found across a wide range of sectors (Gabrielsson, Lindholm-Dahlstrand and Politis, 2012).

The skills of entrepreneurs and management teams are particularly important for growth in young and small firms, whereas formal management structures are more relevant in older and larger enterprises (Wennberg, 2013). Business accelerator programmes have emerged to help develop necessary skills for entrepreneurs with the ambition and potential to grow. Business accelerators provide growth-oriented entrepreneurs with a range of skills-related services such as mentoring, formal training, informal peer learning and business advice (e.g. consulting or help with recruitment). Business accelerators can be either geographically based "physical" locations with associated services; programmes run on a virtual basis, independent of geography; or a hybrid. In rarer cases, the managers of business accelerators take an equity position in the assisted companies. Box 6.2 presents three examples of business accelerators and their different operational models.

The OECD has recently benchmarked business accelerators across OECD and non-OECD countries (OECD, 2013d). The analysis covers varied aspects of the programmes, such as governance, programme activities and the profiles of participating companies. The main findings are:

- **Governance:** Business accelerators tend to take the form of public-private partnerships in which the public sector subsidises part of the activities commonly delivered by private-sector organisations. However, the extent to which programme activities are outsourced varies.

Figure 6.4. **High-growth firms at the local level, 2010**

Percentage of high-growth firms (in employment terms) in total

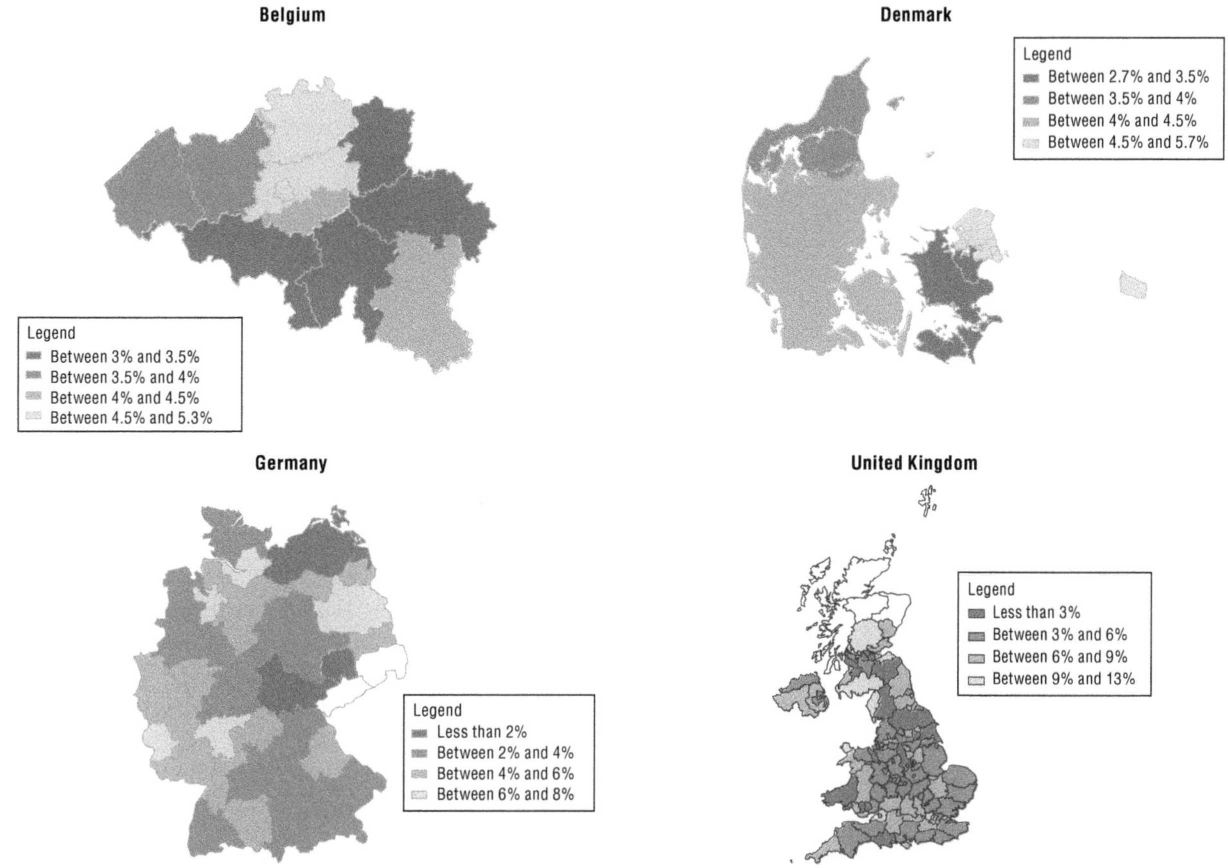

Note: The charts are based on an analysis carried out using the ORBIS database.

Source: OECD based on Hart M. and Temouri Y. (2013), "High-growth firm localities and determinants: Evidence from OECD countries".

- **Programme activities:** A diagnostic tool is commonly used to identify the strengths and weaknesses of participants. Tools usually examine themes such as the business concept, the business organisation, operations and customer relations. After the diagnostic stage, programme activities usually centre on entrepreneurial and managerial skills, through formal training, mentoring, networking events and innovative sessions of peer-to-peer learning. Participants often report that peer-to-peer learning is one of the most important sources of learning in the programme (but that such sessions should not include business owners who are competitors in the same industry). Some programmes also provide consulting-type advice, e.g. on how to utilise other government support schemes or how to procure business support services (e.g. accounting, recruitment).
- **Duration:** Some schemes limit themselves to the business diagnosis and guidance on where relevant service providers can be found. Others provide much more intensive support for periods of up to five years (e.g. the Scottish Companies of Scale programme). Because high-growth firms are a small fraction of the total business population, some degree of resource concentration over longer periods appears justified.

- **Participant companies:** Business accelerators also tend to select firms based on their growth potential, which is a mix of past performance and qualitative criteria such as the ambition of the entrepreneur, the management structure of the firm and the innovativeness of products and services. This makes business accelerators different from other policy programmes where selection is based only on quantitative metrics. Participant companies tend to come from the most affluent regions of the country. This is not surprising given that high-growth firms are primarily found in large urban areas. However, this finding does signal that innovation policies might inadvertently favour leading regions over peripheral. Participant firms come from a diverse mix of sectors.

Box 6.2. **Examples of successful practices for business accelerators**

Through a consortium of five private business consultancies, the Netherlands runs an intensive business accelerator programme (Growth Accelerator) aimed at entrepreneurs who have the ambition to increase their annual turnover from EUR 3-5 million to EUR 20 million. Participating entrepreneurs receive a diagnosis of their growth potential, followed by weekly or bi-monthly sessions of bespoke advice and training for five years. Participants co-finance programme activities.

A Growth Accelerator has also been launched in England. The initial business diagnosis is carried out through an online assessment and phone interviews. Entrepreneurs then meet with a programme manager who sets out a package of possible support and appoints a coach to work with the company. The coaches identify business development service providers for the entrepreneurs. England's programme, launched in 2012 with a budget of GBP 200 million, aims to assist 26 000 firms over a period of three years. An interim evaluation (UK government, 2014) showed that participating companies improved their business on a number of dimensions.

Finland runs a different form of business accelerator (Vigo). Vigo accelerators are privately held profit-seeking companies established and managed by experienced entrepreneurs and executives who offer managerial support to, and take equity stakes in, the assisted firms (with stakes of no less than EUR 30 000 in each company). The Finnish government selects the accelerators, pays accelerator management fees, co-invests in the companies and has a veto on the companies selected by Vigo managers.

Sources: OECD (2013c), "Key Findings of the Work of the OECD LEED Programme on High-Growth Firms", *www.oecd.org/employment/leed/high-growthreport.htm*, and Autio (2013), "Promoting leadership development in high-growth firms", *www.oecd.org/cfe/leed/leadership-development-HGFs.pdf*.

Main policy messages for establishing and operating business accelerators

- Involve private-sector organisations – such as business consultancies, IP and marketing specialists, and business lawyers – in design and implementation.
- Consider co-funding of business accelerators by participating growth-oriented entrepreneurs so as to ensure their commitment to programme activities.
- If the business accelerator takes an equity stake in the assisted companies, ensure that co-ownership does not discourage promising companies from participating. One way to do this is to take a symbolic ownership share in the participant company or make equity participation convertible to debt in the case of success.

Main policy messages for establishing and operating business accelerators *(cont.)*

- Develop a heterogeneous set of skills development activities that use interactive and experiential learning. Make sure that peer-to-peer learning is part of the programme activities.
- Do not scatter resources too thinly across many firms. Rather, prioritise working with firms that have the ambition and potential to grow. Avoid extending public support for too long and without milestones.
- Select firms by using a mix of quantitative criteria (e.g. recent turnover and employment growth rates, export sales, and investment in innovation) and qualitative criteria (e.g. new products or services, personal ambition) with the former being more prominent.
- Ensure that business accelerators do not focus only on technology-based sectors and that growth-oriented companies from less wealthy regions are given special consideration owing to their potential impact on job creation.

Source: OECD (2013d), "An International Benchmarking Analysis of Public Programmes for High-Growth Firms", *www.oecd.org/employment/leed/high-growthreport.htm*.

Based on OECD (2013d), the main messages for policy makers interested in setting up business accelerators are listed below.

6.4. New industrial policy and smart specialisation

The discussion on specific innovation policies above is also affected by the renewed interest in industrial policy around the world (O'Sullivan et al, 2013; Warwick, 2013; Stiglitz, Lin and Monga, 2013). However, the nature of such policies has changed from the traditional focus on government subsidies and national champions. Advocates of new industrial policy see government as a facilitator in the face of complexity and uncertainty, enabling closer co-ordination between individual economic agents as well as greater experimentation in the economy. By comparison with the earlier historical experience of industrial policy, so-called "new industrial policy" might be characterised as exhibiting some or all of the following characteristics (Warwick, 2013; Warwick and Nolan, 2014):

- greater emphasis on building networks, improving co-ordination and promoting awareness
- less reliance on direct support in the form of state aids and (market-failure correcting) subsidies
- greater emphasis on strategic (rather than defensive) industrial policy
- a shift away from sector-based strategies and towards certain technologies and activities.

The two latter points, in particular, have brought industrial policy closer to innovation policy, due to the perceived links between technological development and structural change in the economy. At the same time there has been greater recognition of the importance of ensuring that government only bears risk that is "proportionate", i.e. enough to matter, not too much to lead to moral hazard. Policy makers are also increasingly cognisant of the need to plan for exit from the policy, and to make these plans known in order to help resist pressures from beneficiaries for the retention of policy.

Two other dimensions of new industrial policy are important, as can be seen in the experiences from countries such as the United Kingdom and the Netherlands. First, industrial strategies are *not* just about the money, i.e. funding or subsidies for specific firms, technologies, activities or capabilities. They are very much also about a set of other complementary policy measures that enable a particular set of firms to be successful and create comparative advantage globally. These other policy measures might include reduced regulation (or tighter regulation driving innovation), more flexible labour market policies, migration policies that enable relevant talent to join the firms, or government procurement that enables/fosters innovation and competition in the relevant sectors. At the same time, and despite being less about sectors, industrial strategy *does*, however, recognise that different sectors and groups of firms have different relationships with the government. Some are almost entirely dependent on government policy (e.g. private-sector health or educational establishments, defence industry) while others are much less impacted (e.g. clothes retailing). A differential approach that recognises the role of government – whether positive or negative – is therefore also an important part of a "smart" industrial policy.

In the conduct of such new industrial policies, it is important to recognise and accept that mistakes and errors are inevitable. Part of the policy challenge is to design governance procedures to detect and correct these errors (Rodrik, 2008). It is for this reason that monitoring and evaluation have become central elements of industrial policy. Fortunately, there has been a trend in the evaluation community towards greater use of more rigorous techniques that have two related characteristics: the identification of a counterfactual (what would have happened in the absence of the policy) and the inference of causality (rather than simple correlation). A further discussion on this issue is contained in Chapter 8 of this report.

One application of the emergence of new industrial policy is smart specialisation, which involves regional (and national) governments encouraging investments in domains that would "complement the country's other productive assets to create future domestic capability and interregional comparative advantage" (Foray et al., 2009). What distinguishes smart specialisation from traditional industrial and innovation policies is mainly the process defined as "entrepreneurial discovery" – an interactive process in which market forces and the private sector are discovering and producing information about new activities, and the government assesses the outcomes and empowers those actors most capable of realising the potential (Foray, 2012; Hausmann and Rodrik, 2003). Hence smart specialisation strategies are much more bottom-up than traditional industrial policies.

Like traditional industrial policy, smart specialisation strategies aim to address market/systems and co-ordination failures. But traditional industrial policies required significant levels of information to justify subsidy support and they tended to be implemented in vertically integrated sectors with stable technological paradigms. In contrast, smart specialisation – as well as new industrial policies – recognises the lack of perfect information, the level of advancement of a given activity, and the relative risks for policy. It thus focuses on helping entrepreneurs identify their knowledge-based strengths at the regional level and in a more exploratory approach in which public decision makers listen to market signals using a range of assessment tools (e.g. "SWOT" analysis of strengths, weaknesses, opportunities and threats; surveys) and mechanisms such as public-private partnerships, technology foresight and road mapping to name a few. A recent OECD report on smart specialisation identified the following policy messages (OECD, 2013e):

- **Policies for entrepreneurial discovery.** The smart specialisation approach calls for an "entrepreneurial selection" of market opportunities (e.g. to minimise failures and to avoid ill-informed policy decisions). While successful companies will constitute the new specialisation of the country/region (self-discovery), the role for policy is to develop a flexible strategy focusing on measurable intermediate goals, identifying bottlenecks and market failures and ensuring feedback into policy learning processes. The approach includes incentives to strengthen entrepreneurship and encourage agglomeration.
- **Promoting general-purpose technology platforms and networks.** Given the range of applications of general-purpose technologies, technology platforms involving public and private actors but also standards-setting organisations can help increase productivity in existing sectors and help identify sectors in which to concentrate resources.
- **Diagnostic and indicator-based tools and infrastructure.** Smart specialisation requires regions and countries to maintain an infrastructure and indicator base to monitor and evaluate performance and policies.
- **Strategic governance for smart specialisation.** Good governance and the development of local capabilities are key to identifying local strengths, aligning policy actions, building critical mass, developing a vision and implementing a sound strategy. Further discussion on governance follows in Chapter 8 of this report.
- **Openness to other regions.** The specialisation strategy of regions should take into account that other regions are also involved in knowledge-creating activities and that duplication might lead to lower effectiveness and finally failure. Hence, co-operation with other regions with complementary capabilities and strategies is important.

6.5. Demand-side policy and grand challenges

Demand-side innovation policy seeks to increase the demand for innovations, or improve the conditions for their uptake and their diffusion. As described in OECD (2011b), the key policy instruments used are innovation-oriented procurement, regulations, standards and information dissemination (particularly to consumers).

The concept of fostering innovation through demand-side policy – particularly public procurement – is not new. Indeed, some countries have pursued active technology procurement policies for decades, most notably in defence, energy and transport. However, policy interest in demand-side initiatives has grown in recent years. In part this reflects an expectation that demand-side policy can steer innovation towards specific grand challenges, such as population ageing or climate change. In addition, owing to constrained public finances in most OECD countries, there is attraction in the possibility that demand-side policies might be less costly than some other forms of support. The renewed interest may also reflect greater awareness of feedback linkages in the innovation process between demand and supply.

Key challenges for innovation-oriented procurement

There are various rationales for using public procurement to promote innovation:

- Because of their purchasing power, governments can shape innovation *directly* (because procurement can help firms recuperate the sunk costs of risky and sometimes large investments) as well as *indirectly* (because as a lead consumer, government can influence the diffusion of an innovation).

- The delivery of some essential public services might become more cost-effective (on a like-for-like basis) if relevant forms of innovation succeed.[6]
- Particularly when procuring from small innovative firms, public-sector demand may help to counter problems of access to finance (for instance by providing some degree of security for third-party lenders).
- Governments may need to create a market for a new technology in order to meet a policy challenge that is time-bound. The search for commercial-scale low-carbon emission technologies is such a case.

For some time public procurement has been used to facilitate the emergence of a number of high-tech sectors in countries such as the United States, Japan and France (where procurement has helped to develop high-speed rail and nuclear energy technologies). However, in recent years countries such as Australia, Finland, Germany, Sweden and the United Kingdom have given new emphasis to public procurement as a tool to promote innovation and meet societal goals. Sixteen out of 34 OECD countries report having a strategy to use procurement to support the delivery of innovative goods or services. However, only six regularly monitor or measure the results of such strategies (OECD, 2013f).

Despite the existence of national strategies, a recent survey of firms in six OECD countries (Austria, Belgium, Finland, Germany, Portugal and Sweden) indicates that innovation requirements are relatively rare in procurement contracts (Appelt and Galindo-Rueda, 2015). While public procurement represents around 16% of GDP in OECD countries, little of this spending explicitly addresses innovation. It is probable that more of the potential innovation benefits of procurement could be realised. But to do so various challenges must be addressed:

- Innovation-oriented public procurement must be efficient, competitive and accountable.
- Expertise in innovation is often lacking in specialised procurement agencies, whose main responsibility is for the efficiency of purchasing. Some OECD countries have issued guidelines for innovation-oriented procurement (e.g. United Kingdom), while others (e.g. Finland) have introduced funding instruments to encourage innovation-oriented procurement.
- Procurement is often fragmented across local, regional and national governments. The fragmentation of public demand limits scale effects that can be helpful to innovative procurement, although small scale may also enable experimentation with new ideas and public services models. However, for many investments in a potential innovation, having a larger public market will improve the risk-return profile.
- Where the procurement system is decentralised and professional procurers are few, the lack of skills for innovative purchasing is also an important challenge. A survey in 2013 of public procuring units in Finland suggests that the capacity of procuring units influences the incidence of innovation procurement. Procurement units with 1 000 or more employees are more likely than smaller units to make purchases as first users. The larger procurement units are also more likely to award contracts that require delivery work (which may entail some level of innovation). Procurement officials sometimes lack guidance on how to take innovation, social or environmental criteria into account in public procurement.
- Procuring innovation entails risks additional to those present in all procurement procedures. These risks, which might act as a disincentive to innovative procurement, include:

- **technological risks:** risks of non-completion stemming from technical features of new goods or services
- **organisational and societal risks:** risks arising within the procuring organisation and/ or risks related to uptake of the good or service by users
- **market risks:** includes the risk that suppliers do not respond to the tender, but also that the prospective market fails to develop.

The risks entailed in innovation-oriented procurement tend to increase with the degree of innovation involved. Pre-commercial public procurement, in particular, can present considerable risks. Mitigation options exist for all of the risks, for instance through contract design and procedural modification. But risk mitigation itself requires skilled and experienced staff. General government procurement can be made more innovation-friendly with little additional risk simply by specifying the goods and services to be procured in terms of functionalities, rather than predetermined technical characteristics. This will provide space for the creativity of markets to propose new products or product designs to meet the specified functions.

Procurement processes should not disadvantage (innovative) SMEs. Procurement processes often favour established enterprises (which have greater capacity to respond to government tenders). Engaging as broad a range of the enterprise population as possible is good for equity and expands the range of ideas proposed. Moreover, procurement should not discriminate against non-national actors. OECD countries must also adhere to international standards and obligations (e.g. the WTO Government Procurement Agreement).

There have been almost no systematic assessments of the impacts of innovation-oriented public procurement (see also Warwick and Nolan, 2014), apart from some evaluation in the context of the United States' Defense Advanced Research Projects Agency (DARPA). However, the OECD is seeking to improve measurement. The aim is to produce policy-relevant indicators of innovation procurement and to analyse the impact of procurement on innovation and other economic outcomes (Appelt and Galindo-Rueda, 2014).

Key challenges for innovation-oriented regulation

Regulation refers to the implementation of rules by public authorities and governmental bodies to influence the behaviours of private actors in the economy. Regulation influences innovation indirectly, since it affects the framework conditions for firms, and involves no direct outlay of public funds. Regulations can affect the performance (e.g. quality, compatibility, emissions standards) or consequences (e.g. on health, safety, the environment) of products or services, thus having a direct impact on demand for innovative goods and services. For example, in Japan, the Top Runner programme uses a dynamic process of setting and revising energy efficiency performance standards by taking the current highest efficiency of products as a benchmark in 23 product groups. Without drawing on public finances, this flexible setting of benchmarks creates positive incentives for manufacturers to quickly improve product performance.

Policy making must be informed by the following considerations:

- Regulation requires deep industry- and technology-specific expertise. The effects of regulation on innovation are complex and can be ambiguous *a priori*. Regulations can both inhibit and stimulate innovation (as already suggested in the discussion on the Porter hypothesis in Chapter 1 of this report).

- The impacts of regulation on innovation are likely to be highly technology- and industry-specific. The time period over which policy yields impact might also vary from one regulation to another, again reflecting industry-specific characteristics.
- To assess the appropriateness of targeted regulatory policy, analysts need to ask whether the market would introduce the right technology in the absence of the regulation. For instance, could vehicle manufacturers be expected to increase engine fuel efficiency regardless of regulations on the performance of engines? Is the market characterised by lock-in in incumbent technologies and platforms? Whether the market is efficient or not will likely reflect industry-specific considerations.
- Close attention is needed to the precise form that regulation takes. For example, uncertainty in the duration of a regulation could reduce its strength or influence on demand conditions.
- OECD analysis suggests that it is helpful to take account of the specific design characteristics of different instruments (market- or regulation-based).[7] Design characteristics that require attention include:
  - **Stringency:** How ambitious is the environmental policy target?
  - **Predictability:** What effect does the policy have on investor uncertainty?
  - **Flexibility:** Does the innovator have the freedom to identify the best way to meet the objective?
  - **Incidence:** Does the policy target the externality directly, or is the point of incidence a "proxy" for the pollutant?
  - **Depth:** Do incentives exist to innovate through a range of potentially ascending objectives?

The ideal policy instrument will be one that is *sufficiently stringent* to encourage an optimal level of innovation, *stable enough* to give investors adequate planning horizons for risky investments, *flexible enough* to encourage innovators to create genuinely novel solutions and *closely targeted* on the policy goal, and that provides incentives for continuous innovation. There is no automatic correspondence between the type of instrument and the critical design attributes. For instance, different environment-related taxes can have varied combinations of these design attributes, and a regulatory standard might have more in common with a tax than a technology-based standard.

Social cost-benefit analysis of regulation – both *ex ante* and *ex post* – is essential. Even when regulation spurs innovation, a regulation-based policy might be cost-ineffective overall. For instance, cost-benefit analysis of some vehicle efficiency regulations in the United States shows that a small increase in the gasoline tax would deliver equivalent savings in fuel consumption but at a much lower cost to society (in part because regulation of engine efficiency can lower the marginal cost of driving and so induce more driving). Such analysis can be complicated in isolating the specific effects of regulation from other influences. This reflects, among other reasons, inherent complexity in the pathways by which regulation can shape innovation.

Key challenges for setting innovation-oriented standards

Standards are documents based on various degrees of consensus (industry-wide, national, regional or international) that lay out rules, practices, metrics or conventions used in technology, trade and society at large. Standards affect innovation and other economic outcomes through multiple routes, and the economic benefits of standards have become clearer to policy makers in recent years.

Unlike regulation, the setting of standards is mainly the responsibility of industry. Government acts as facilitator or co-ordinator. The public sector's role with respect to standards largely involves measures to include under-represented groups in their development (in particular the research community), along with support for establishing international standards. Indeed, there is a clear trend towards standardisation work being conducted internationally. This is because, in a globalised economy, compatibility and interface across borders are increasingly important. (Today, for example, producers of energy-efficient light bulbs face different performance standards in different markets.) Countries and firms that play primary roles in setting international standards can enjoy advantages from doing so, to the extent that the new standards align with their own national standards and/or features of their productive base.

If standardisation is brought into effect too early, it could preclude and shut out better technologies. But if standardisation occurs too late, then the costs of transition to the new standard could be high enough to slow or prevent diffusion. If product life cycles are shortening, issues of timing are likely to increase in importance. Unfortunately, however, there is no generally applicable rule by which to judge if a standard is premature, because future innovation is always likely to have unforeseen features. Finally, standards set by different agencies should not conflict or hinder each other. An example of how this question might be addressed is the United Kingdom's decision to set up a government-wide committee to discuss and decide on standardisation policy matters.

Main policy messages: Demand-side policy and grand challenges

Innovation-oriented public procurement

- As with traditional procurement, innovation-oriented public procurement must be efficient, competitive and accountable. It is probable that in most countries there is potential to benefit from innovation in procurement, both through lower costs over time and through more innovation activity.
- Clear guidance, tools and support for capacity building should be made available if procurement is to be used to foster innovation. Innovation-oriented procurement entails risks additional to those present in all procurement procedures. Risk-mitigation options exist but require skilled and experienced staff for their implementation.
- Procurement processes should not disadvantage (innovative) SMEs. For example, the process for participation must not be overly costly or bureaucratic. Engaging as wide a range of the enterprise population as possible is positive for equity and for capturing as many good ideas as possible.

Innovation-oriented regulation

- Regulation requires deep industry- and technology-specific expertise. The effects of regulation on innovation are complex and of uncertain duration, and can be ambiguous *a priori*.
- Close attention must be given to the precise form that regulation takes. Even minor features of a regulation can give rise to unwanted behaviours. Social cost-benefit analysis is essential. This is because regulation can lead to innovation but not be cost-effective overall.

Main policy messages: Demand-side policy and grand challenges *(cont.)*

- Ideally, a regulatory instrument will be sufficiently stringent to encourage optimal innovation, stable enough to give investors adequate planning horizons, flexible enough to encourage genuinely novel solutions and closely targeted on the policy goal, and will provide incentives for continuous innovation.

Innovation-oriented standards

- The public sector's role with respect to standards largely involves measures to include under-represented groups (in particular the research community) in the process of developing standards, as well as support for establishing international standards.
- Care should also be taken that standards set by different agencies do not conflict with or hinder one another.

Key policy messages from the OECD's work on demand-side policies are presented below.

6.6. Consumer policy and innovation

Consumers are relevant for firms' innovation not only in regard to the way they shape market demand for products but also in regard to how they may shape firms' innovation activities. Consumers can be directly involved in firms' innovation processes. Policy can influence consumers' demand for and contributions to innovation by establishing regulations and standards in a way that reduces the risks for consumers in adopting innovative products and services; ensuring efficient ICT infrastructure that allows collaboration between firms and end users; stimulating user-driven innovation through grants; using public procurement for innovations that will ultimately be sold to private end users; and providing information, organising awareness campaigns, and promoting education initiatives to inform and educate consumers about innovative products and services.

The impact of consumers and consumer policy on innovation has attracted renewed attention in recent years. This has occurred as advances in ICTs, changing consumer behaviour and increased competition have fostered consumer-driven innovation and raised new challenges for consumers engaging in more complex, data-driven, service-based and global markets. In 2014, recognising the vital role of well-protected and empowered consumers in economic performance and innovation, the OECD recommended that policy makers revisit their approaches to consumer issues arising in today's dynamic market environment (OECD, 2014b). The resulting OECD Council Recommendation was based on the OECD's *Consumer Policy Toolkit* (OECD, 2010b). Using a six-step process, the *Toolkit* provides a framework for policy makers and enforcement bodies to determine whether, when and how to intervene to effectively address market failures.

The *Toolkit* highlights that when consumers make well-reasoned and informed decisions, they represent powerful drivers of innovation, productivity growth and competition. Consumers have had a notably important role as drivers of innovation and productivity in the ICT and Internet environments. Here, the increasing ability to access and compare products and transaction information through a range of innovative platforms (such as social media and price and product comparison websites) as well as interactive tools (such as product ratings and reviews) has boosted consumer interest in a growing array of innovative physical and digital products. Such interactions, and the demand for innovative products, have spurred businesses to develop new business models, improve products, develop new products, and provide consumers with more personalised offers in a timely and cost-effective way. Businesses have been helped to do this by using consumer data and data analytics (OECD, 2015).

However, despite the many benefits for both businesses and consumers, stakeholders have emphasised the need for a better balance between innovation on one side and consumer protection and empowerment on the other. In this regard policy makers in a number of countries have focused not only on reviewing consumer policy, but also reviewing policies in related areas such as telecommunications, privacy and competition.[8] Along these lines, the OECD has explored ways to strengthen the effectiveness of disclosures and consumer protection – against fraudulent, misleading and deceptive practices – in the areas of communication services and e-commerce.

Consumers in communication services

In many OECD countries, deregulation and privatisation in the telecoms sector have provided consumers with many benefits, including a continuous supply of new and sophisticated products. At the same time, consumers have faced challenges in understanding and comparing complex telecom packages, pricing plans, and contract terms and conditions. Telecoms regulators and enforcement agencies have pointed to a number of misleading and fraudulent marketing techniques and commercial practices that affect consumer trust (OECD, 2011c). These have included, for example, advertised broadband speeds that reflect hypothetical speeds, and advertising that emphasises the minimum cost of a service without sufficient attention to additional charges that could apply. Many policy and enforcement options have been implemented to address such issues. These range from industry codes and other self-regulatory tools (regarded by most authorities as the preferred approach), to laws and regulation (for example, in information disclosure) as well as criminal and civil sanctions.

An example of a flexible approach may be found in the United Kingdom. There, the Office of Communications (Ofcom) contracted with a company to provide a platform through which consumers could measure the broadband speed they were actually getting, as compared with the advertised speed. This initiative also helped the Ofcom to better understand the differences in broadband speed across different geographical locations. Another useful example is found in Colombia. Based on a resolution issued in 2011 by the country's Commission for Communications Regulation (CRC), communication service providers are obligated to publish service quality data on their websites. In the light of such information, the CRC issues information about quality of service by supplier and region, which in turn helps consumers to make more informed decisions (OECD, 2014c).

Consumers in e-commerce

As a part of its review of the OECD's 1999 *Guidelines for Consumer Protection in the Context of Electronic Commerce*, the OECD has recently explored three areas where innovation has benefited consumers and businesses, and SMEs in particular. These are:

- **Mobile and online payments** (OECD, 2014d): Access to more convenient and easy-to-use e-commerce payment mechanisms have simplified the acquisition and use of products. Consumers benefit from low-cost payment options, which are increasingly processed by non-traditional financial organisations such as mobile operators and social media.

- **Digital content products:** Continued innovation and enhanced competition have provided consumers with a greater supply of high-quality products at competitive prices.
- **Participative e-commerce:** E-consumers have directly engaged in innovation through, for example, the sharing of information about their experiences with products and brands, or through crowdfunding.

With the growing complexity of products and transactions, combined with constant technological changes, consumers' ability to absorb and analyse information has become more limited. As pointed out in the *Toolkit*, the biases and misperceptions that often shape consumer decisions in the offline environment are particularly harmful in the more distant online context. This problem seems to be particularly pronounced when consumers acquire products via mobile devices while on the move (OECD, 2010b).

A number of policy and enforcement initiatives have been implemented to address the above issues and strengthen trust. For example, in 2014, with a view to enabling informed consumer decisions, the European Commission published a (voluntary) model for the provision of information to consumers acquiring digital content products.[9] In addition, in some countries, businesses have co-operated with governments to implement programmes to enable convenient consumer access to the consumption data held on them by businesses in the energy, financial and retail sectors. Furthermore, taking into account the fact that consumers tend to follow default settings online and end up unnecessarily paying extra charges, the European Union's Consumer Rights Directive introduced a ban on pre-ticked boxes in online offers.[10] To enhance transparency in e-commerce interactions, public authorities and the private sector have also developed guidelines on endorsements and testimonials, on trust mark schemes and on price and product comparisons. For example, in 2012, following a review of price comparison sites, the United Kingdom's Office of Fair Trading requested leading website operators to provide clearer information to consumers on: 1) the way search results were presented; 2) how they were ranked; and 3) the identity of the business operating the scheme. Since 2006, Ofcom has run an accreditation scheme for platforms that compare prices for services, such as mobile and broadband services. The main policy messages of the OECD's work on consumer policy and innovation are below.

Main policy messages: Consumer policy and innovation in ICT markets

- **Consumer policies in the rapidly evolving, innovative and complex ICT market should aim to create a trustworthy environment for businesses and consumers** (OECD, 2014b). Consumer protection and empowerment initiatives should keep in mind, and be balanced with, the benefits provided by product and service innovation. To enable informed decisions, consumers should be empowered with choices, dynamic pricing options, personalised offers, and strategic information about products and transactions. Enhancing protection against fraudulent and misleading commercial practices is also needed to strengthen trust and stimulate demand for new and innovative products.
- **The interface among consumer policy and related policies that increasingly impact consumers in ICT markets, such as telecommunications, privacy and competition policy should be further explored.**

6.7. The policy mix for innovation policies

The mix of instruments that is used to support innovation differs considerably across countries. Recent years have seen increased interest in this policy mix. Whereas much emphasis was previously placed on the design and evaluation of individual instruments of innovation policy, there is now greater interest in understanding the effectiveness of the larger portfolio of policy instruments used to improve a country's innovative potential and capabilities. This view of the policy landscape reflects a growing appreciation of the interdependence of policy measures and an understanding that the performance or behaviour of innovation systems requires a more holistic perspective. Yet, while there is evidence that the complementarities and trade-offs among policy instruments are significant for assessing a country's STI policy and its impact on innovative and economic performance, they remain poorly understood.

For the policy mix concept to be useful in policy making and analysis, individual policy instruments and the interactions among them need to be defined. Policy instruments can be characterised in several ways: by their target groups, their desired outcomes, or their mode of intervention (e.g. funding, regulation). Some of the most popular characterisations are binary in nature, e.g. supply-side versus demand-side instruments. They should not be interpreted as alternatives but as possible complements. In fact, a key challenge is to strike an appropriate balance, taking into account the current state of the innovation system concerned and a vision for the future. An analysis of the "bottlenecks" in a particular innovation system is an important diagnostic tool for identifying the areas for intervention that are required to complement existing public- and private-sector activities. Much of the empirical work on innovation policy mixes has been concerned, for the most part, with discussing balances (and by extension, policy gaps). Far less attention has been paid to interactions, no doubt on account of the conceptual and practical challenges involved. Yet the effectiveness of a policy instrument almost always depends upon its interaction with other instruments, sometimes at different times and for different purposes.

Countries' instrument mixes will differ, as they will have developed over time and will have been adapted to the country's specific political and socio-economic circumstances. This is confirmed by countries' replies to the OECD *Science, Technology and Innovation Outlook* policy questionnaire 2014 (OECD, 2014a). This questionnaire invited countries to rate the balance in the policy mix for business R&D and innovation over time (ten years ago, today and in the next five years) for five sets of policy instruments: population-targeted versus generic instruments; sector- or technology-targeted versus generic instruments; financial versus non-financial instruments; competitive versus non-competitive instruments; and supply-side versus demand-side instruments. The main findings are as follows:

- **Population-targeted versus generic (non-population-targeted) instruments:** Population-targeted instruments are those targeted towards specific types of firms, especially SMEs or new-technology based firms. The questionnaire results indicate that many countries have moved towards more population-targeted instruments over the last decade and that this will continue in the next five years. There are, however, important exceptions: Poland's policy instrument mix has been and will remain predominantly generic, while those of France, Germany, Sweden and the United Kingdom have increasingly moved away from population-targeted instruments, a trend that is set to continue over the coming years. A notable trend among practitioners is a gradual move away from using firm size as a segmentation criterion, towards differentiators more to do with firm age and, most recently, growth rate or potential.

- **Sector- and technology-targeted versus generic (non-technology-targeted) instruments:** Sector- and technology-targeted instruments support specific fields of R&D and innovation or specific industry sectors. Countries vary markedly in the balance of sector- and technology-oriented and non-sector/non-technology-oriented instruments. Close to half of all countries claim that their policy instrument mix is becoming more sector- and technology-oriented than previously, owing, perhaps, to an interest in "new industrial policy". A few OECD countries are moving in the opposite direction. Sweden expects policy to move from what was a strong sector and technology orientation a decade ago to a strong generic orientation in the next five years; over the same period, Finland and Germany expect to move from a policy mix that was slightly more sector- and technology-oriented to one that will be slightly more generic. Outside of the OECD, China expects to move from a policy mix with a strong sector and technology orientation ten years ago to one that is equally balanced in the next five years.
- **Financial versus non-financial instruments:** Financial instruments include both direct (e.g. credit loans and guarantees, repayable advances, competitive grants, innovation vouchers) and indirect funding (e.g. R&D tax incentives), while non-financial instruments include a variety of tools, including business innovation services, organisation of events, and information campaigns that promote business innovation. The bulk of support to business R&D and innovation has been financial in nature. While there has been some movement towards more non-financial instruments in about half of the countries answering this question, the balance in about three-quarters remains at the financial instrument end of the spectrum.
- **Competitive versus non-competitive instruments:** Competitive policy instruments selectively allocate funding on the basis of criteria such as expected performance and relevance. Non-competitive policy instruments may be granted universally or after a selection process based on eligibility criteria. Countries show a strong preference for competitive instruments. Close to half of the countries answering this question indicated a shift towards more competitive instruments. However, among OECD countries, Canada, the Netherlands and, to a lesser extent, the United Kingdom indicate that their policy mix is and will remain more non-competitive, which may partly reflect the strong reliance on R&D tax credits in their support for business innovation.
- **Supply-side versus demand-side instruments:** Supply-side instruments aim to boost knowledge production and supply, with a view to accelerating knowledge spillovers and externalities. As discussed earlier in this chapter, demand-side instruments focus on boosting market opportunities and demand for innovation, as well as on encouraging suppliers to meet expressed user needs. The results confirm the long-standing focus on supply-side instruments but also the recent emergence of demand-side policy to stimulate and articulate public demand for innovative solutions and products from firms. Many countries indicate that the next five years will see increased emphasis on demand-side instruments, though the majority expect supply-side instruments to remain dominant.

In summary, based on countries' self-assessments, it is evident that the balance of their policy mixes differs and that these balances change over time. Overall, more countries have been moving towards more targeted policy mixes, involving more competition and mobilising a broader diversity of instruments. A further discussion on the policy mix, its evaluation and the practical implementation of policies follows in Chapter 8 of this report.

Notes

1. Government funding of science and basic research is discussed in Chapter 5 of the report.
2. All of these regimes, including the proposed Italian IP box, will need to be assessed against the substantial activities requirement set out by the OECD Forum on Harmful Tax Practices. The UK government has already announced that its current regime will be closed to new entrants as of June 2016 and phased out by 2021, and the current regime is likely to be replaced by a regime that complies with the OECD requirements.
3. Knowledge spillovers across a geography are typically measured by the way that patent citations in a given technology area decay with distance (i.e. beyond a particular distance, commonly a 150-200 km radius, citations are significantly less likely) (OECD, 2013b, provides a literature review).
4. Section 7.2 on inclusive innovation will return to this question.
5. The OECD defines high-growth firms for statistical purposes as "enterprises with an average annualised growth greater than 20% a year, over a three-year period, and with ten or more employees at the beginning of the observation period". Growth is measured by turnover or employment.
6. At the same time, innovation can also contribute to more costly policies, as citizens and politicians choose the higher-quality, more expensive option – e.g. in the health sector.
7. These characteristics pertain specifically to the OECD's work on environmental regulation, but most apply also to other areas of regulation.
8. In Canada, in 2013, the Canadian Radio-Television and Telecommunications Commission (CRTC) established a national mandatory code of conduct for wireless service providers. The code is intended to make it easier for consumers to get information about their contracts and about their associated rights and responsibilities, as well as to establish standards for industry behaviour (*www.crtc.gc.ca/eng/info_sht/t14.htm*).
9. See: *http://ec.europa.eu/justice/consumer-marketing/files/model_digital_products_info_complete_en.pdf*.
10. See: *http://ec.europa.eu/justice/consumer-marketing/rights-contracts/directive*.

References

Akcigit, U., D. Hanley and N. Serrano-Velarde (2014), "Back to basics: Basic research spillovers, innovation and growth", *NBER Working Paper Series*, No. 19473.

Andrews, D. and C. Criscuolo (2013), "Knowledge-based capital, innovation and resource allocation", *OECD Economics Department Working Papers*, No. 1046, OECD Publishing, Paris, *http://dx.doi.org/10.1787/5k46bh92lr35-en*.

Appelt, S. and F. Galindo-Rueda (2015), «Measuring the link between public procurement and innovation», *OECD Science, Technology and Industry Working Papers* (forthcoming), OECD Publishing, Paris.

Autio E. (2013), "Promoting leadership development in high-growth firms", paper prepared for the OECD workshop *Management and Leadership skills in High-Growth Firms*, Warsaw, 6 May 2013, *www.oecd.org/cfe/leed/leadership-development-HGFs.pdf*.

Bravo-Biosca, A., C. Criscuolo and C. Menon (2013), "What drives the dynamics of business growth?", *OECD Science, Technology and Industry Policy Papers*, No. 1, OECD Publishing, Paris, *http://dx.doi.org/10.1787/5k486qtttq46-en*.

Brown, R. and S. Mawson (forthcoming), "Trigger points and high growth firms: A conceptualisation and review of public policy implications", *Journal of Small Business and Enterprise Development*.

Carlino, G. and W. Kerr (2014), "Agglomeration and innovation", *Harvard Business School Working Paper*, 15-007.

Foray, D., P. David and B. Hall, (2009) "Smart Specialisation – The Concept", Knowledge Economists Policy Brief No 9, June 2009, *http://ec.europa.eu/invest-inresearch/pdf/download_en/kfg_policy_brief_no9.pdf?11111* .

Foray, D. (2012, forthcoming) "Economic Fundamentals of Smart Specialisation", *Ekonomiaz*, special issue.

Gabrielsson, J., A. Lindholm-Dahlstrand and D. Politis (2012), "Sustainable high-growth entrepreneurship: A study of rapidly growing firms in the Scania region", paper presented at the OECD-DBA (Danish Business Authority) workshop on *High-Employment-Growth Firms: Local Policies and Local Determinants*, Copenhagen, 28 March 2012, *www.oecd.org/cfe/leed/growthfirmsworkshop.htm*.

Griffith, R., H. Miller and M. O'Connell (2010), "Corporate Taxes and Intellectual Property: Simulating the Effect of Patent Boxes", IFS Briefing Note 112, Institute for Fiscal Studies, London, *www.ifs.org.uk/bns/bn112.pdf*

Hart, M. and Y. Temouri (2013), "High-growth firm localities and determinants: Evidence from OECD countries", paper prepared for the OECD-DBA workshop on *High-Employment-Growth Firms: Local Policies and Local Determinants*, Copenhagen, 28 March 2012.

Hausmann, R. and D. Rodrik (2003), "Economic Development as a Self-Discovery", *Journal of Development Economics*, Volume 72, No. 2.

Henrekson, M. and D. Johansson (2010), "Gazelles as job creators: A survey and interpretation of the evidence", *Small Business Economics*, Vol. 35, pp. 227-244.

Martin, P., T. Mayer and F. Mayneris (2011), "Spatial concentration and plant-level productivity in France", *Journal of Urban Economics*, Vol. 69/2, pp. 182-195.

OECD (2015), *Digital Economy Outlook 2015*, OECD Publishing, Paris, *http://dx.doi.org/10.1787/9789264232440-en*.

OECD (2014a), *OECD Science, Technology and Industry Outlook 2014*, OECD Publishing, Paris, *http://dx.doi.org/10.1787/sti_outlook-2014-en*.

OECD (2014b), "OECD Recommendation on Consumer Policy Decision Making" (booklet), OECD Publishing, Paris, *www.oecd.org/sti/consumer/Toolkit-recommendation-booklet.pdf*.

OECD (2014c), *OECD Review of Telecommunication Policy and Regulation in Colombia*, OECD Publishing, Paris, *http://dx.doi.org/10.1787/9789264208131-en*.

OECD (2014d), "Consumer policy guidance on mobile and online payments", *OECD Digital Economy Papers*, No. 236, OECD Publishing, Paris, *http://dx.doi.org/10.1787/5jz432cl1ns7-en*.

OECD (2013a), *Regions at a Glance 2013*, OECD Publishing, Paris, *http://dx.doi.org/10.1787/reg_glance-2013-en*.

OECD (2013b), *Regions and Innovation: Collaborating Across Borders*, OECD Publishing, Paris, *http://dx.doi.org/10.1787/9789264205307-en*.

OECD (2013c), "Key Findings of the Work of the OECD LEED Programme on High-Growth Firms", OECD Publishing, Paris, *www.oecd.org/employment/leed/high-growthreport.htm*.

OECD (2013d), "An International Benchmarking Analysis of Public Programmes for High-Growth Firms", OECD Publishing, Paris, *www.oecd.org/employment/leed/high-growthreport.htm*.

OECD (2013e), *Innovation-driven Growth in Regions: The Role of Smart Specialisation*, OECD Publishing, Paris, *www.oecd.org/sti/inno/smart-specialisation.pdf*.

OECD (2013f), "Strategic public procurement", in *Government at a Glance 2013*, OECD Publishing, Paris, *http://dx.doi.org/10.1787/gov_glance-2013-46-en*.

OECD (2011a), *Financing High-Growth Firms: The Role of Angel Investors*, OECD Publishing, Paris, *http://dx.doi.org/10.1787/9789264118782-en*.

OECD (2011b), *Demand-Side Innovation Policies*, OECD Publishing, Paris, *http://dx.doi.org/10.1787/9789264098886-en*.

OECD (2011c), "OECD consumer policy toolkit workshop on communication services: Summary of proceedings", *OECD Digital Economy Papers*, No. 221, OECD Publishing, Paris, *http://dx.doi.org/10.1787/5k480t1g546j-en*.

OECD (2010a), *The OECD Innovation Strategy: Getting a Head Start on Tomorrow*, OECD Publishing, Paris, *http://dx.doi.org/10.1787/9789264083479-en*.

OECD (2010b), *Consumer Policy Toolkit*, OECD Publishing, Paris, *www.oecd.org/sti/consumer/consumerpolicytoolkit.htm*.

OECD (2007), *Competitive Regional Clusters: National Policy Approaches*, OECD Publishing, Paris, *http://dx.doi.org/10.1787/9789264031838-en*.

O'Sullivan, E. et al. (2013), "What is new in the new industrial policy? A manufacturing systems perspective", *Oxford Review of Economic Policy*, Vol. 29/2, pp. 432-462.

Rodrik, D. (2008), "Normalizing industrial policy", *Commission on Growth and Development Working Paper*, No. 3, Commission on Growth and Development, Washington, DC.

Rosenthal, S.S. and W.C. Strange (2004), "Evidence on the nature and sources of agglomeration economies", in Henderson, V. and J.F. Thisse (eds.), *Handbook of Regional and Urban Economics*, Vol. 4, North-Holland, Amsterdam.

Stiglitz, J.E., J.Y. Lin and C. Monga (2013), "The rejuvenation of industrial policy", *World Bank Policy Research Working Paper*, No. 6628.

UK Government (2014), Interim Evaluation of Growth Accelerator, BIS Research Paper No. 187, Department for Business, Innovation and Skills, London, *https://www.gov.uk/government/uploads/system/uploads/attachment_data/file/375059/bis-14-1204-interim-evaluation-of-growth-accelerator-2014.pdf*

Uyarra, E. and R. Ramlogan (2012), "The effects of cluster policy on innovation", *Nesta Working Paper*, No. 12/05.

Warwick, K.S. (2013), "Beyond industrial policy: Emerging issues and new trends", *OECD Science, Technology and Industry Policy Papers*, No. 2, OECD Publishing, Paris.

Warwick, K.S. and A. Nolan (2014), "Evaluation of industrial policy: Methodological issues and policy lessons", *OECD Science, Technology and Industry Policy Papers*, No. 16, OECD Publishing, Paris, *http://dx.doi.org/10.1787/5jz181jh0j5k-en*.

Wennberg, K. (2013), "Managing high-growth firms: A literature review", paper prepared for the OECD workshop on *Management and Leadership Skills in High-Growth Firms*, Warsaw, 6 May 2013, *www.oecd.org/cfe/leed/Wennberg_Managing%20a%20HGF.pdf*.

Westmore, B. (2013), "R&D, Patenting and Productivity: The Role of Public Policy", *OECD Economics Department Working Paper*, No. 1046.

Wilson, K. and F. Silva (2013), "Policies for seed and early stage finance: Summary of the 2012 OECD Financing Questionnaire", *OECD Science, Technology and Industry Policy Papers*, No. 9, OECD Publishing, Paris, *http://dx.doi.org/10.1787/5k3xqsf00j33-en*.

The Innovation Imperative
Contributing to Productivity, Growth and Well-Being

Chapter 7

Applying the framework for innovation

This chapter explores some of the applications of the policies for innovation. The applications differ according to the national context, but may also be affected by the sector or technology concerned, and by the specific objectives of innovation. It first discusses the national agenda for innovation, and then turns to a number of specific policy challenges, namely the role of innovation for inclusive growth, health innovation and the role of innovation in the green growth agenda. Particular attention is also paid to innovation in the public sector, which is important to improve effectiveness and efficiency within the public sector, but which can also help support innovation throughout the economy.

The statistical data for Israel are supplied by and under the responsibility of the relevant Israeli authorities. The use of such data by the OECD is without prejudice to the status of the Golan Heights, East Jerusalem and Israeli settlements in the West Bank under the terms of international law.

This chapter explores some of the applications of the policies for innovation discussed in Chapters 3 to 6. This application differs according to the national context, but may also be affected by the sector or technology concerned, and by the specific objectives of innovation. Chapter 7 first discusses the national agenda for innovation (Section 7.1), and then turns to a number of specific policy challenges, namely the role of innovation for inclusive growth (Section 7.2); health innovation (Section 7.3); and the role of innovation in the green growth agenda (Section 7.4.). Particular attention will also be paid to innovation in the public sector (Section 7.5), which is important to improve effectiveness and efficiency within the public sector, but which can also help support innovation throughout the economy. Chapter 8 will turn to some of the specific challenges in implementing these policies in practice, and turning a strategy for innovation into effective action and outcomes.

7.1. The national agenda for innovation

Countries differ substantially in the basic conditions for innovation, such as the level of economic development, the structural make-up and trade specialisation of the economy, and geography, but also in their institutional characteristics and approaches to policy, e.g. in regard to the role of government and different private and public actors in the economy. As a result, countries face different constraints under which policy responses to their challenges and opportunities need to be developed and different barriers that provide a rationale for policy action (as discussed in Chapter 1). Moreover, depending on these conditions, global developments in the area of science, technology and innovation – such as the internationalisation of research and development (R&D) or the changing nature of innovation (as discussed in Chapter 2) – may translate into different sets of challenges and opportunities. For example, the emergence of new players in global R&D affects countries with a strong R&D base in a different manner than it affects countries with more limited R&D capabilities. This section explores some of the key issues in setting a national agenda for innovation and how initial conditions – in terms of economic performance and structural features – affect this agenda.

Many countries today share the overarching objective to move towards more innovation-led sustainable growth, and many have become more aware of the role of innovation for achieving this goal. Innovation is considered important for growth and competitiveness, but also in helping address social and global challenges. As a result, innovation policy is becoming an integral part of economic policy in a wider range of countries, both advanced and emerging, resulting in some convergence in the policy agenda for innovation across countries.

Yet the precise nature of the challenges that innovation policy makers need to address and the approaches and set of policies they need to adopt in order to contribute effectively to sustainable economic growth differ across countries, depending on their stage of development, economic structure, capabilities of firms, etc. To be effective, innovation policy and the related governance system need to be adapted to the specific challenges. Moreover, the specific choice and combination of innovation policies and related governance need to be aligned with the capabilities of each country in terms of policy making and policy implementation. A number of stylised policy agendas can be distinguished in this respect, although many countries will be confronted with their own particular mix of challenges (OECD, 2010a, 2014a, 2014b).

Key agendas for innovation policy

Boosting productivity growth in high-income countries

A first policy agenda where innovation plays an important role involves strengthening productivity growth, typically mainly in advanced economies. In many countries, some sources of per capita income growth that have played an important role in the past are now reaching their limits. Because of demographic developments in advanced economies, growth is bound to become increasingly dependent on rising multifactor productivity (MFP). OECD long-term projections suggest that the MFP contribution to gross domestic product (GDP) per capita could rise from around 54% to 88% in OECD countries between 2010 and 2060 (Figure 1.3; Braconier, Bloom and Davis, 2014). Policies that can help boost productivity growth therefore become crucial for achieving sustainable growth of per capita income (OECD, 2014b). Fostering innovation is an important and intrinsically unlimited way to achieving this goal. Innovative capacity boosts productivity by advancing the technology frontier (mainly in advanced economies), by speeding up the adoption of existing technology (in both advanced and emerging economies), and by diffusion of technologies, processes and practices within economies.

Many high-income countries already have a well-functioning innovation system. Yet even in this group of countries there are large differences in regard to innovation performance, the role different types of innovation play in driving growth and the type of innovation system, including the systems of innovation policy governance. This also implies considerable variation in some of the key challenges for innovation policies. Some examples of the agenda for innovation in advanced economies are discussed in Box 7.1.

Closing the development gap of low- and middle-income countries

A closely related objective to which innovation can contribute is reducing development and productivity gaps. Many emerging-market countries have grown rapidly in recent years, but the pace of catching-up is likely to slow sharply as GDP per capita in these countries converges towards higher levels. Nevertheless, important gaps in productivity levels remain with the most advanced economies (Figure 7.1; OECD, 2014e) that are primarily linked to differences in MFP. Stimulated by successful examples, many low- and especially middle-income countries are therefore paying increased attention to fostering their innovation capabilities as a means to continue their catch-up with more advanced countries.

Developing and emerging economies typically share certain types of weaknesses in their innovation system, including a lack of skilled human resources, weak innovation capabilities in business firms, and a disconnect between industry and universities and public research organisations, in regard to both research and education. Moreover, many are confronted with fragmented systems of innovation, where the institutional frameworks and capabilities for innovation policy still require development, and governance of innovation policy is often in its infancy. Challenges related to governance and implementation, as discussed in Chapter 8, are therefore particularly important, as is public-sector innovation, discussed in Section 7.5. In many cases, serious shortcomings in the framework conditions for innovation, such as lack of competition or poorly functioning product and financial markets, also need to be addressed in order to make policies targeted towards innovation activities more effective.

Box 7.1. The agenda in advanced OECD countries: France, the Netherlands and Sweden

The national context of innovation agendas: Shared features and differences

France, the Netherlands and Sweden are high-income countries with a strong record in science and technology. France is the second-largest economy in the euro area. The Netherlands and Sweden are smaller but highly internationalised, and both are home to enterprises of global reach. France has been more reliant on its much larger domestic market but also has a core of global firms although its small and medium-sized enterprises (SMEs) tend to be more inward-looking. The Netherlands is a key European trade and logistics hub and has manufacturing strengths, including in processing industries, some of which have sprung from agriculture. Sweden, too, has a large services sector but retains a strong and varied manufacturing base that has undergone major changes in ownership, restructuring and "servitisation". The declining share of manufacturing has been a cause of concern in France.

The three countries are home to some of the world's most innovative multinational enterprises (MNEs) and have historically had strong public research and education systems. Business research is heavily concentrated in large MNEs. Increasing and broadening business research has proven difficult despite long-standing efforts in all three countries. A sluggish macroeconomic environment has subdued business innovation investments in the France and the Netherlands. In the case of France, this is compounded by a complex and insufficiently selective system of business support. The efficiency of its tax incentives is in particular need of improvement. In recent years, the Netherlands has relied heavily on tax incentives to support business innovation, which to a great extent benefit SMEs. The Netherlands features a comparatively low R&D intensity.

Sweden and the Netherlands have two of the most prolific public research systems. France has witnessed an erosion of its global position overall but remains one of the world's pre-eminent research powerhouses with pockets of leading research in numerous areas. There is a renewed policy emphasis on increasing the economic and societal relevance of public research, in the form of valorisation agendas (in the Netherlands), new organisational models (Sweden and the Netherlands) and ongoing discussions about governance reform (France). Research funding regimes favour R&D collaboration between the public and private sectors. This involves, for example, calls for collaborative research project proposals and the establishment of more strategic, long-term competence centres/centres of excellence.

In both Sweden and the Netherlands, there appears to be some mismatch between the frontier-shifting orientation of university research and the less ambitious types of innovation that are relevant to some parts of the business sector, including many SMEs. Public research is being extended to traditionally non-research performing institutes – in universities of applied science in the Netherlands and the community universities in Sweden – with a view to improving learning and better responding to regional economic and social needs.

In both France and Sweden, middle-level agencies play a key, if not leading, role in the governance of the system, whereas in the Netherlands policy programming is contained within the relevant ministries. There are 'horizontal' co-ordination issues in all three countries, including tensions between ministries of education or research and ministries of economy. In the Netherlands these tensions are managed in a bilateral way, whereas in Sweden they are managed at the agency level, essentially between the Swedish Research Council (VR) and the country's innovation agency (VINNOVA). In Sweden, innovation policy is weak relative to other policy areas (e.g. higher education), which has frustrated a more holistic approach to innovation policy. The Netherlands and France have a long experience with co-ordination that cuts across government ministries, which despite some notable successes, can be further improved.

Sweden and France have an uneven record on evaluation, which has been resisted in parts of the French public research system. Certain specialised functions of the new public research actors that do not typically lead to publications in high-ranking journals pose a challenge to national evaluation frameworks. In the Netherlands there are efforts under way to better recognise and monitor the usefulness of this type of research in its own terms.

Box 7.1. **The agenda in advanced OECD countries: France, the Netherlands and Sweden** *(cont.)*

Challenges and opportunities

Globalisation has increased opportunities, but also competition. Emerging economies are increasingly competing in the higher end of value chains. The Netherlands and Sweden embraced globalisation early on, and more comprehensively than most countries. France, until this day, has remained more inward-looking, which also reflects its large size. Globalisation challenges some long-standing features of their economic models, with significant transfers of ownership and corporate restructuring and some offshoring in the case of Sweden and some difficulty in benefiting from the growth of emerging markets in the case of the Netherlands. France has concerns about international competitiveness.

Raising productivity and competitiveness is a challenge for all three countries. All of them have high levels of labour productivity, and the Netherlands and France are among the leading OECD countries in this respect. While there is still scope to raise participation rates, notably in France, productivity growth is the key for further improving living standards in the longer term. In advanced countries, sustained productivity growth requires more innovation. Through greater efficiency in production and development of new products, innovation improves the competitiveness of firms and helps them sustain high wage and income levels.

Despite the high levels of human capital, education performance is an issue in France (in large segments of the population and in the supply of doctorates), Sweden (decline in scores on the OECD's Programme for International Student Assessment [PISA] test) and, to some extent, the Netherlands (falling tertiary completion rates and mismatches). Universities are under strain in the face of budget limitations, widening participation, and pressure to be more responsive to the needs of the economy and society. Public research institutes (PRIs) too are under pressure to demonstrate and improve the economic impact of their activities. The challenge is reflected in contemporary policy attention to university spin-offs, licensing income and research-inspired start-ups. PRIs are facing difficulty in adapting to the changing role of the state and the types of R&D it commissions as well as shifts in the corporate strategies of previous national champions that have become much more internationalised and typically rely less on PRIs for R&D.

All three countries face major societal challenges such as an impending demographic shift, energy security and environmental sustainability. Innovation will play a key role in addressing these challenges. Responses may open up opportunities for industrial diversification, in areas such as health, the bio-economy and renewable energy. Emerging sciences, enabling technologies and changes in the way global innovation is organised (e.g. a move towards openness and sharing) provide many more opportunities besides.

National agendas and reform processes

Formalised, government-backed processes play a key role in agenda setting and co-ordination in the Netherlands and in France. Sectoral plans drawing on increasingly interactive public-private relationships set much of the agenda in both countries. In the Netherlands, planning happens largely within the top sectors, a new form of industrial policy meant to set agendas, facilitate co-ordination between businesses and government, pool resources, leverage public research capacities, improve the relevance of skills, and encourage a whole-of-government approach. In France, in addition to the sectoral plans, national strategies for research and innovation were introduced in 2009 and 2014. Their preparation involved various stakeholders including research agencies, researchers, businesses and users (e.g. patient associations), but stopped short of proposing a resource plan for implementation of their objectives, which remained rather general. Social capital, trust in broadly effective and autonomous government agencies, and excellent research institutions help facilitate effective co-ordination in Sweden. The national strategy seems to play a less prominent role in co-ordination. The research and innovation policy bills introduced every four years help agencies with midterm planning and strategies while providing a framework for the government to set priorities.

Box 7.1. **The agenda in advanced OECD countries: France, the Netherlands and Sweden** *(cont.)*

Societal challenges are addressed in both national plans and in the context of European efforts under Horizon 2020. In the Netherlands, societal challenges are increasingly being reflected in the priorities of the top sectors, while in France research strategies have been set for ten societal challenges. Emerging technologies are addressed as horizontal themes in the Dutch top sectors, whereas in France efforts on major impact innovations are under Innovation 2030.

Source: OECD *Innovation Reviews* for France (OECD, 2014c), the Netherlands (OECD, 2014d) and Sweden (OECD, 2013a).

Figure 7.1. **Total factor productivity as a percentage of the US level, 2000 and 2008 (in %)**

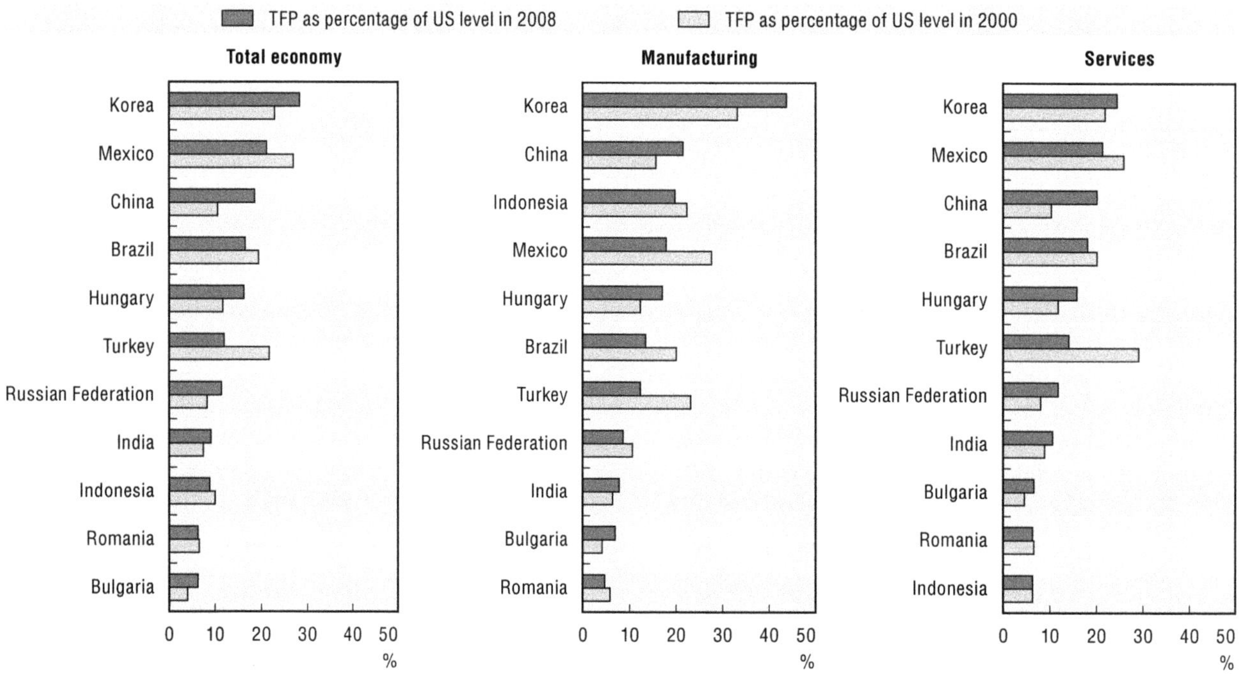

Note: TFP or total factor productivity as shown in the graph is the same as MFP as discussed in the text.

Source: OECD (2014e), *Perspectives on Global Development 2014, http://dx.doi.org/10.1787/persp_glob_dev-2014-en.*

Middle-income countries can consider four key areas when developing their strategies to improve innovation and productivity (OECD, 2014e). These areas are not mutually exclusive, so countries can make improvements in different areas at the same time and they are often interlinked. Moreover, some countries have greater opportunities and possibilities than others in some areas depending on their specific conditions and capabilities. The four key areas include:

- **Diversifying continuously into higher value-added activities within the economy.** This can imply a move from primary sectors to industry and services, but also changes within agriculture, industry and services to diversify into higher value-added activities. This would also boost productivity and is a requirement to remain competitive in global markets at the middle-income level.
- **Innovating through the adoption of existing knowledge elsewhere in the world and increasingly through the development of local technological capabilities.** Middle-income countries often still have significant room for technological catch-up, as seen by their still

very low labour productivity and MFP compared with advanced countries. Besides better international integration through trade and foreign direct investment (FDI), countries can make effective use of technology licensing; getting technology, designs, production and management assistance from foreign buyers, consulting firms and technical experts; and foreign education and training, among others. However, countries also need to innovate new products, processes, services and forms of organisation that are better suited to their needs than what they can get from abroad, as well as to develop their own frontier shifting innovations to create competitive edges. To adopt technologies from abroad, and to develop local strengths, boosting the absorptive capacity of the economy is particularly important.

- **Reforming product, labour and financial markets as well as skills development schemes.** In many middle-income countries, the development of productive, innovative businesses is often constrained by an inadequate regulatory environment or unavailability of appropriate skills. Chapter 4 discussed the importance of such framework policies in detail.
- **Fostering competitive and innovative service sectors.** The domestic services sector can grow to meet the demand of the growing middle classes. Services can also increase the competitiveness of manufacturing and be a source of export earnings. Strengthening innovation in services is important in this context.

Making improvements in the four key areas above can be a focus for both low- and middle-income countries, and also high-income countries, but getting these areas right becomes more crucial in the upper-middle-income bracket. At the same time, most middle-income countries (including in the upper bracket) can and should continue exploiting the "old" drivers of growth (OECD, 2014e), including shifting labour from lower productivity (agricultural) sectors to higher productivity (agricultural, industry and service) sectors and fully reaping the benefits from growth led by factor accumulation, which includes the better utilisation of labour and the accumulation of human and physical capital. Most middle-income countries still have room to make improvements in these areas.

Box 7.2 illustrates the innovation agenda for three large emerging economies, namely the People's Republic of China (hereafter "China"), Mexico and the Russian Federation, and shows that while there are important similarities in the agenda, there are also important differences in challenges, opportunities and the conditions for innovation more generally. Some large economies are shifting towards a new growth model that increasingly relies on growing domestic demand as an additional source of growth. Therefore, these economies are on the one hand trying to identify new forms of partnership with foreign companies to enhance technology transfer and linkages with domestic companies; on the other hand, they are investing in supporting the development of SMEs and innovation to better adapt to the changing economic landscape.

Box 7.2 also illustrates that innovation policy – and reforms in policy governance – in countries as diverse as China, Mexico and the Russian Federation, but also in Chile, Korea and South Africa, are informed by a change in the understanding of the role of and interplay between the creation and diffusion of technology, and typically include strategies aimed at fostering the emergence of new comparative advantage. The outstanding economic development of Korea, as well as the historically preceding example of Japan, is very instructive in this respect. The case of Korea shows that a significant level of S&T capabilities is required for successfully implementing imitation strategies, and all the more for moving up the value chain to accelerate catching up (OECD, 2009b, 2014f).

Box 7.2. **Key emerging economies: China, Mexico, Russian Federation**

The national context of innovation agendas: Shared features and differences

China, Mexico and the Russian Federation are major emerging economies, and members of the G20 group of countries. All three are economies in the middle-income range, while their development trajectory and specialisation vary significantly. China and the Russian Federation have both been undergoing a transformation from a centrally planned economy to a market-based one, but have chosen very different approaches in this process. While China has risen to become the world's leading export nation, supplying a wide range of export products, the Russian Federation has remained highly dependent on the export of raw materials, including oil and gas. Mexico has been undergoing important economic reforms and has benefited from its proximity to the large North American market integrated by the North American Free Trade Agreement (NAFTA). All three countries are facing challenges to their respective growth model, and innovation is an important factor in all strategies to tackle them. They face the challenge of making a transformation into industries that thrive on innovation. China is making efforts to rebalance and reorient the economy, tapping into new sources of growth, including innovation. Starting at very modest levels in the 1990s, China has made a persistent effort to invest in science, technology and innovation. It has also been skilful in leveraging the large size and dynamism of its market to get access to technology. China's success in increasing its innovation capabilities contrasts with the much slower pace of change in the Russian Federation and Mexico. While all three countries are making some progress in upgrading human capital, only China has managed to rebalance its innovation system so it is centred on the business sector and has substantially improved its performance.

Innovation systems performance, capabilities and governance

The three countries have very different innovation capabilities. China has made major strides in upgrading its innovation system, whereas in the Russian Federation and Mexico efforts to upgrade innovation capabilities have faltered. Expenditure on R&D as a percentage of GDP has remained below 0.5% in Mexico; it is slightly above 1% in the Russian Federation, while China – after persistent increases since the 1990s – is now close to 2%, approaching the OECD average. In Mexico and the Russian Federation, many of the standard innovation output indicators (trade, scientific output, patents, etc.) suggest rather modest innovation performance. By contrast China has made leaps in innovation capabilities both in science and, especially, in the business sector. To some extent, upgraded capabilities and performance are reflected in China's strong MFP growth, which unlike many other countries' picked up further in the second half of the 2000s. Notable productivity improvements in the Russian Federation in the late 1990s and early 2000s were likely driven by efficiency gains linked to market reforms rather than innovation. On the opposite end, Mexico experienced poor productivity growth. The pronounced regional concentration of innovation capabilities characterising all three countries is to some extent a reflection of their size, but also signifies their inability to use effectively their full potential.

Available indicators suggest that the majority of Mexican firms do not systematically innovate. Russian Federation manufacturing firms are much less likely to engage in innovation activities than their counterparts in OECD countries. Of the three countries, only China has managed to shift the balance of its innovation system, with 76% of R&D performed by the business sector, as opposed to just under 40% in Mexico. The bulk of Russian Federation R&D is still performed in state-owned branch research institutes, which are mostly separate from industrial firms and universities. In Mexico and particularly in the Russian Federation, links between universities and public research institutions and businesses remain weak, with incentives on both sides working against closer co-operation. An important obstacle lies with firms themselves, which have too few capabilities to innovate, little absorptive capacity for innovation, weak links to public research institutes and universities, and, above all else, easy access to economic rents that provide few incentives to innovate.

Improving innovation performance will require sustained investments and qualitative improvements in human resources. Starting from low levels, Mexico has developed a significant pool of engineers over two decades, but levels remain below most other OECD countries, discouraged by low effective demand for such

Box 7.2. **Key emerging economies: China, Mexico, Russian Federation** *(cont.)*

skills. The majority of the Mexican labour force, however, has a low level of skills. Although China has the world's largest pool of human resources for science and technology (S&T), the tertiary-qualified share of the population is still very low. Furthermore, China still lacks world-class researchers, although there are signs that this is changing. The Russian Federation stands out in terms of formal education attainment. The share of the working adult population with a higher education qualification is similar to some of the most advanced OECD countries. In terms of the quality of education, parts of China show an impressive performance in the OECD PISA assessment of secondary education students, while the Russian Federation and Mexico lag behind most OECD countries.

In the Russian Federation, much of the science base remains in a poor state following a long period of underfunding. Judging by the number of scientific publications, the Russian Federation has a level of scientific output comparable to that of Brazil or the Netherlands. Its output, however, is of much lower scientific impact than Mexico's or China's. Mexico's "national system of researchers" has made public research more efficient over time and allowed some top-quality universities and public research centres to emerge. The system has suffered from resource constraints, however, and an emphasis on predefined outputs appears to have discouraged more long-term research. Over the past decade the performance of public research has improved dramatically in China. Its aggregate scientific output is now second only to the United States, though the citation impact of publications is just above that of Mexico, lagging considerably behind most OECD countries.

Challenges and opportunities

All three countries face challenges to their current sources of economic growth. Cost pressures and environmental degradation are gradually placing a brake on China's low-cost manufacturing growth "engine". In the face of deteriorating framework conditions and mounting social challenges, export-led manufacturing growth has stagnated in Mexico, and primary product exports remain important. Russian Federation growth performance has been dependent on transitory factors in the past, especially terms-of-trade gains reflecting world oil and gas prices. Despite the nuances in the challenges, the solution invariably includes productivity improvements and innovation-driven economic diversification and upgrading.

Improving the framework conditions for business innovation is a major challenge, particularly for the Russian Federation and Mexico. Overly restrictive regulation, exceptions to the rule of law and a lack of competition remain major disincentives to innovation and entrepreneurship. Flawed legal frameworks, corruption, and weak physical and information and communications technology (ICT) infrastructures are problems shared by all three countries.

The three countries face significant social and environmental challenges that could benefit from innovation. These include environmental degradation, public health and the energy system. Food security is an issue for China and Mexico, while the Russian Federation and, uncharacteristic for a country of its income level, China face the challenge of rapidly ageing populations with pervasive consequences for the economy and society, including the innovation system. A regional rebalancing of innovation capabilities across their vast territories will be important for meeting a multitude of social, environmental and economic development challenges. Upgrading the capabilities of their scientific institutions and kick-starting a virtuous cycle of capability accumulation and innovation in the business sector, while aligning their activities towards national goals in a more direct and effective manner than has been the case so far, are of utmost necessity.

National agendas and reform processes

National innovation agendas differ in prominence and scope across the three countries. China's innovation agenda is quite wide-ranging, incorporating environmental sustainability, societal challenges and inclusiveness perspectives beyond the traditional focus on economic growth. In this respect, it resembles the innovation agendas of many advanced OECD countries. Thus, the National S&T Major Projects focus strongly on public health, ageing, food and drug safety, and disaster prevention, while the Innovation 2020 Programme of the Chinese Academy of Sciences includes energy and health among its

Box 7.2. **Key emerging economies: China, Mexico, Russian Federation** *(cont.)*

four sectoral focuses. The Russian Federation's innovation agenda continues to have a high-tech bias and favours strategic industries related to the country's large military-industrial complex, as well as energy and ICTs. Mexico's innovation agenda is somewhat narrower and less prominent and is strongly associated with activities of the public research system, which is comparatively small. It, too, has recently given greater prominence to societal challenges, such as climate change and food security.

Compared with OECD countries, public R&D funding in China is heavily oriented towards applied research and development, suggesting that some rebalancing is needed towards cutting-edge fundamental research essential for China to go beyond "catch up". The Russian Federation's public R&D funding has increased sharply over the last seven or eight years, with much earmarked for developing research infrastructure (including in universities) and for frontier science programmes in areas such as materials. While growth in Mexico's public R&D funding has outpaced the OECD average in recent years, spending remains low overall and is used largely to fund researchers' salaries.

In all three countries, the public sector continues to play a prominent role in national innovation systems. Even in China, where perhaps most progress has been made towards establishing a firm-centred innovation system, PRIs and state-owned enterprises (SOEs) remain dominant. Both China and the Russian Federation have sought to reform Soviet-era PRIs. Radical transformations took place in the 1990s in China, and reform processes continue with a view to improving PRIs' alignment with new social and economic development challenges. Reforms in the Russian Federation have been slower and more piecemeal, though have recently accelerated with a major reorganisation of the Russian Academy of Sciences in 2013 and ongoing consolidation of the state-owned, nominally privatised, industry branch institutes.

The innovation agendas of all three countries aim to encourage innovation in large firms and to support entrepreneurship and SMEs. In China, various policy instruments aim to foster a firm-centred innovation system and emphasise the indigenous innovation capacity of Chinese firms. However, the dominance of SOEs, especially in public utilities, tends to mitigate the pressures to innovate that normally arise from competition. The Russian Federation has a number of policies targeted at firms, including measures that seek to relieve administrative barriers and improve framework conditions (including taxation and customs regimes). There are also measures that target major SOEs, notably the Innovation Development Programme (IDP), which charges them with developing innovation strategies and co-operating with universities and research institutes. Much of Mexico's support to business innovation is tied to public-private partnerships with PRIs and universities.

All three countries give prominence to human resources and skills in their innovation agendas. In China, various talents programmes aim to attract and retain top-tier academics (including from overseas) and to address the business sector's need for innovative personnel, for example through mobility programmes. In the Russian Federation, initiatives have been put in place to improve the qualifications of engineers in strategic industries through new training programmes and internships in leading research and engineering centres in the Russian Federation and abroad. Much of Mexico's efforts are focused on support to high-end human resources in science and technology (HRST) and researchers. More resources have been mobilised for government-sponsored fellowships and to improve the quality of graduate programmes offered by universities and PRIs.

All three countries have initiatives to address regional imbalances in innovation activities. In China, to boost the development of the western region, the Great Western Exploration Strategy supports investments in research infrastructure, research collaboration, and human resource mobility between the eastern and western regions. In the Russian Federation, the government is supporting the development of regional clusters and injecting funds into selected regional universities to build their research capabilities and linkages with local firms. In Mexico, the federal government is collaborating with state governments to provide modest funding for applied research in the regions.

Source: OECD Reviews of Innovation Policy, China (OECD, 2008a), Mexico (OECD, 2009a and OECD, 2013b), Russian Federation (OECD, 2011a).

However, there are pending issues that are difficult to resolve and need attention. For example, the required combination of dedicated effort, openness and flexibility that is necessary to avoid "lock-in" in "underperforming trajectories" or capture by special interests may be difficult to achieve. Moreover, the governance system needs to ensure a fairly long time horizon for innovation policies, which is often difficult to maintain given the incentives of policy makers and the presence of short-term demands ("time inconsistency") and high levels of governance capabilities. Approaches to such high-level governance are further discussed in Chapter 8.

Moreover, there is a high degree of interdependence of the innovation agendas of different countries. For example, the success of major players is significantly changing the environment under which other catch-up economies will operate. Many countries will need to strengthen their R&D capabilities and related infrastructure and also improve the links between foreign companies and the innovation system, e.g. in order to stay internationally competitive as a location of R&D-related activities (OECD, 2008b).

Diversifying the economy and moving up the value chain

A closely related policy objective that can be supported by innovation policies relates to competitiveness and the opportunities for countries to upgrade their position in global or regional value chains. Many countries perceive a need to diversify their economy and to move up the value chain, e.g. by fostering a shift to higher value-added businesses and activities. This perception is reinforced by the fact that many – particularly smaller – countries are exposed to the impact of new competition from emerging economies such as China in labour-intensive activities. Examples of the innovation agenda for such countries are included in Box 7.3.

Hungary is another example of a country that is seeking to move up the value chain.[1] It has developed a significant manufacturing base (mainly operating as a platform for exports to other EU countries) while at the same time showing a high degree of similarity in its structure of exports with that of Chinese exports. In order to foster competitiveness and to move up the value chain, firms need to boost productivity and thus "competitiveness" through innovation. FDI by MNEs in Hungary has created a significant manufacturing sector, producing and exporting goods that are conventionally classified as medium- or even high-tech (although Hungarian-based firms are relatively low in terms of R&D intensity). A major task of policy is to embed this sector better into the national innovation system – including universities and public research organisations – and to upgrade the absorptive capabilities of the domestic SME sector to overcome the dual structure of the economy characterised by a highly productive and internationalised multinational sector and a much less dynamic domestic SME sector.

Innovation in natural resource-based economies

Diversification is also a major challenge for resource-based economies. Chile is a major world exporter of copper, for example, and Norway of oil and gas. In some areas, such as salmon farming, the two countries are competitors for world markets, though closely linked through FDI and knowledge flows (e.g., Norwegian engagement in Chilean aquaculture). South Africa is another country known for its strong resource base. Several of the countries reviewed by the OECD, including Chile, Norway, the Russian Federation and South Africa, have benefited from strong global demand for raw materials. New Zealand benefited from increased global demand for food, although it did not benefit from globalisation to the full extent. However, in export-oriented industries such as dairy products and packaged meat, New Zealand is realising economies of scale that, given its market size and location, are difficult to achieve in many other industries.

Box 7.3. **Smaller emerging economies: Colombia, Peru, Viet Nam**

The national context of innovation agendas: Shared features and differences

Colombia, Peru and Viet Nam are middle-income countries that have each enjoyed periods of favourable economic performance and social development in the past one or two decades. Colombia and Peru made breakthroughs in overcoming long-standing internal conflicts that had seriously hampered their development. Viet Nam looks back at impressive socio-economic achievements, starting with the *doi moi* reforms in the late 1980s. All three economies showed resilience to the crisis, though income per capita growth has been slower in the less buoyant international environment of recent years. Progressively opening the economy, Peru and notably Colombia have moved from the paradigm of import substitution to a more open model geared to reap the benefits of globalisation. The economies of Colombia and Peru largely benefited from the surge in demand for raw materials that boosted activity of their mining sectors as well as non-tradables, while exerting pressure on the development of domestic manufacturing, tradable services and agriculture. Viet Nam, located in one of the world's most dynamic regions, has been inspired by East Asian successes pioneered by Japan, and emulated and selectively adapted by cohorts of Asian "tigers" and "dragons", and lately, China. Export-led, manufacturing-based growth invariably was a cornerstone for their rise, while they differed with respect to the role of FDI. Investment by East Asian and Western firms has enabled producers in Viet Nam to link to buyer-driven global value chains (GVCs) (e.g. in clothing, furniture and electronics), which now account for a large share of international trade in many standardised commodities. While Viet Nam has succeeded in developing export strengths across a broad range of products, exports have concentrated on items such as petroleum, rice, coffee and seafood, together with garments and footwear. More recently there has been a surge in production activity in the electrical and electronics industry (e.g. the assembling of mobile phones by Korean firms).

Innovation systems performance, capabilities and governance

Available indicators point to low capabilities and resource commitments to science, technology and innovation (STI). While there is evidence of considerable new-to-the-country innovation, overall knowledge intensity of production remains low. In the case of Viet Nam and Peru, current R&D commitments are impossible to ascertain as key statistics are a decade old. Fragmentary evidence suggests that businesses fund and perform a minor share of gross domestic expenditures on R&D (GERD). Patenting activity of international relevance is very low, and data on productivity suggest that innovation activity is having only limited economic impact.

The three countries have made considerable achievements in upgrading skills, the outcome of long-standing efforts to improve and expand secondary and tertiary education. Significant issues remain, however, both in terms of quantity and in quality. Peru and Colombia rank low in the 2012 OECD PISA assessment of the performance of secondary students. In contrast, Viet Nam ranked similar to countries with much higher income. The liberalisation of tertiary education in Peru was not accompanied by appropriate regulation, which has placed a strain on quality. Moreover the education systems of Colombia and Peru fail to take full advantage of available human resources, as they provide fewer opportunities to disadvantaged students and thus contribute to already high levels of inequality. In all three countries there is still much room for improving the relevance of skills to the needs of industry.

The performance of universities and PRIs has improved somewhat, as reflected in the increasing trends on the number of scientific publications and citations. However, despite a set of reforms, important bottlenecks persist in their governance. Viet Nam did not overhaul its PRI sector as thoroughly as China did. Neither Peru or Colombia has comprehensively addressed the role of tertiary education and public research in supporting the accumulation of innovation capabilities within firms, e.g. by better aligning the

Box 7.3. **Smaller emerging economies: Colombia, Peru, Viet Nam** *(cont.)*

curricula to the needs of innovating firms or by moving closer to "learning by doing" modes of instruction. PRIs seem insufficiently responsive to changing market and societal demands and tend to be tied to narrow sectoral interests, which does not make them well suited to supporting the national effort for economic diversification.

In Colombia and Peru, institutional rigidities and concerns about corruption limit the options for public support to business sector innovation and discourage experimentation with instruments and modes of governance. Peru, Viet Nam and, to a lesser extent, Colombia do not yet possess an information base for STI policy that is comparable in terms of definitions, coverage and timeliness to OECD countries. This complicates the design of appropriate policies to kick-start a much-needed virtuous cycle of innovation capability accumulation in the business sector. Lack of strategic intelligence is an obstacle to identifying key constraints and improving the design and performance of policy instruments.

Challenges and opportunities

Economic diversification and upgrading towards activities of higher value added is a key challenge for all three countries. Colombia's and Peru's success depends heavily on the export of raw materials. Innovation can help develop new economic activities, boost productivity to sustain income and employment growth for a rising urban population, foster agricultural diversification to improve rural livelihoods, and raise the environmental sustainability of growth. A major challenge for Viet Nam – a more diversified economy – is to strengthen its position in GVCs by building innovation capabilities. Investing more in innovation capacity building could help Viet Nam ease some of the constraints on technology absorption and foster sustained productivity growth. For example, technological spillovers from FDI seem to have been rather low so far, as MNEs were mainly attracted by low labour cost. Better embedding their activities in the domestic economy – which would also be desirable for Colombia and Peru – would help put in place a virtuous cycle of upgrading.

National innovation performance crucially depends on the in-house capabilities of businesses to innovate. In this respect all three countries need to step up efforts to achieve sustained improvements. Businesses are the driving force of all successful innovation systems. A well-performing, market-driven innovation system requires framework conditions that encourage businesses to innovate. Though there have been improvements with respect to the rule of law, regulatory regimes, increasing openness, more vigorous competition, financial development and a strengthening of intellectual property rights (IPRs), there remains much scope for improvements. The dividends of such improvements can be high. While framework conditions are a prerequisite, they need to be complemented by dedicated policy focused on encouraging the systematic accumulation of innovation capabilities in a broad range of businesses.

Raising the profile of innovation policy and its credibility as a driver of sustainable economic development is a formidable challenge of its own. In Viet Nam and Peru, this is reflected not only in the modest resources devoted to innovation. In Colombia there are some encouraging signs of improvement in the form of the considerable public resources channelled to STI from a share of mineral royalties and some notable achievements in promoting the digital economy. All three countries face major societal challenges. These include challenges related to security (notably Colombia and Peru), the environment (e.g. in mining districts), urbanisation and health. Innovation plays a key role in addressing these challenges and provides opportunities for economic diversification.

National agendas, reform processes

Innovation has gained much ground in national policy agendas in the three countries. Contrary to a decade ago, innovation is now reflected in key strategy documents. Viet Nam and Colombia have placed a strong policy focus on education and allocated considerable public resources. Viet Nam has set ambitious goals to increase public S&T expenditure. Current efforts are focusing on improving the innovation capacities of public institutions and their transition towards a market orientation. Agendas also signal efforts to use public actors and funds to leverage innovation more broadly, albeit with limited success so far. Viet Nam

Box 7.3. **Smaller emerging economies: Colombia, Peru, Viet Nam** *(cont.)*

has been attempting to use SOEs to drive upgrading. Colombia has devised a complex mechanism for the regional redistribution of mineral-royalty-derived STI expenditure while Peru developed a similar but much more limited scheme, focused on the development of regional universities.

To develop innovation systems that can take on their role as drivers of sustainable social and economic development, much needs to be done to put ambitions into practice. In the past, ambitious plans were not always matched by either adequate resources or reforms to bring about the necessary implementation capacity, co-ordination and steering towards national goals. Despite progress made, much scope remains to tackle shortcomings in institutions and governance arrangements in all three countries: lack of accountability and evaluation, unclear demarcation of institutional functions, lack of cross-government co-ordination, and low administrative capacities limit the efficiency and effectiveness of policy.

Moreover, policy has often been based on a concept of the national innovation system that is centred on universities and PRIs and consequently emphasises science, technology and R&D at the expense of more broad-based innovation. In Colombia, similar to other countries with a strong "legalistic" tradition such as Peru, this is formalised in a legal definition of the national innovation system that includes only public research actors. This contrasts sharply with the reality of all successful innovation systems which are driven by a dynamic business core. In Viet Nam, the legacy of the Soviet-style research system with its characteristic separation of R&D and production is only gradually being overcome. Narrow concepts of innovation provide little guidance for addressing the needs of business sector innovators and often neglects non-R&D-based innovation, including innovation in processes, design, organisation and marketing, which are increasingly relevant to (among others) the growing services sectors of all three countries. Drawing directly on experiences in highly advanced innovation ecosystems, innovation policy efforts are in some cases (e.g., in Colombia) overly focused on the supply of PhDs, R&D-based start-ups, etc. covering only a rather narrow segment of innovation.

Source: OECD Reviews of Innovation Policy, Colombia (OECD, 2014g), Peru (OECD, 2013c) and Viet Nam (OECD, 2014h).

The economic success of most of these countries, where it occurred, has not been based simply on the exploitation of natural resources, but on how these resources and associated revenues were managed, and how new opportunities related to resource extraction were seized. This required strengthening capabilities to assess such opportunities and specific policies to seize them, as well as conducive framework conditions.

Those countries that derive a large share of their income from the extraction of natural resources – for example, Chile from copper, Norway and Mexico from the hydrocarbon sector, South Africa from a range of commodities, including metals – face specific challenges. First, prices of raw material show a high degree of volatility, as illustrated by the recent fluctuations in the oil price, with repercussions to the real economy. Nevertheless, some countries have profited from the rise in revenues from natural resource extraction to establish new mechanisms to finance industrial and regional development. At the same time, there is a potentially negative impact of the exports of natural resources on other sectors of the economy. In addition, countries may be vulnerable because natural resources are limited, or may be substituted as a consequence of innovation.

Natural resource-based countries therefore need to manage a number of tasks – which require adequate governance mechanisms – to avoid being afflicted by the "resource curse". This includes stabilisation through smoothing disposable revenues from raw materials (such as oil and gas, metals), while avoiding the risk of an adverse impact on other sectors of the economy ("Dutch disease"). Also, it is important to facilitate and foster structural change towards new economic activities, to prepare for a future of declining revenue from natural resources. And policy needs to provide adequate incentives for long-term change towards more innovation-driven growth. This can be a difficult task when rents have become a major source of income for major parts of the economy.

A number of resource-based economies are making efforts to promote a diversification of their economy, and in some cases making use of the revenues derived from the exploitation of natural resources and of the technological capabilities built in and around industries linked to their resource base. Countries' experience and performance vary significantly. Norway has been a prime example of how to manage large revenues from its petroleum sector prudently, and with a long-term perspective. It has combined good management of its oil and gas revenues with seizing opportunities for knowledge-intensive activities in and around this sector, using it as a platform for developing its own technological capabilities leading to the development of marketable goods and services.

Chile is also actively managing its revenues, and allocates a proportion to long-term investment in innovation. The management of these funds based on revenues from raw material extraction present specific challenges in relation to the role of regions and states within the economy and reinforce the need to strengthen the dialogue between levels of government in matters related to innovation policy (OECD, 2013d). Chile has created a range of good framework conditions and has been very successful in entering new export-oriented businesses in traditional industries, notably in the agro-food sector. It has made a persistent effort to move towards more innovation-based development. In order to achieve this goal, new governance structures and funding mechanisms for innovation have been put in place. These include the Innovation Fund (FIC), and the associated levy on mining revenues, the creation of a National Innovation Council for Competitiveness and an Inter-ministerial Committee for innovation. The new National Innovation Council, together with the Production Development Corporation (CORFO), has implemented cluster initiatives that are expected to strengthen and broaden the scope of the country's competitive advantage in different areas, from mining to agro-food and services. In recent years, the government has also been focusing more on smart specialisation strategies, aiming to transform the structure of production, increase productivity and diversify the economy, and encourage sustainable development. It promotes a public policy that aims – through private-public co-operation – to enhance strategic sectors where Chile has a high growth potential, providing the resources to remove obstacles for the development and scaling of these industries. New Zealand is also seeking to move towards a more innovation-driven growth path and has made innovation part of its growth strategy.

Creating a positive feedback loop between a high level of adaptability and innovation helps an economy to diversify and to move up the value chain. Opportunities can also arise from concerted effort covering areas of strength in science and the economy. For example, more than many other countries, Norway has nurtured strong social support for action to contribute to solving problems of global relevance, such as sustainable development, and related issues, such as clean energy. Large-scale programmes to address such topics could be very efficient focusing devices for public support to innovation and have potentially widespread impact on industries and fields of science and technology.

One natural resource sector that offers opportunities for innovation in many countries, both advanced and emerging, is agriculture. The agricultural sector is not only important for several countries as a source of growth and competitiveness, but also as the provider of food for a growing and increasingly wealthy world population. Innovation in food and agriculture is therefore a particularly important policy challenge (Box 7.4).

Box 7.4. **Innovation in food and agriculture**

The food and agriculture sector is expected to provide healthy, safe and nutritious food for a growing and wealthier world population, feed for increasing farm animal populations, and fibre and fuel for a growing range of industrial uses, without depleting available land, water and biodiversity resources.

Governments and the international community recognise that innovation is essential to achieve the productivity growth required to meet these goals, while responding to sustainability and climate change challenges. In the last two decades, MFP growth, driven by the adoption of innovation and structural adjustment, has been the main source of increases in agricultural production. Evidence from a large range of econometric studies shows that the estimated benefits of agricultural R&D far exceed its costs with annual rates of return ranging from 20% to 80% (Alston, 2010). At the microeconomic level, it is clear that the adoption of innovation leads to better productivity performance (OECD, 2014i). Innovations in farm inputs and farming practices have allowed improvements in sustainability performance in most OECD countries (OECD, 2013e). A number of technologies and practices, such as reduced tillage, crop rotation, soil cover and improved varieties, already allow for "sustainable production intensification". While large improvements could be realised with greater adoption of current technologies, in particular by smallholders, future challenges such as climate change require the creation of innovative solutions that are better adapted to evolving and diverse demands.

Drawing on the OECD Innovation Strategy, OECD work on innovation in food and agriculture has developed a framework to review policy incentives and disincentives to innovation in the sector. This framework has been applied to pilot country reviews, which consider the extent to which the general policy environment facilitates investment, and whether incentives to food and agriculture ensure that the agriculture innovation systems align the supply of innovation with sector demand and facilitate the adoption of innovation at farm and firm levels (OECD, 2013f, 2014j, 2014k, 2014l).

Innovation in agriculture is heavily influenced by policies that provide support to farmers, and that currently account for 18% of gross farm receipts on average for the OECD area (OECD, 2014m). Some countries continue to rely heavily on measures that distort production and trade, and tend to discourage innovation. Others provide more neutral income support, which improves producers' investment capacity, but does not encourage adaptation. Incentives to improve sustainable use of resources often target the adoption of specific production practices rather than encouraging more flexible approaches to attaining environmental outcomes (OECD, 2012a, 2013f).

Agricultural innovation systems often have their specific funding and specialised institutions and governance, although in most countries there are institutional linkages with the general innovation system. Public R&D intensity is generally higher than for non-agricultural activities (see figure below). Private investment is lower in the agricultural sector, possibly due to the small scale of firms and farms. Private investment is concentrated in large input and food processing companies, and in areas such as farm equipment and seeds. In many countries, agricultural education fails to meet the changing needs of the sector. Technical assistance is provided by both public and private actors and is often subsidised. Adoption of innovation remains, however unequal.

Policy guidance on innovation in food and agriculture (OECD, 2014m, 2013f)

- Move away from farm income support to invest in knowledge, education and strategic infrastructure that can help improve the long-term productivity, sustainability and profitability of the sector.
- Strengthen the governance of innovation in agriculture to improve the strategic orientation on long-term issues. Make systematic evaluation an integral part of public funding mechanisms for innovation.
- Strengthen co-ordination between agricultural innovation actors and policies to better connect supply and demand.
- Clarify public and private roles in innovation, identify areas for partnerships, and design well-working governance systems around public-private partnerships.

Box 7.4. **Innovation in food and agriculture** *(cont.)*

Public R&D intensity in selected countries, agriculture and all activities

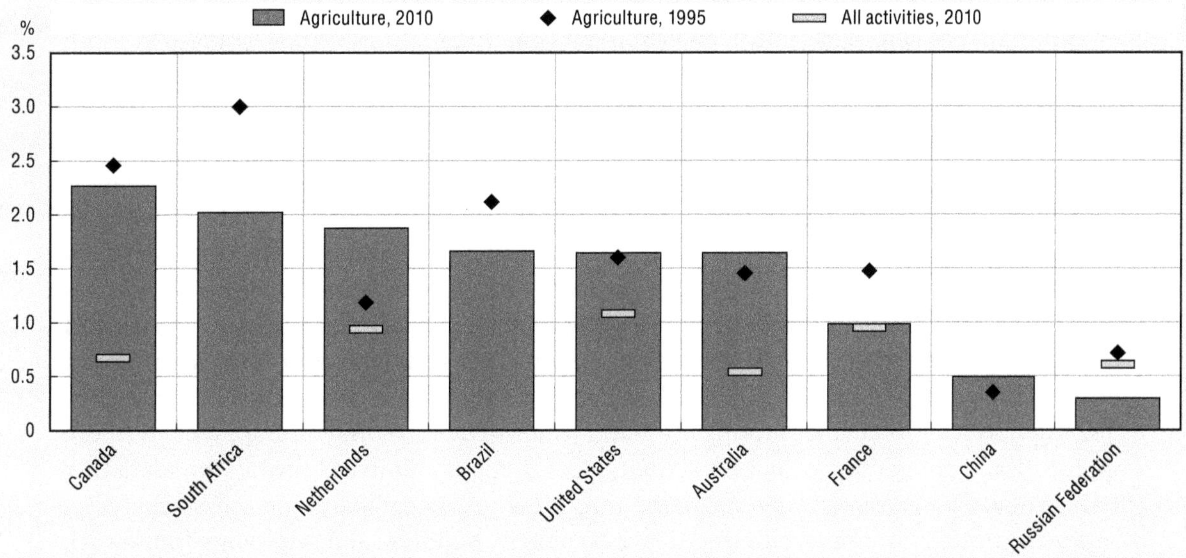

Source: OECD Research and Development Statistics, 2014, *http://www.oecd.org/sti/inno/researchanddevelopmentstatisticsrds.htm*; ASTI (Agricultural Science and Technology Indicators), IFPRI, 2014, *http://www.asti.cgiar.org/*.

- Ensure that farmers have access to independent extension and advisory services to improve technical knowledge as well as professional skills.
- Strengthen co-operation through international, regional and subregional research networks to increase R&D spillovers and to enhance the efficiency of national innovation systems.
- Facilitate access to information systems, such as genetic information and soil data.

Sources: Alston, J. (2010), "The benefits from agricultural research and development, innovation, and productivity growth", *http://dx.doi.org/10.1787/5km91nfsnkwg-en*; OECD (2014i), *Dynamics of Dairy Farm Productivity Growth*; OECD (2013e), *OECD Compendium of Agri-environmental Indicators*, *http://dx.doi.org/10.1787/9789264186217-en*; OECD (2013f), *Agricultural Innovation Systems*, *http://dx.doi.org/10.1787/9789264200593-en*; OECD (2014j), *Innovation for Agriculture Productivity and Sustainability: Review of Australian Policies*; OECD (2014k), *Innovation for Agriculture Productivity and Sustainability: Review of Brazilian Policies*; OECD (2014l), *Innovation for Agriculture Productivity and Sustainability: Review of Canadian Policies*; OECD (2014m), *Agricultural Policy Monitoring and Evaluation 2014: OECD Countries*; OECD (2012a), *Agricultural Policy Monitoring and Evaluation 2012: OECD Countries*, *http://dx.doi.org/10.1787/agr_pol-2012-en*.

Innovation for inclusive and green growth

As noted in Chapter 1, in recent years, the policy agenda for innovation has broadened to goals beyond growth, productivity and competitiveness. As innovation is an important driver of growth, the benefits of that growth should ultimately trickle down throughout the economy, sometimes supported by social and redistributive policies that would help spread the benefits of innovation-led growth across the wider population. Strong growth in major emerging economies, notably China and India, has contributed to a significant reduction in global poverty over the past decade. At the same time, inequalities have been growing in many countries, leading to a call for more inclusive growth. In several emerging economies, the policy agenda for innovation has now broadened to a specific focus on fostering inclusive innovation, i.e. innovations that are specifically aimed at improving the welfare of low-income and excluded groups. China, Colombia, South Africa, India and Indonesia are among a growing group of countries where inclusive innovation is explicitly considered as part of the broader approach to innovation. This will be explored further in Section 7.2 of this Chapter.

Another goal that is increasingly part of the innovation agenda is green growth and sustainability, often as part of a broader effort to focus innovation more on addressing social and global challenges. A few countries have made this their key goal for innovation, e.g. Korea during its focus on green growth, but it many countries this has now become an important consideration in the development of horizontal innovation policies. A more extensive discussion of what this implies in practice follows in Section 7.4.

Innovation strategies today

Reflecting these various objectives, starting points and approaches to innovation policy development and implementation, innovation strategies differ considerably across countries. Country responses to the OECD Science, Technology and Industry Outlook policy questionnaire 2014, covering all 34 OECD countries and more than 10 emerging economies, reveal both similarities and differences in goals and policy priorities across countries and also point to some international features in national STI strategies as well as some broad cross-country policy patterns (OECD, 2014a). A first similarity is that almost all countries give high priority to business innovation and innovative entrepreneurship, whatever the approach and modalities of public action. Second, most countries aim at consolidating their innovation ecosystem by strengthening public R&D capacity and infrastructures; improving overall human resources, skills and capacity building; and improving framework conditions for innovation (including competitiveness). Third, while countries at different stages of socio-economic development share some STI policy priorities, other priorities are specific to certain countries. This is reflected in the relative concentration of countries in strategic STI policy fields according to the intensity of their GERD; see Figure 7.2.

Typically, for countries that already rank high in terms of business R&D and innovation, there is a focus on investing in the science base, both public research and human resources, to strengthen the basis for future innovation. These high-performing countries are also prioritising their research and innovation support to gain competitive advantage for future growth areas such as green technologies and health and to help address global challenges. Some recent examples of strategic approaches for Finland, Germany and Korea are listed in Box 7.5.

Countries that identified the contribution of innovation to sustainable and green growth as a major STI policy priority in 2014 tend to be more R&D-intensive. For OECD countries in which innovation performance lags, there is a particular focus on building the institutional capacity to steer or "govern" STI policies, to strengthen the links between public research and industry, and to improve the quality of higher education and research (OECD, 2010a). Small open OECD countries with high exposure to trade and FDI are also more likely to consider challenges raised by STI globalisation and increasing international co-operation as major policy priorities.

For their part, catching-up and emerging economies are seeking to include STI strategies in their longer-term economic development strategies. Emerging and middle-income economies (e.g. Argentina, Colombia, Costa Rica, Malaysia, Viet Nam) are developing strategies to diversify their economies and mobilise innovation to improve their competitiveness, move up GVCs and escape the "middle-income trap". Less R&D-intensive countries tend to set a priority on the contribution of innovation to structural adjustment and a new approach to growth, on improving the returns to and impact of science, and on increasing the skills base. For example, China's Medium- and Long-Term Plan for S&T Development (2006-20) aims to use innovation as a tool for restructuring Chinese industry and shift from investment-driven to innovation-driven growth.

Figure 7.2. **Major national STI policy priorities and patterns by level of R&D intensity, 2014**

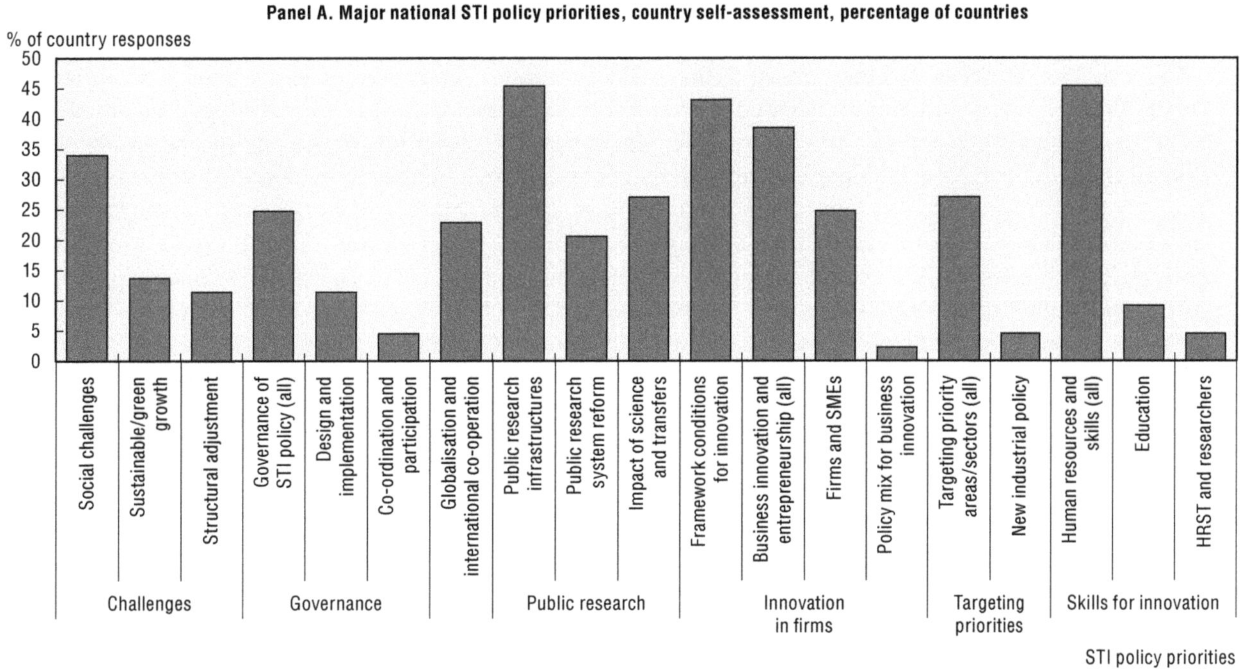

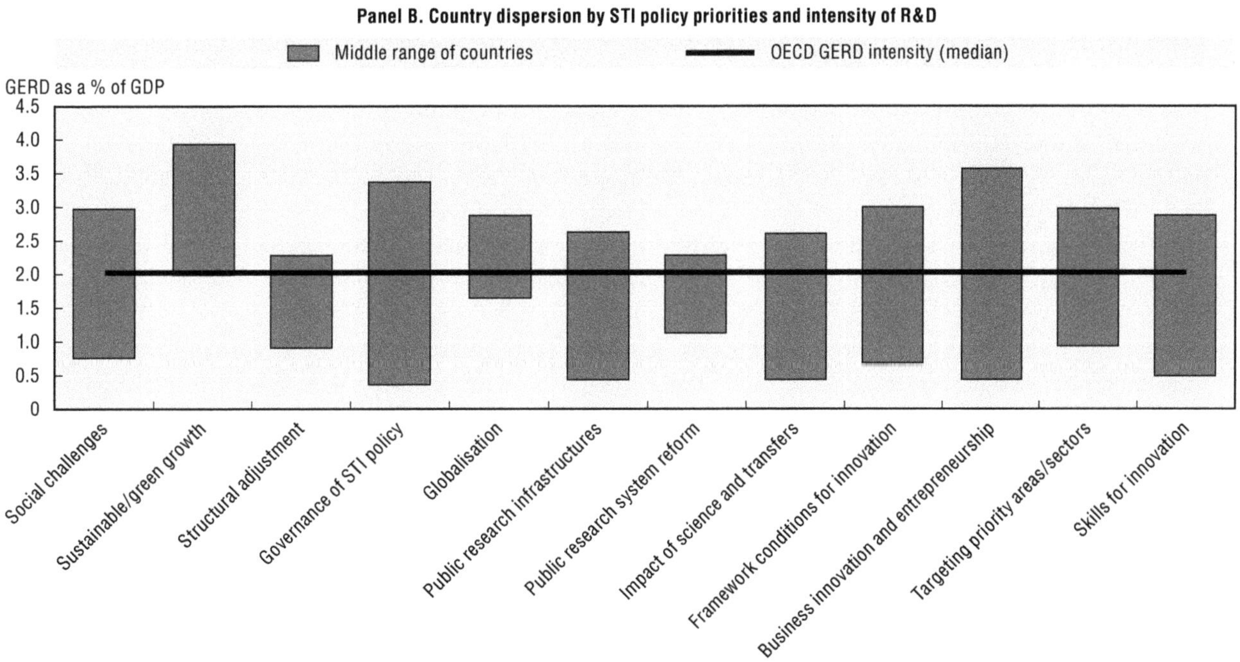

Notes: STI policy priorities are defined by country self-assessment answers to the question: "What are the major STI policy priorities in your country? Please select three (maximum five) STI policy priorities in the drop-down lists below and describe briefly 'in your words' (one sentence) these major policy priorities." Responses are provided by delegates to the OECD Committee for Scientific and Technological Policy.

Panel B illustrates the extent to which national STI priorities can be linked to the degree of advancement of R&D system. It shows the countries that reported each STI policy priority as a major issue according to their GERD intensity. The middle range of countries includes all OECD countries and non-OECD economies to the exclusion of the two most R&D-intensive and two least R&D-intensive countries. For the policy priorities related to sustainable/green growth and structural adjustment, however, the middle range of countries includes the top two and bottom two to compensate for the small number of countries in these two policy categories. The intensity of GERD is expressed as a percentage of GDP.

Sources: OECD (2014a), *Science, Technology and Industry Outlook 2014, http://dx.doi.org/10.1787/sti_outlook-2014-en.*

Box 7.5. Examples of strategic approaches to innovation: Finland, Germany and Korea

Finland: The SUUNTA strategy. In 2013 the main innovation policy agencies (Academy of Finland, Tekes, Sitra, Finnvera and Finpro) decided to join forces and develop a new joint strategy (the so-called SUUNTA strategy) with the aim to shift the focus of support for research, development and innovation beyond individual projects, companies and sectors towards broader business ecosystems in areas of key importance to Finland. This was preceded by extensive analysis and foresight work to explore megatrends, significant business opportunities and challenges that Finland is facing, and aims to create innovative new ways to implement policy. Increasing uncertainty and complexity in the global economy, coupled with rapid digitalisation and tightening competition, implies that companies increasingly build their competitiveness through symbiotic relationships with collaborators, competitors and other actors within business ecosystems. This is challenging current innovation policy thinking, implementation and impact assessment.

The SUUNTA strategy identified natural resources and resource efficiency, digitalisation, well-being and health as areas where there is the most potential for new business ecosystems to emerge, and also considers ways through which these ecosystems can connect to global hubs and attract investments to Finland. The emphasis of the SUUNTA strategy is on identifying emerging business ecosystems and boosting these through better co-ordination among the main innovation policy agencies, for example by mandating ecosystem orchestrators, creating physical or virtual platforms for ecosystem extension, identifying and addressing barriers for developments, securing political support where needed, and developing policy instruments so that the agencies can work more seamlessly together.

Germany: High-Tech Strategy. The High-Tech Strategy is a comprehensive, interdepartmental innovation strategy. The goal is for good ideas to be translated quickly into innovative products and services. The strategy aims to create favourable conditions for innovation, consolidate resources and promote transfer. New initiatives regarding skilled personnel are to be launched including initiatives in science, technology, engineering and mathematics (STEM), and efforts are also made to enhance the attractiveness and permeability of vocational training in Germany. In addition, technical regulations and standards will be further harmonised to improve the business environment. Moreover, Germany is to become more internationally attractive as a centre for venture capital investments. Germany plans to develop an open-access strategy that will improve the framework for effective, continued access to publicly financed publications.

Regarding the system for knowledge diffusion, new measures are adopted under the High-Tech Strategy to strategically expand universities' options for co-operation with industry and society, to close gaps in commercialisation, and to advance internationalisation of leading-edge clusters and other comparable networks. Furthermore, the development of a competitive, employment-strong industry is promoted. The plan is to expand the number of SMEs participating in innovation funding schemes by making the funding conditions even more user-friendly. Moreover, the number of innovative start-ups in Germany is to be increased by improving the existing policy instruments and by connecting start-ups to global centres of growth and value creation. Via innovative public procurement, new incentives for innovation will be provided for the benefit of industry.

Korea: Creative Economy strategy. To face challenges resulting from changing global circumstances and limitations to Korea's old catching-up strategy, the Park Administration introduced "Creative Economy" as a presidential pledge in 2013. Creative Economy is a strategy to shift Korea's growth paradigm from industrial economy to knowledge economy based on the use of science, technology, innovation and creative ideas. By putting the highest priority on creativity for strengthening Korean S&T and ICT capabilities, the Park Administration aims to create new markets, new jobs and new opportunities for sustainable growth.

Box 7.5. **Examples of strategic approaches to innovation: Finland, Germany and Korea** *(cont.)*

To enable the effective implementation of the Creative Economy strategy, the Ministry of Science, ICT and Future Planning (MSIP), released an Action Plan for the Creative Economy in June 2013. The plan sets six strategic actions: 1) building an ecosystem that rewards creativity and promotes start-ups; 2) supporting SMEs to lead the creative economy; 3) generating new growth engines to pioneer new industries and markets; 4) nurturing global creative leaders with dreams, talent and a challenger spirit; 5) strengthening S&T and ICT innovation capacities as the basis of the creative economy; and 6) building a creative economy culture where people and the government work together. Each strategy involves three to five specific tasks including establishment of the "Creative Economy Town," an online open platform for idea development, and the "Creative Economy Innovation Center," an offline innovation platform for public-private and/or central-local partnership.

Developing countries are also exploring new ways to promote innovation. Some natural resource-rich countries, such as Chile, Colombia and Peru, are implementing new schemes to finance innovation, including sectoral technology funds, sometimes in the context of smart specialisation strategies, as in Chile. Public procurement is also seen as a tool to promote innovation and production upgrading in priority areas, as in Brazil, China, India and South Africa. In addition, developing countries, such as Brazil, Morocco and India, are increasingly using FDI as a tool to foster innovation and industrial upgrading by promoting linkages and technology transfer. Some developing countries are also prioritising sustainable development: they are investing in developing new technologies and in fostering the creation of new environmentally friendly business models. Development banks are often playing a key role in financing green innovations (OECD, 2013d). South-South co-operation is also increasing as a way to promote innovation in developing countries. For example, the Brazilian Agricultural Research Corporation (EMBRAPA) has strengthened its partnership with Africa, increased collaboration with the Forum for Agricultural Research in Africa (FARA) and opened a local branch in Ghana. EMBRAPA Africa is in charge of knowledge sharing and technology transfer to improve the competitiveness of agricultural products in Africa and their access to global markets. It carries out research programmes and offers technical assistance. It also participates in the Africa-Brazil Agricultural Innovation Marketplace, an initiative targeting smallholder producers aiming to enhance agricultural innovation for development in Africa. A growing number of developing economies also have increased their focus on start-ups (Box 7.6).

At the same time, OECD countries and emerging economies share certain concerns and priorities in regard to the governance of their innovation system and policy; support to innovation in firms, entrepreneurship and SMEs; and the contribution of innovation to meeting social challenges (including inclusiveness). Most countries have also adopted quantitative targets to benchmark their performance and progress, especially through targets for R&D spending (OECD, 2014a). The volume of GERD to be achieved is often expressed as a percentage of GDP, and in some cases, the relative contribution of the business or the public sector is specified as well. China and the Russian Federation target S&T output in terms of patents, citations and publications. New Zealand takes into account economic performance as reflected in the increase in exports, while Korea looks at S&T-related job creation. Denmark and Switzerland monitor educational outcomes and the share of a youth cohort completing upper secondary or higher education programmes.

Box 7.6. **Policies for start-ups in developing economies**

Chapters 2 and 4 already called attention to the important role of young firms and start-ups for innovation. The spread of ICTs and the transformation of the organisation of production throughout the world, where firms increasingly work in networks, have helped generate growing interest in start-ups both in OECD and in developing countries. Start-ups can contribute to structural change by introducing new knowledge-intensive products and services. They can also help sustain innovation, drive productivity and create opportunities for good-quality jobs. Although there is growing interest in supporting the creation of start-ups, there is no universal definition for this type of firm. They can be defined on the basis of their performance, i.e. their growth potential, or on their innovative and technological focus. These firms are innovative and tend to provide solutions to emerging problems or create new demands by developing new types of businesses. For example, Argentina and Brazil define them as technology-based firms; Chile as high-growth firms; and Kenya, Nigeria and Peru as ICT-based firms.

Start-ups can be determinant in inducing innovation and in boosting and diversifying the economy. But they must overcome major hurdles to get up and running and to expand, especially in developing countries. Knowledge-based companies flourish in dense innovative environments, where new entrepreneurs can interact with a high-quality and vibrant science base and where access to technologies and finance is easy and the regulatory framework is business-friendly (OECD, 2010b, 2013g). The diffusion of ICT and the upsurge of new areas including biotechnology and renewable energy have prompted the creation of those firms, especially in OECD countries. However, even though to a much lesser extent, this phenomenon is rising in several countries in Africa, Asia and Latin America. This is due to three main factors:

- High GDP growth over the last decade has contributed to creating opportunities for those firms in developing economies.
- The rise in talent mobility and global production unbundling has helped workers from developing economies to acquire professional skills in foreign universities and companies, favouring the development of an entrepreneurial culture.
- The spread of ICT has created opportunities for knowledge exchange, making start-up companies a feasible business option in developing economies as well. In addition, governments in developing countries have also started to look at those firms as a potential new source of growth (OECD, 2013d, 2013g).

Innovative start-ups face greater challenges than traditional firms because their business is risky and uncertain, particularly during the early phases. Public policies can support them through a wide policy mix at each stage of development, offering finance, advice and training, in addition to promoting a favourable legal framework. In natural resource-rich countries, incentives to promote start-ups help create an innovative cluster around traditional activities. In countries with a good record of FDI attraction in high value-added activities, promoting start-ups can be an effective way to strengthen competitiveness and to increase the impact of FDI on the local economy. Often, start-ups develop in sectors related to ICT; therefore they require adequate digital infrastructure (i.e. fast and reliable Internet access) as a precondition for their creation and expansion. In Africa, technology-based start-ups operate mostly in the software and telecommunications sectors, according to the *Forbes Africa* ranking of the top African start-ups (*Forbes Africa*, 2012). Most of these companies are quite young and offer a variety of innovative services for the African continent, as for example a start-up from Kenya specialising in systems for online retailing in Africa.

Developing countries are starting to put in place instruments to support the creation of start-ups. Much attention has been paid to access to finance; however, although fundamental, financial resources alone are not enough to foster the creation and expansion of those firms. A legal framework that makes it easier

Box 7.6. **Policies for start-ups in developing economies** *(cont.)*

to start and run technological businesses is paramount. Targeted programmes to foster the development of technical and managerial skills are also important. The experience of OECD countries shows that what matters is not only the design of one specific instrument, but the implementation of a policy mix that accompanies the firms in all the stages of development. For example, venture capital is effective when early-stage financing (e.g. seed and angel investors) is available and when measures to support the development of ideas into business plans are in place, including, for example, incubators and accelerator programmes. Latin American countries, for example, support start-ups in various ways, with a range of approaches under differing institutional arrangements. Some, like Brazil and Chile, have been accumulating experience since the 1990s; others, like Colombia and Peru, have only recently begun introducing programmes devoted to start-ups.

Sources: OECD (2010b), *SMEs, Entrepreneurship and Innovation, http://dx.doi.org/10.1787/9789264080355-en;* OECD (2013g), *Start-up Latin America, http://dx.doi.org/10.1787/9789264202306-en;* OECD (2013d), *Perspectives on Global Development 2013; Forbes Africa* (2012), "Ranking of top 20 African startups".

National strategies for STI typically follow a vision and are designed on the basis of data-driven evidence, tools such as scenarios and strengths-weaknesses-opportunities-threats (SWOT) analyses. The process of making an innovation strategy is perhaps more important than the document, as it helps reveal problems, barriers and hidden opportunities and promotes a learning process. National strategies typically offer a vision of the main challenges that the national innovation system is confronted with and the directions that should be taken to address these challenges. Innovation strategies have many possible uses: they can trigger an exchange of views among stakeholders and help reach a common strategic vision of research and innovation, foster the convergence of views among stakeholders and decision makers, and help in setting priorities and planning resources.

In practice, national innovation strategies vary considerably across countries (in terms of their ambitions, time horizon, scope, how operational they are, etc.). This generates a range of possible design and implementation issues, thus providing ample scope for learning and the exchange of good practices. National strategies are not always effective, however, which can result from inadequate design, including a lack of realism in the choice of objectives, but also from an inadequate design process, including lack of consultation with stakeholders, or from the implementation process itself, including lack of monitoring and evaluation. Obstacles may also result from a lack of motivation or even the resistance of certain actors, whose concerns and agendas have not been sufficiently integrated. They may also be due to institutional settings, which may not be sufficiently amenable to the possible reorientation of resources required by new strategies. Moreover, as innovation becomes a tool to achieve a wide range of policy objectives, and as the range of ministries, agencies and stakeholders involved expands, developing a coherent strategy and implementing it in a consistent way becomes increasingly challenging. The challenges this raises will be further discussed in Chapter 8 of this report. A number of key policy learnings from the OECD's work on national innovation strategies are below.

Main policy messages: Developing innovation strategies at the national level

- Countries differ substantially in the basic conditions for innovation, and face different constraints that need to be overcome in strengthening innovation performance. Nevertheless, many advanced, emerging and developing countries today share a common objective to move towards more innovation-led sustainable growth and many have become more aware of the role that innovation can play in achieving this goal.
- Most countries consider innovation important for growth and competitiveness, but also in helping address social and global challenges. As a result, policies for innovation are becoming integral to economic policy in a wide range of countries, both advanced and emerging, resulting in some convergence in the policy agenda for innovation across countries.
- To be effective, policies for innovation and the related governance system need to be adapted to the specific challenges faced by each country. Moreover, the specific choice and combination of innovation policies and related governance need to be aligned with the capabilities of each country in terms of policy making and policy implementation.
- At the same time, there is much scope to exchange experiences and good practices across countries, even though the institutional context for policy making will differ by country.
- The design and implementation of national innovation strategies are critical to their success. The process of developing a national strategy is particularly important and requires early and adequate involvement of key stakeholders.

7.2. Innovation and inclusive growth

Chapter 1 already noted that economic growth, as measured by GDP, can no longer be the only goal for government policy, as socio-economic well-being will not depend only on achieving growth. This section briefly discusses two aspects of the relationship between innovation and inclusive growth, namely 1) the extent to which innovation policies are inclusive and should be made more inclusive; 2) innovation policies that are specifically aimed at improving the well-being of lower-income and excluded social groups. OECD work is looking at the link between innovation and inclusive growth in today's knowledge-based economies. This will also examine the extent to which new and emerging technological changes might lead to job-less growth, which could provide a serious challenge to inclusive growth.

Innovation policies and inclusive growth

The OECD (2014n) defines inclusive growth as "economic growth that creates opportunity for all segments of the population and distributes the dividends of increased prosperity, both in monetary and non-monetary terms, fairly across society". This report uses the term "social inclusiveness" to denote processes that create opportunities for all segments of the population, particularly people in the lower deciles of income distribution and people who are excluded for other reasons. Social inclusiveness is characterised by three dimensions: 1) multidimensionality; 2) an emphasis on distribution; and 3) policy relevance (Box 7.7).

Box 7.7. **Characterising the OECD Initiative on Inclusive Growth**

The following dimensions characterise the approach adopted by the OECD Initiative on Inclusive Growth:

- **Multidimensionality.** There is widespread recognition that GDP captures only part of economic welfare and excludes other dimensions that also matter for well-being, such as jobs, skills and education, health status, environment, civic participation, and social connections (Stiglitz et al., 2009).
- **Emphasis on distribution.** "Inclusive growth" means that people, independently of their socio-economic background, gender, place of residence or ethnic origin, should have fair opportunities to contribute to growth (i.e. they are part of the growth process), and that their contribution should yield equitable benefits (i.e. they benefit from the process outcomes). The specific emphasis on the "target" group to be "included" is very much a policy question specifically reflecting countries' socio-economic characteristics.
- **Policy relevance.** Inclusive growth should be policy-actionable and make a link between policy instruments and the relevant monetary and non-monetary dimensions, taking into account distributional impacts. This requires in particular assessing the impact of policies and institutions on the different dimensions of inclusiveness, as well as the trade-offs and complementarities that are expected to exist between pro-growth and pro-inclusiveness policies.

Source: OECD (2014n), *All on Board: Making Inclusive Growth Happen*, *http://dx.doi.org/10.1787/9789264218512-en.*

Inclusive growth starts with the premise that economic growth is critical to well-being, as it provides the resources that help create the conditions for more inclusive growth. For example, economic growth strongly correlates with a decline in poverty (OECD, 2015a). According to World Bank data, 972 million people were living on less than USD 37.50 per month in China in 1981; this number dropped to 157 million by 2009. India has also made substantial advances in reducing poverty: the official poverty ratio declined from 45% in 1994 to 37% in 2005. Between 2005 and 2012 – a period during which India achieved the fastest rate of economic growth in its history and also implemented a number of policies aimed at helping the poor – extreme poverty declined to 22% of the population, or some 270 million people (Gupta et al., 2014).

However, the link between innovation and social inclusiveness is often more complex than can be gleaned from such aggregate data. Evidence from regional data on innovation shows a non-linear relationship between R&D investment and regional inequality. On average, the regions investing very heavily (more than 2% of their GDP) in R&D in 2004-06 had the highest levels of income inequality in 2010. Inequality was lower in regions with intermediate levels of R&D investment in 2004-06, but rose again at the bottom of the range (less than 0.8% of regional GDP spent on R&D). It is therefore important to look at detailed dimensions of inclusive growth, such as industrial inclusiveness.

Industrial inclusiveness

Industrial inclusiveness refers to the degree of similarity of innovation capacities across firms, sectors, regions, universities and PRIs within countries. Its opposite is the concentration of leading innovation capacities in firms, sectors, regions or universities of international standing that are highly advanced compared to others in the economy. For example, the concentration of excellence among firms, especially within very narrowly defined sectors and regions, is characterised by the co-existence of very productive firms with weakly productive firms, productivity being closely related to innovation capacities. Microeconomic data point to substantial dispersion: in the United States, for instance, Syverson (2004) shows that "the plant at the 90th percentile of the productivity distribution makes almost twice as much output with the same measured inputs as the 10th percentile plant." Hsieh and Klenow (2009) find the dispersion among firms in China and India to be even more substantial. New OECD work (OECD, 2015b) also points to a big gap between frontier firms and laggards in terms of productivity performance.

The front runners are typically globally competitive and include the top R&D investors in the world, as well as some firms from emerging and developing countries (European Commission, 2013). Another dimension of innovation concentration becomes apparent when observing the differences across firm size. In most countries represented in Figure 7.3, SMEs account for less than 40% of total business expenditures on R&D (BERD). Estonia and New Zealand – where SMEs account for over two-thirds of total BERD – are the exceptions.

Figure 7.3. **BERD by size class of firms, 2011**

As a percentage of R&D performed in the business sector

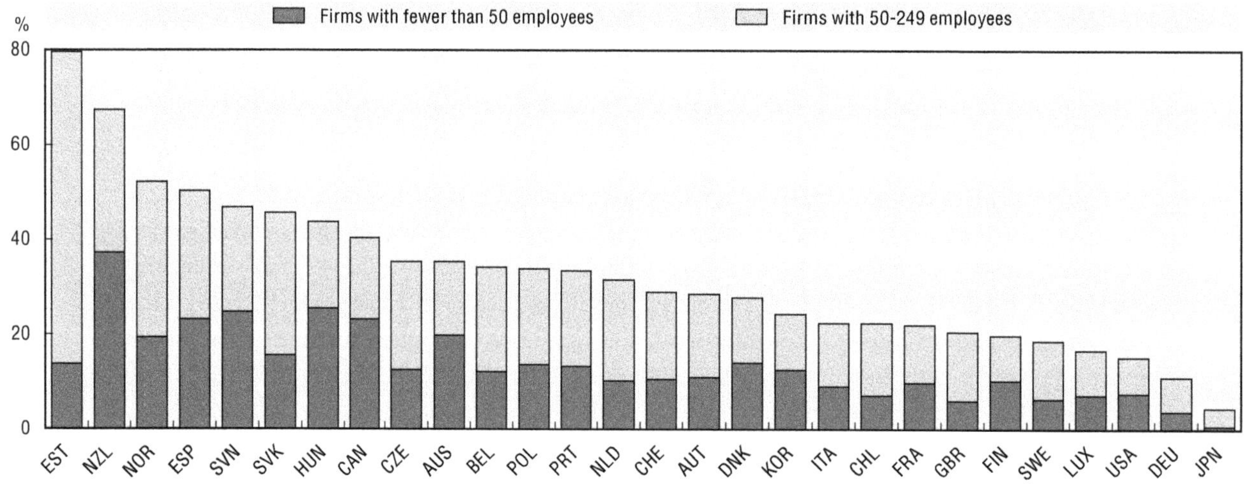

Source: OECD (2013h), *OECD Science, Technology and Industry Scoreboard 2013: Innovation for Growth*, *http://dx.doi.org/10.1787/sti_scoreboard-2013-en*.

A few leading universities and PRIs also concentrate the contribution of public research to innovation. Their leadership often correlates with other high-performance indicators, notably teaching quality. Rankings such as the Academic Ranking of World Universities (known as the "Shanghai Ranking", introduced in 2003) and the Times Higher Education World University Ranking (launched in 2004) illustrate the heightened importance of competing for excellence among leading universities. Becoming a world-class research university generally requires long-term investments (Salmi, 2013).

In OECD countries, a limited number of regional innovation hubs concentrate resources: over 33% of R&D takes place in the top 10% of large OECD regions – which also concentrate around one-fourth of skilled employment – and the top 10% of small OECD regions apply for 58% of patents (OECD, 2013e). Concentration is even more pronounced in non-OECD countries: in China, the top three regions – Guangdong (46%), Beijing (14%) and Shanghai (13%) – account for almost three-quarters of all patenting activity. In India, the top three regions – Maharashtra, which includes the capital, Mumbai (26%), Delhi (24%)

and Andhra Pradesh (13%) – account for almost two-thirds of national patenting activity (Creszenci et al., 2012).

Frontier research often takes place in regional centres of excellence, which by their very nature are tied to a place and (often) embedded in a local network. These centres create opportunities locally, but not evenly across the nation. Indeed, a closer look at the distribution of patenting activity – measured here by the number of patents filed under the Patent Cooperation Treaty – shows large regional differences. In China, for example, the majority of patents are filed in regions along the coastline. In England or Germany, the southern regions are more active than the northern regions. In France and Spain, the regions around the capital cities of Paris and Madrid concentrate patenting activities.

Other regional indicators, such as the share of regional scientific publications per 1 000 inhabitants, also show evidence of strong concentration. In 2010, the top 40 OECD regions (out of almost 1 700 regions with data) represented one-third of all scientific publications (OECD, 2013h). A similar picture of concentration emerges for regional shares of R&D expenditure (OECD, 2013i).

Factors determining the concentration of innovation

Innovation concentration is not a new phenomenon: it reflects the substantial economies of scale and scope resulting from agglomeration effects. However, the significant differences in the various dimensions of concentration require different approaches to addressing them. The co-existence of high-performing and low-performing firms in some markets is somewhat surprising, as competition could be expected to force bad performers out of the market. Weak competitive pressures, combined with lack of market integration, likely feature among the reasons that technology and productivity gaps across firms are greater in developing countries. In addition, framework conditions may particularly affect small firms and younger businesses, and thus generate a skewed distribution of innovative firms.

Other factors, this time related to the heterogeneous distribution of knowledge-based capital (KBC), contribute to the skewed distribution of innovations. Evidence from two knowledge outputs – patents and/or publications – shows that only a very small share of ideas have high value, as measured by the number of citations they receive or other criteria. A major factor why ideas translate into skewed value distribution relates to the non-rival and non-excludable nature of knowledge – marginal costs are low and therefore successful ideas easily capture entire markets, replacing all others. This occurs even more as markets become increasingly global.

These dynamics may in turn lead to a stronger concentration of innovation capacities among actors, since agglomeration and reputation benefits reward those generating winning ideas. Success often attracts not only talent, but also resources investing in the future generation of ideas. Chances are high that more leading innovations will emerge in areas where capacities are concentrated. This will reinforce certain actors' dominant role as generators of leading innovations, given that synergies arise from concentrating the best resources. These dynamics apply to both firms and universities. While they are particularly important in KBC-based industries – particularly the software industry – the important transformations of other sectors make them increasingly relevant to a broader group.

The benefits of agglomeration also have implications for industrial inclusiveness, the proximity of innovation capacities across different regions within countries. Inventors' need to access specific infrastructure and production infrastructure also contribute to concentration. Imperfect financial market conditions outside of the leading agglomerations can hinder innovation efforts by other regions to match the same scale of innovation activities. Cities tend to produce the largest agglomeration benefits – the positive externalities arising from bringing many firms, workers and customers together in the same place. These benefits make firms and workers in large cities generally more productive than in small cities or rural areas. This means the same amount of inputs produces more output – which in turn results in higher GDP per capita – in larger cities. The importance of spatial proximity for certain forms of innovation collaboration can also reinforce agglomeration forces. The propensity of inventors to co-patent with partners from the same region is higher than the propensity to co-patent with co-inventors from different regions within the same country or from abroad.

By contrast, the creative destruction that characterises innovation-based growth can challenge leaders' position in the global economy, lessening the concentration of innovative activities among owners of "winning ideas". Given the assets they can rely on, however, leaders may have an edge in maintaining their leadership – unless too many of their assets are stranded in existing facilities. At the same time, many governments have tended to concentrate their support on top actors to foster national competitiveness, further intensifying the concentration of innovation capacities.

The democratisation of innovation

The "democratisation of innovation" refers to the widening of the group of successful innovators to include actors who did not previously participate in innovation processes – particularly smaller entities, i.e. individuals, firms and entrepreneurs from a variety of backgrounds that are typically considered outsiders – and have opportunities to succeed with bottom-up initiatives. The extent to which these outsiders succeed in innovation is closely related to their ability to reach a sufficient scale – which is therefore at the heart of opportunities for more democratic innovation dynamics. While much innovation is highly concentrated, evidence also points to the reverse: Figure 7.4 shows that the share of young enterprises in innovation can be substantial, particularly when it comes to business-sector services in certain countries.

Figure 7.4. **Patenting activity of young firms by sector, 2009-11**

Share of young patenting firms and share of patents filed by young patenting firms

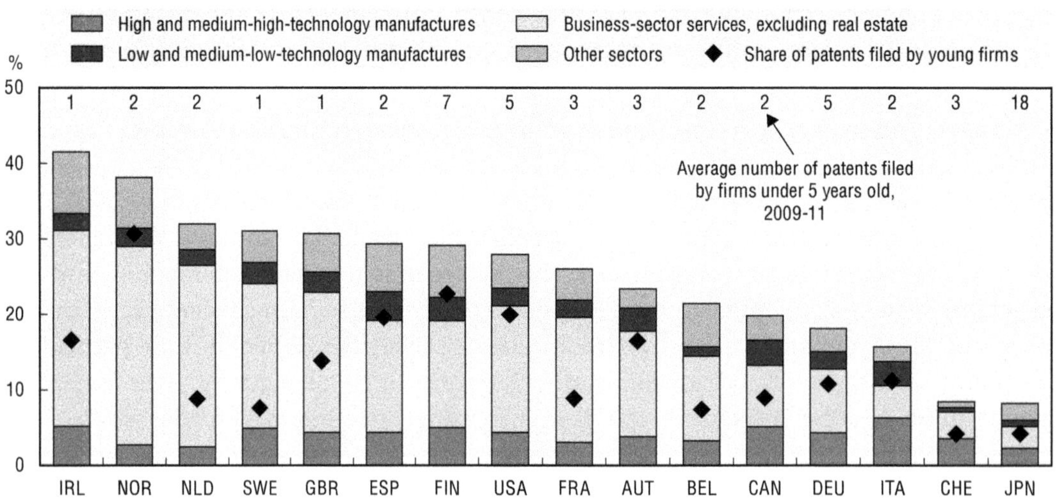

Source: OECD (2013h), *OECD Science, Technology and Industry Scoreboard 2013: Innovation for Growth, http://dx.doi.org/10.1787/sti_scoreboard-2013-en.*

Among the ICT-empowered trends that have that have helped democratise innovation, the following have played an important role:

- **Product distribution has become cheaper, reducing the cost of launching innovations.** This cost reduction stems from the fact that accessing customers on the Internet is less expensive than through brick-and-mortar stores. Dedicated digital platforms, such as Apple's App Store for mobile apps, allow producers of mobile applications to sell their products directly to their main target audience. Similarly, Amazon offers third parties the opportunity to sell their products on its platform, while Facebook and Twitter (among others) facilitate targeted product marketing.
- **Innovation-related production costs have decreased in some sectors.** Software has helped reduce the costs of making high-quality products in a number of fields. Cloud computing services – such as Amazon Web Services – provide high-quality data processing capacities without requiring large upfront investments, spurring a new dynamism of digital start-ups. To date, such developments appear to have mainly benefited larger firms, but there are indications that smaller firms are also starting to derive advantages.
- **The risks and time span between product development and market launch have decreased for certain innovations.** Software start-ups, in particular, can open nearly instantaneously – and if unsuccessful, wind their activities down rather easily. Market observers also point to opportunities for a more experimental approach to innovation, where innovators offer a multiplicity of products to consumers and adjust them based on information provided by test users. If convergence is achieved, they scale up the products fully, sometimes even globally. A number of online sites, such as UserTesting.com, also help companies experiment.
- **The demand for innovation can be assessed more easily.** An increasing amount of data is being collected about consumption behaviour, allowing firms to better understand demand for their products. Moreover, platforms such as InnoCentive, which allow the posting of innovation challenges, can facilitate the innovation "crowdsourcing" of innovation. Segmenting product innovation also allows for wider participation, deepening technological markets. Moreover, the opportunity to consult more systematically with users, and even the option of user-driven innovation, can arguably lower the costs associated with identifying demand.
- **Using the Internet facilitates access to knowledge for innovation.** Evidence shows these benefits arise particularly for businesses in developing countries, which were previously at a disadvantage in accessing formal or informal information. Platforms such as TechShop provide support for small-scale entrepreneurship, reducing the barriers to engaging in innovation.

In order for the Internet to play as critical a role in innovation activities across developed and developing economies alike, the development of backhaul and cross-border networks, which enable local networks to connect to the wider Internet, is critical. In several developing countries, communication networks often resemble rivers, with small branches of regional networks delivering their traffic to a central national backbone that ends at one submarine fibre, making cable cuts a greater risk to the functioning of the economy. Such infrastructure shortcomings prevent Internet-based business development. The presence of data centres or other local facilities that can host Internet exchange points (IXPs) and servers is also essential (OECD, 2014o).

The increased demand for more customised products – often including an important service component – may also favour small, agile entrepreneurs with a smaller-scale innovation approach and the capacity to adjust to shifting demand. In other words, small firms might benefit from their ability to reduce information asymmetries between users and producers (Von Hippel, 2005). Adopting bottom-up and improvised approaches to innovation may also be more beneficial to competitive innovators than more inflexible innovation approaches centred on R&D departments' contributions to innovation (Radjou et al., 2013). Effectively small businesses do not have the constraint of a legacy business model that may limit large businesses in changing their operations.

Incremental and non-technological innovations for growth (compared with more technological innovations) also offer wider opportunities for people who are not part of the professional elite to engage in innovation. At the same time, the growing number of highly skilled individuals has contributed to raising the pool of potentially successful small-scale innovators. This is relevant to the democratisation of innovation, which seeks to involve excluded groups in innovation processes. Moreover, higher levels of skills among groups of product "lead users" (possibly coupled with wider opportunities for developing products) allow consumers themselves to be more active as innovation producers (Von Hippel, 2005). This development can abolish some of the challenges posed by possible information asymmetries between producers and users, and sometimes even effectively stimulate entrepreneurship.

Despite these developments, a number of trends point, on the contrary, to further concentration:

- The growing importance of GVCs might result in stronger concentration on a specific set of tasks – i.e. those in which a country's firms have a comparative advantage. Depending on the governance structures of GVCs, this can lead to an increasing concentration of innovation capacities among national actors.
- Industries characterised by "innovation contests" may also experience increased concentration. Larger firms might find it easier to engage in such contests, since they do not risk their very survival by engaging in innovation – particularly as they have diversified their investments in innovation, can rely on other sources of market income and do not rely on returns from their newly introduced innovations to survive (Fernandes and Paunov, forthcoming).
- Evidence shows that smaller firms, notably in developing economies, are much less efficient. This points to the potential benefits of concentrating resources further (Hsieh and Olken, 2014); in other words, efficiency in these economies would be better served by greater concentration than by the opposite.

At the same time, concentrating innovation activities and democratising innovation are not opposites. In many cases, innovation leaders are connected to – or even included in – broader innovation ecosystems comprising large and small, universities, etc. within such ecosystems. Democratisation can facilitate access to a wider community, which will share in the rewards of the wider ecosystem if it wins the race. Moreover, differences in scientific and/or technical domains influence both the opportunities for democratisation and the needs for concentration. The increasing costs of developing and deploying innovations in a number of fields – e.g. pharmaceuticals – may also lead to greater concentration in such sectors. By contrast, services innovations – particularly those in marketing or organisation – often require lower investments.

Trickle-down dynamics

The distance between innovation leaders and the remainder of the economy depends on the ease with which leading-edge technologies are diffused across the economy. The extent to which such processes take place influences the gaps in innovation capabilities between insiders and outsiders. While wider diffusion fosters industrial inclusiveness – as well as social inclusiveness – some exclusive rewards to the innovator are needed to encourage innovation. This is the very essence of the intellectual property (IP) system, which provides inventors with exclusive rights to the rewards from their invention for a period of time. However, facilitating diffusion is also essential to fostering the innovation process. Innovation and technical change depend on new knowledge because unlike physical property, knowledge grows over time. New knowledge expands based on the existing stock of knowledge ("standing on the shoulders of giants"), new discoveries rely on the existing level of science, and new ideas originate from past experiences. Thus, to the extent that innovation ultimately depends on connecting to diverse sources of knowledge, its increased availability can provide wider opportunities for corporate innovation (Arthur, 2007).

Opportunities for leapfrogging point to a different approach to diffusion. Such opportunities often arise in more excluded or laggard regions or groups in emerging economies that lack core infrastructure (including electricity or fixed-line telephone networks). New developments, such as mobile telephony, can allow leapfrogging. Thus, adoption processes need not necessarily be linear and follow the same path as in other countries. In China, solar thermal heating – developed by Tsinghua University – has allowed bypassing gas- or electricity-based heating in a number of rural areas and provided new opportunities for competitiveness (Lee, 2014).

Knowledge spillovers are a critical contributor to diffusing innovation from insiders to outsiders, thereby enhancing overall performance. Wider opportunities for knowledge spillovers have been shown to have positive impacts on firms' performance. Bloom et al. (2013) find evidence in the United States of positive technology spillovers from R&D investments and show that the social returns from R&D are at least twice as high as the private returns. Knowledge lends itself to such spillovers: once created, it can be replicated and disseminated at virtually no cost, and benefits more firms than its original creator (Arrow, 1962).

Geographic proximity plays an important role in knowledge spillovers (Krugman, 1991; Audretsch and Feldman, 1996). Even for knowledge that is codified in the form of patents, there is a rich literature documenting that patent citations are geographically localised, a fact that holds true when controlling for the pre-existing concentration of technologically related activities (Jaffe et al., 1993). In some countries, the success of regional innovation hubs contributes to innovation-related or economic benefits in nearby regions. However, this is not always the case and depends in part on regional absorption capacity and agglomeration dynamics as illustrated by the cases of India and China. Evidence in OECD countries notes that inter-regional spillovers from R&D investment depend in part on the characteristics of the neighbouring region; notably rural region neighbours appear to benefit more economically than neighbouring urban regions, the latter being more likely to compete for key resources (Lembcke, Ahrend and Maguire, 2015).

The increased opportunities provided by the widespread uptake of ICTs can reduce the barriers to transmitting increasingly sophisticated pieces of knowledge. Videoconferencing is one means of transferring ever-larger amounts of information in ways that match the concept of "proximity". Firms with weaker access to "offline" knowledge networks (e.g. firms in remote locations) may have more to gain from Internet-enabled knowledge spillovers. This points to the possible benefits of the Internet in helping to democratise innovation by enhancing lagging performers' opportunities to compete with top performers. The evidence also confirms earlier studies showing that smaller firms, rather than larger ones, benefit more from spillover effects (see for example Acs, Audretsch and Feldman, 1994). The Internet also facilitates spillover benefits for researchers and their universities. Ding et al. (2010) show that the Internet facilitated the inclusion of women scientists, as well as the overall research output of people working at non-elite institutions, by providing increased access to the knowledge of others and larger opportunities for collaboration.

Several case studies illustrate how informal and grass-roots innovators derived advantages from the Internet and mobile networks. In their study on Uganda, Muto and Yamano (2009) show that farmers located farther away from the country's centre gain more from these networks – independently of whether they themselves own mobile phones – effectively finding evidence of spillover effects from such infrastructure. Studies have also shown that micro-enterprises – including those operating in the informal sector – tend to benefit from ICTs, notably through mobile phones (OECD, 2015a).

Such opportunities have caveats, in that knowledge networks alone often do not guarantee corporate performance – which is instead also driven by firms' own "absorptive capacities". The limitation of knowledge flows in the context of low internal capacities has been a core theme in the literature on knowledge spillovers (e.g. Görg and Greenaway, 2004). Firms need the capacity to deal with the knowledge they access, otherwise they have little to gain (Hu et al., 2005; Kokko et al., 1996). This is because knowledge often has a "tacit" component that cannot be easily transferred, or might be inappropriate in specific firm contexts requiring adjustments. Moreover, framework conditions might affect businesses differentially. The heterogeneous impacts of framework policies on firms have been a core theme of the OECD DynEmp Project (Criscuolo et al., 2014). One of its findings has been that in many countries, policies are not well geared towards the growth of young innovative firms, sometimes explicitly favouring incumbents.

Innovation policies have different outcomes on industrial and territorial inclusiveness depending on how they interact with other policy measures and framework conditions. Several design/procedural aspects can ultimately shape the impacts of innovation policies on exclusion. Irrespective of objectives, policies can have different outcomes and may contribute to excluding certain individuals/groups by virtue of their design, which may feature: 1) lengthy or costly application procedures before rewards are provided, hindering start-ups; 2) complex application procedures requiring expertise possessed only by selected firms; 3) rewards for past performance in subsequent application procedures, advantaging incumbents; 4) insufficient focus on advertising the existence of policy programmes to outsiders, potentially reducing the share of external participants; and 5) budgetary cuts affecting the amounts of available funding, potentially resulting in greater applicant selectivity.

The same issue applies to territorial inclusiveness, as innovation policies that target specific sectors, social challenges or types of institutions will have a *de facto* place-based dimension that can add to, or detract from, inclusiveness. Certain programme rules, such as matching regional funds requirements, can also reinforce the flow of public innovation funding to the leading regions. While policy discussions often disregard these policy aspects, they are critical to how the policies will serve industrial or territorial inclusiveness, since they tend to aggravate "exclusiveness".

This challenge affects advanced, emerging and developing economies alike. It is often much easier and more straightforward to identify the largest contributors to innovative potential than the smallest or latest. Moreover, past performance (e.g. using previous publications to select research excellence initiatives) is a simple selection criterion, as predicting potential future success is more challenging and involves greater risk. The challenges can be greater in developing and emerging economies, since selection criteria often exclude informal-sector participants.

Complementary policies supporting the policy environment in which firms operate can be critical in creating conditions for democratising innovation. The example of policies pertaining to IPRs illustrates this well. Even though IPRs provide opportunities for different actors, large businesses often use them more intensively. One reason is that enforcement costs are a significant hurdle for small companies, since the costs are not proportional to firm size; attorney fees, management costs and the time required to deal with litigation issues can be substantial. The size of firms' patent portfolios can help avoid costly litigation by using cross-licensing strategies, meaning that small firms are at a disadvantage compared with larger firms – which can also reach agreements more easily thanks to repeated interaction with their competitors (Lanjouw and Schankerman, 2004). To make things worse, the fact that smaller companies are less prepared to withstand litigation increases their risks of facing further litigation. Meanwhile, the lack of capacities to manage and negotiate IP portfolios imposes a sunk cost that hinders smaller firms' ownership of IPRs. Corrupt business environments have also been shown to affect smaller firms' ownership of IP titles (Paunov, 2014). Moreover, IP rights are useful to businesses only if they can use IP-protected inventions to generate innovations; this requires financial resources small firms might not possess. Thus, IP may only serve the largest firms, unless complementary policies are in place. The importance of the interaction effects highlights the critical role of implementing a whole-of-government approach. This is insufficient in a scenario where policy interactions matter regarding who will benefit from their implementation.

Another example of a policy instrument that may have biases is R&D tax credits (see Chapter 6 of this report). Governments can choose among various instruments to promote business R&D. In addition to giving grants or loans and procuring R&D services, many also provide fiscal incentives. Tax incentives for business R&D expenditures include allowances and credits, as well as other forms of advantageous tax treatment such as allowing for the accelerated depreciation of R&D capital expenditures. Today, 27 of the 34 OECD countries and a number of non-OECD economies give preferential tax treatment to R&D expenditures and do so in many different ways. MNEs benefit the most, as they can use tax planning strategies to maximise their support for innovation. This can distort the playing field and lead to a disadvantage for purely domestic and young firms. In response, Australia, Canada, France, Korea, the Netherlands and Portugal give more generous treatment to SMEs than to large firms. Well-designed direct subsidies may also support small businesses.

Policies also have differential impacts depending on the company's local environment, and particularly on whether firms have access to critical ingredients for innovation (including finance, human capital, knowledge and infrastructure). Access to these critical ingredients can vary within a metropolitan area or region within the same country. Access to sources of both finance and knowledge is a key requirement for innovators. Larger innovators have the opportunity to internalise some of these sources (e.g. by creating their own R&D labs and relying on internal resources to support innovation investments). By contrast, smaller firms rely on external sources, as they do not have sufficient own resources to internalise them. Especially in developing and emerging countries, business framework conditions can constitute stumbling blocks for companies' innovation performance, particularly that of smaller and catching-up firms (Tybout, 2000).

Liberalisation efforts have also fuelled the use of mobile phones in India. Africa, like India, has experienced high growth in the number of mobile subscribers. Yet calls to Africa have not increased in the same manner as for India. International inbound traffic to India (measured by minutes or calls) was less than Africa's in 2003 but grew to ten times higher by 2011. At the same time, the rates to call India decreased tenfold. The difference lies in whether governments let the market set the rates for incoming calls or impose a single rate through an official cartel. Between 2003 and 2011, for example, the termination charges paid by telecommunication operators carrying traffic from the United States to the rest of the world halved on a per minute basis (from around USD 0.09 to USD 0.04). For the highly competitive India market, rates dropped from more than USD 0.14 to less than USD 0.02 over the same period. In Africa on average, rates increased, suppressing demand for calls to people on that continent (OECD, 2014p).

Regional development policy and innovation policy can be mutually reinforcing to promote territorial inclusiveness. Historically, regional development policy focused on simply transferring resources from wealthy to poor regions. However, a more growth-oriented approach to regional development policy has taken hold across OECD countries, spurred by the objective of strengthening the overall domestic innovation capacity, including in less developed regions (OECD, 2011b). Regional development policies can thus complement innovation policies to better support territorial inclusiveness. Furthermore, regional and local level governments themselves can take important complementary actions to improve the impact of national innovation policy instruments, such as providing innovation advisory services to firms in a nationally financed technology park.

Reaching this objective requires stronger regional capacity for innovation policy in both OECD and non-OECD countries. The European Union has promoted the development of regional innovation strategies for many years. Most recently, it financed a platform dedicated to developing such "smart specialisation" strategies. In fact, possessing a strategy is now a condition for receiving EU Structural Funds, since a significant share of those funds – particularly in the most advanced EU regions – is spent on innovation and business development. Another means of building sub-national capacity and improving the use of innovation funds is instituting a regional level council or forum for innovation. From South Africa to Denmark, such entities are used to drive greater innovation success of national policies and sub-national initiatives.

The discussion above focused on the impacts of policy contexts on industrial inclusiveness, but the policies' very design might equally have impacts on industrial and territorial inclusiveness, with potential effects on social inclusiveness. Further investigation should explore whether concentrating excellence is increasingly important for growth and inclusiveness – and, conversely, whether opportunities to democratise innovation, i.e. broaden individuals' and small companies' access to innovation activities and markets, support growth and inclusiveness. Ongoing OECD work is analysing the link between these market characteristics of knowledge-based production processes and rising inequalities many OECD economies experienced over the past decades, as documented in OECD (2014e).

Inclusive innovation policies

As noted already in Chapter 1, both developed and emerging economies have to boost growth and make it more socially inclusive, by offering opportunities for different groups in society (OECD, 2014q). In developing and emerging countries, addressing high levels of poverty remains critical. Innovation can contribute to this objective as a driver of income growth and job creation, which under certain conditions will benefit all in society, be they directly or indirectly affected. In addition, innovations that are specifically aimed at lower-income and excluded groups ("inclusive innovations") can substantially improve their welfare.

Inclusive innovations are innovations that improve the welfare of lower-income and excluded groups. Inclusive innovations often modify existing technologies, products or services to better meet the needs of lower- and middle-income groups. One way consists in lowering unit product prices by preserving only critical functionalities while keeping core quality. The lower price allows lower-income groups to purchase those innovations. Examples of inclusive innovations include the Tata Nano (for goods), a low-cost car produced in India based on a no-frills strategy, and the Narayana Health Cardiac Care Centre (for services), which provides heart surgery at a much lower price due to standardisation and the use of less skilled labour. Table 7.1 provides examples. Inclusive innovations that involve excluded groups not only as consumers but also in the innovation process are referred to as grass-roots innovations.

Table 7.1. **Examples of inclusive innovations**

Service innovation		Product innovation
Empresas Públicas de Medellín (EPM)	**Narayana Hrudayalaya (NH) cardiac hospital**	**MoneyMaker irrigation pump**
EPM is a utility company providing energy and water services. Low-income users can pay using prepaid cards, so as to use the energy service according to their cash flows. No fixed installation costs have to be paid by households. **Innovation:** Pay-per-use method **Operator:** Public utility company **Sector:** Energy and water **Country:** Colombia **Scale:** 43 000 low-income users have been reconnected since implementation in 2007.	NH provides cardiac surgeries and other healthcare services at low cost to the poor. It also caters to isolated communities via telemedicine. **Innovation:** *Standardisation* to decrease costs. *Use of ICT* to set up healthcare centres in remote locations for poor rural communities. **Operator:** Private corporation **Sector:** Healthcare **Country:** India **Scale:** 6 200 beds spread across 23 hospitals in 14 cities (up from an initial 300 beds in 2001).	Low-cost handpowered irrigation pumps created and commercialised **Innovation:** No electricity or fuel to function are required and operating cost is lower **Operator:** US-based NGO (KickStart). **Sector:** Agriculture **Country:** Kenya, Mali, Tanzania **Scale:** The pumps are distributed in local shops and sold to other NGOs for wider diffusion in the three countries.

Sources: Suárez Franco, C.F. (2010) for Empresas Públicas de Medellín; Kothandaraman, P. and S. Mookerjee (2008) and *www.narayanahealth.org* for Narayana Health; OECD (2013) and *www.kickstart.org* for the MoneyMaker irrigation pump.

Opportunities for the wider development of inclusive innovations are larger than before: the growing wealth of emerging economies is generating an increase in the demand for inclusive innovations as the purchasing power of larger segments of society is rising but insufficient to access the same products available to households in developed countries. The BRIICS countries – Brazil, the Russian Federation, India, Indonesia, China and South Africa – and China and India in particular, are increasingly important as they offer a growing market relative to the overall size of the world economy (Figure 7.5).

Several large multinational companies have placed an emphasis on developing inclusive innovations, particularly to serve the growing middle class in emerging economies. Companies such as Siemens, for instance, have set up research projects under the SMART

Figure 7.5. **Non-OECD countries' share in the global economy has been steadily rising**

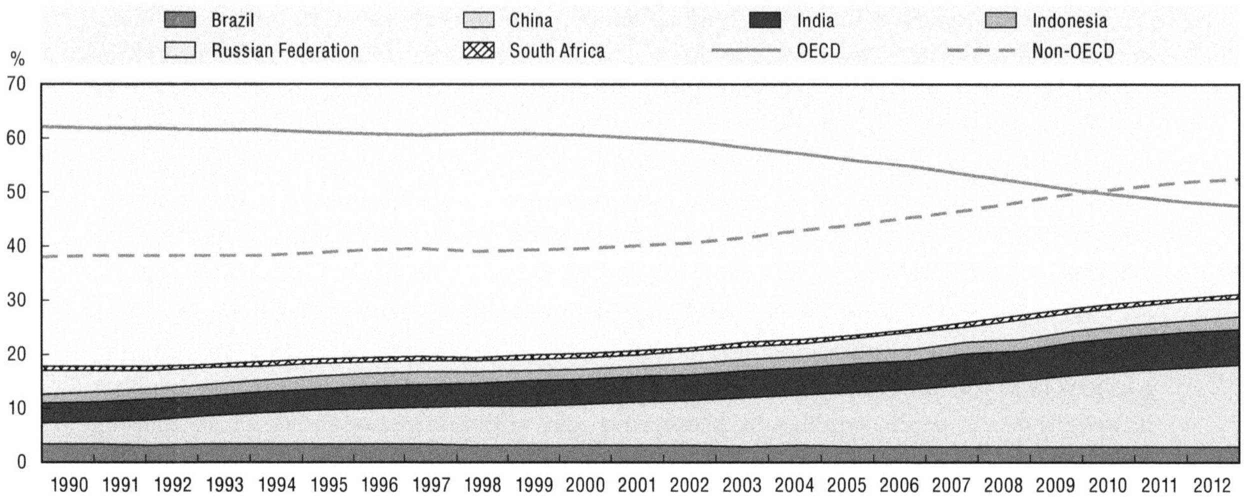

Source: OECD (2014l), *Perspectives on Global Development 2014, http://dx.doi.org/10.1787/persp_glob_dev-2014-en.*

objective (that is "simple," "maintenance-friendly," "affordable", "reliable," and "timely to market.") to introduce low-cost, high-quality products. These include, for instance, a foetal heart monitor (FHM) that does not rely on costly ultrasound technology (Siemens, 2011). Prahalad and Hart (2002) popularised the notion of market opportunities at the "bottom of the pyramid". While it is subject to debate whether new market dynamics will really affect the poorest, the market opportunities for the rising middle classes are certainly a reality. Innovations aimed at this market might also serve the poorest in OECD economies.

Inclusive innovation approaches have a sound rationale for policy action because innovators in the field are often exposed to various types of market failures. Barriers to information about customer needs, infrastructure challenges and shortcomings in access to credit can contribute to a "poverty penalty", i.e. a higher cost of supplying the poor than supplying higher-income markets (Mendoza, 2011). Moreover, many inclusive innovations involve public services (education, health, transportation, etc.) from which the poor often find themselves excluded. The government is, therefore, a core stakeholder as it not only provides but also regulates public goods.

An important challenge for inclusive innovation is achieving a sufficient scale. An innovation's scale depends on market segmentation or consumer location. Localisation can be critical (e.g. for agricultural activities) not only to improve local production techniques, but also to adapt them to specific rural contexts. Given their potential consumers' income and numbers, standard innovators may have better opportunities than inclusive innovators to attain production scale and product standardisation (since agriculture plays a lesser role and local specificities have less impact on products not typically required by the poor). Inclusive innovators, on the other hand, may face cost-based challenges, which ICT-based services (among others) can help address. This is because economies of scale for expanding ICT-based services are often very low.

In the absence of representative statistics, the evidence to date suggests that few inclusive innovations have reached scale. The type of innovation is very much a factor when it comes to scaling up. Reaching maximum scale depends strongly on demand – which will be quite low for localised products, but may involve millions of customers for broader-based services, e.g. mobile banking. Furthermore, product-level scaling is not an absolute necessity: the very process of designing local innovations to serve local needs may support an inherently small-scale market, while also contributing to poverty alleviation.

Inclusive innovations that have scaled up successfully include mobile phones and some mobile services (such as M-PESA), several microfinance initiatives, and Jaipur Foot, Fuel from the Fields and Narayana Health. This success has occurred for several reasons.

- The product responded to **strong demand**, as demonstrated by the poor's willingness to pay for such services. Mobile phones, for instance, were taken up even where electricity supply was a challenge, because communication needs were substantial. In 2013, mobile phone subscriptions per 100 inhabitants amounted to 89.4% in developing countries (ITU, 2014).
- Successful innovators **invested in gaining a deep understanding of the requirements of the poor**, which can be achieved by involving them directly in innovation processes.
- **Developing profitable business models** was a priority. This process often involves multiple iterations, aimed at identifying opportunities for success, which might be described as "**thinking out of the box**". Innovative **pricing and financing strategies**, as well as modified business processes, have also proved critical. While cost reduction was generally the main criterion, other factors (including ensuring product quality and the application's usefulness) were critical too. Cost-effectiveness and profit-driven objectives have often underpinned successful initiatives.
- **Favourable regulatory conditions** and experimentation with different approaches were often critical. For instance, public-private partnerships (e.g. the Aashkar tablet in India) were used to support outreach to poorer communities in India or South Africa. In Kenya, the success of M-PESA would not have been possible without regulations enabling this type of service development.
- **Private entrepreneurial initiative** was a driver of scale. Private companies (e.g. Nokia and Motorola) have adapted handsets for the developing world, while microfinance institutions (MFIs) have received a substantial boost from participation by commercial banks. Other actors – notably non-governmental organisations (NGOs), not-for-profit organisations, and universities and/or PRIs – have often contributed adjustments to supply a wider market with a better product.
- **Open access to information infrastructures, including data**, enables the development of innovative goods and services. Therefore, equal and non-discriminatory access can maximise the economic and social value of information infrastructures.
- Building on **existing infrastructures helped achieve scale** by overcoming obstacles through relying on existing delivery networks for the poor (e.g. using small community-based shops) and existing knowledge sources (e.g. NGOs operating in the field). Fuel from the Fields, a grass-roots entrepreneurship initiative that allows production of charcoal from agricultural waste, relies on partner institutions to disseminate the technology and know-how to diverse communities (Paunov and Lavison, forthcoming).

Microcredit – the granting of small loans rarely amounting to more than a few hundred US dollars – is an interesting case because unlike other inclusive innovations, it is a more mature product that has undergone substantial experimentation and managed to reach significant scale. According to estimates, about 200 million people worldwide took out loans at an MFI in December 2010, of which over 130 million were living in extreme poverty – i.e. on less than USD 1.25 per day, or less than half the national poverty line (Maes and Reed, 2012). The microfinance market, estimated at USD 60 billion to USD 100 billion in 2013, caters to about 20% of demand for credit by the poor worldwide (IFC, 2013). Among the various MFIs, the Grameen model is quite widespread, with the Grameen Bank numbering over 8.37 million members in 2012 (Grameen Bank, 2013). Microfinance is also interesting because it facilitates the uptake of inclusive innovations.

The success of the microfinance model is based on constant efforts to provide sustainable credit services to geographically scattered and remote poor clients. Unlike higher-income groups, these people often have neither collateral nor a credit history and may even sometimes lack verifiable identities. To avoid moral hazard, MFIs needed to find alternatives to traditional approaches (e.g. collateral-based loans to ensure borrowers do not have incentives to default). Providing low-income groups with access to credit with joint liability has been one critical solution. Another solution has been to provide dynamic incentives – e.g. the promise of larger future loans conditional on timely repayment of the initial smaller loans.

Inclusive innovations demonstrate that innovation can effectively improve the welfare of lower-income and excluded groups. New technologies, in particular ICTs, have heightened opportunities to develop inclusive innovations. The private sector's interest in serving the growing middle-income groups in emerging economies in particular offers opportunities for inclusive innovations to successfully reach scale despite the many challenges they face – from the lack of financing and technical expertise for grass-roots innovators to limited information about actual consumer needs for pro-inclusive innovators. Policy plays a role in creating a favourable environment for inclusive innovations to develop scale, effectively leveraging market-based creativity to tackle these development challenges more efficiently.

Co-operation across various dimensions is critically important for policies to successfully support inclusive innovations. The public services nature of products requires cross-institutional collaboration within governments. Moreover, involving a variety of actors including PRIs and universities, the private sector, banks and other financing institutions, and non-governmental institutions is crucial to tackle market failures. Finally, it is critical that lower-income and excluded groups are involved in the innovation process and that their participation is not marginalised. This not only reduces risks of low product uptake, but also supports ownership by the communities involved. This is often critical in the diffusion of inclusive innovation. Schemes such as Colombia's Policy on Social Innovation aim at involving a wider community (see Paunov and Lavison, forthcoming, for further detail).

Moreover, alleviating financial constraints is critical as constraints remain large despite the considerable growth of the industry of "impact investing" in the last years (Koh, Karamchandani and Katz, 2012). Several countries including India and South Africa have looked into the possibility of creating specific financial support for inclusive innovation initiatives. While China does not have a specific fund for inclusive innovations, the "Special Fund" dedicated to its S&T Programme for Public Well-Being has supported 23 projects that include various inclusive innovations in the areas of health, ecology, public safety, etc. (CASTED, 2014).

Providing access to knowledge and technical expertise can be a critical tool to support grass-roots innovations. One way to promote greater access is for governments to incentivise universities and PRIs to support grass-roots innovators. Another is to foster intermediary institutions that build bridges between formal innovation facilities (PRIs, universities) and people at the grass-roots level. It is also essential to stimulate linkages between innovators and private-sector companies for scaling up and between grass-roots innovators themselves. The most well-known example of a successful intermediary institution is the Honey Bee Network in India. Also, as early as 1986 the Chinese Ministry of Science and Technology (MOST) initiated the Spark programme, which aims at transferring and diffusing science and technology over the vast rural areas of China through grant funds, technology training to farmers and solving local technology problems using the know-how from research institutes.

Finally, prizes and related instruments can be particularly relevant as a means of drawing attention to inclusive innovation initiatives and might in that way play a specific role. Examples include the G20 Challenge on Inclusive Business Innovation, which is a global competition aimed at identifying "business with innovative, scalable, replicable and commercially viable ways of reaching low-income people in developing countries". This competition received 167 applications between November 2011 and February 2012, with 50% of applications from the agricultural sector and the remainder from the retail, housing, health and education sectors. Also, an India-EU Prize for Affordable and Inclusive Innovation is currently being developed in collaboration between the Indian National Innovation Council, the Indian Department of Science and Technology and the EU delegation. The prize will reward inclusive innovation resulting from collaboration between Indian and European individuals or organisations, and support the development of awarded innovations (via incubation or scaling up, depending on maturity) (NInC, 2013).

Below are some policy conclusions about policies that can help foster inclusive innovation.

Main policy messages: Supporting inclusive innovation

Governments can approach the topic of inclusive innovation through multiple channels. The following factors have proven particularly pertinent:

- making use of advanced technologies and ICTs – notably but not only mobile telephony – because they can be used as platforms for multiple services
- involving intermediary institutions – including NGOs or universities – that connect to the user group and "translate" inclusive innovations
- developing financing mechanisms in support of inclusive innovation and introducing prizes to incentivise such efforts
- involving ministries beyond those in charge of innovation. Inclusive innovation often falls within the responsibilities of other ministries that focus on poverty alleviation, rural development and specific issues in education, health or infrastructure
- addressing regulatory challenges that emerge when entrepreneurs that address the needs of low-income groups follow a perspective that is neither purely for-profit nor purely social
- firmly inserting inclusive innovation policies in the overall innovation policy agenda, thereby ensuring the joint objective of achieving growth and inclusiveness.

7.3. Fostering health innovation

The challenge of health innovation

Chapter 1 already pointed to the important role that health innovation has played in improving well-being and living standards. Health innovation is an interactive and distributed process that involves several phases, including: 1) identification of need; 2) R&D; 3) commercialisation; 4) delivery; and 5) diffusion. These stages are increasingly understood to be circular, iterative and highly interconnected – unlike the traditional notion of a linear step-by-step process. Health innovation is tightly connected to the provision, uptake and use of new treatments: feedback from purchasers, providers and patients is essential in shaping the innovation process.

Health innovation is a complex and expensive undertaking that involves multiple actors from the public and private sectors. At each stage of the innovation cycle, many social and economic factors can affect the development, uptake or diffusion of new health technologies. Across the OECD, a number of studies have identified major health innovation bottlenecks in the cycle and strategies for removing them. While the details vary, the roadblocks share similar characteristics across countries. Innovation in prevention, therapy and healthcare relies on (uncertain) technical and scientific advances; involves multiple players; requires large financial commitments, high risks and long time frames; and is highly regulated. Moreover, the provision of healthcare is often considered a public good and in some countries a right.

The government has an important role in funding both basic and clinical research (Figure 7.6); regulating safety and efficacy; determining availability, equity and access; and often pricing and/or reimbursing health products and services. Important questions for policy makers are whether the innovation system is delivering the innovations that are most needed by patients, and what the costs associated with those innovations are. Public opinion matters greatly for health innovation, and there are numerous examples of public pressure on government health policies (to provide greater access, to maintain affordability of products/services, to increase/relax the burden of regulation).

This section does not provide a full overview of all the challenges and policies related to health innovation. Rather, it points to findings from some recent OECD work focused on the contribution of innovation to better health outcomes, and the policies than can help ensure such outcomes. The section touches on five areas where recent work has been undertaken, namely: 1) innovation and global health challenges; 2) new approaches to optimising the role of modern technologies for drug development; 3) the role of big data for both care and the development of better treatments; 4) the role of ICT technologies; and 5) issues related to payment for innovation.

Meeting the health challenge

Improving the health of the world's population is an enormous policy challenge that requires both national and international policy action. Over the coming decades, innovation – technical, social and organisational – will play a major role in the delivery of more personal, predictive and preventive healthcare products and services and will radically change how medicine is practised and healthcare is delivered.

Health challenges in terms of the incidence of chronic diseases such as diabetes, obesity and dementia are on the rise in developed and developing countries alike. This has led governments to seek both to contain the costs of treatment and to find new approaches to

Figure 7.6. **Government funding of health R&D, 2012**

(as a percentage of GDP)

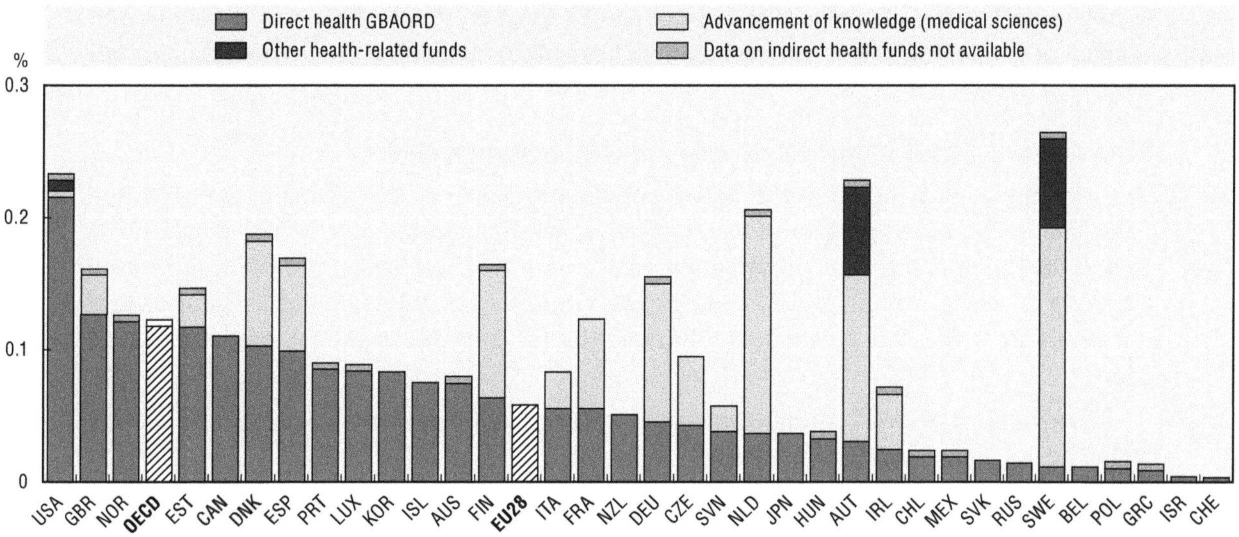

Notes: GBAORD = government budget appropriations or outlays for R&D. *Direct health* GBAORD comprises government R&D budgets primarily committed to the socio-economic objective of protecting and improving human health. Funds related to the *advancement of knowledge* are non-oriented research funds and general university funds, i.e. the estimated R&D content of government block grants to universities, for which it is possible to identify the amount dedicated to R&D in the medical sciences. *Other health-related funds* are ad hoc OECD estimates based on national sources that cover general support for R&D in hospitals and related areas that are excluded from GBAORD estimates.

Source: OECD (2013h), *Science, Technology and Industry Scoreboard 2013*, *http://dx.doi.org/10.1787/sti_scoreboard-2013-en.*

prediction and prevention. Ageing societies have led to an increase in chronic diseases and are also putting cost and delivery pressure on public health systems, forcing governments to consider new ways to deal with long-term care. At the same time, infectious diseases such as HIV/AIDS, malaria, tuberculosis and lower respiratory infections continue to threaten significant portions of the world's population. New global health threats, such as the rise of antibiotic-resistant bacteria, highly contagious strains of the influenza (flu) viruses, and the recent Ebola crisis, coupled with a highly mobile population, raise the spectre of potential pandemics. Amid all these challenges, government healthcare costs continue to rise and consume an increasing share of GDP, creating a situation that may be neither politically nor economically sustainable over the long term.

There are significant opportunities to improve the efficiency of our health systems and the quality of human health dramatically by harnessing the results of current scientific discoveries, building on the public investment in health R&D (Figure 7.6). The era of genomic medicine, launched by the sequencing of the human genome, promises a major shift in health innovation, with life increasingly understood, and medicine increasingly practised on a molecular level. Sequencing the human genome took a decade to complete and cost USD 3 billion. Since then, the price of sequencing has dropped rapidly to where it stands now at approximately USD 1 000 per genome, and whole genomes can be sequenced in a day. Understanding genomes and their interaction with environmental factors, as well as the use of emerging technologies in, for example, pharmacogenetics and systems biology, can lead to earlier, more accurate diagnosis and therapeutic intervention, i.e. the practice of precision medicine.

Electronic health records and biomedical data collections (e.g. human biobanks and genetic research databases) are examples of innovations designed to reduce costs, increase efficiency and optimise the use of research outcomes. Entirely new therapies and diagnostics are being developed via technologies based on stem cells, nanotechnology and synthetic biology.

New models of health innovation and knowledge management

The rapid pace of scientific and technological advances in the life sciences, the complexity and heterogeneity of knowledge relevant to health innovation across fields and subfields, and the need to integrate vast amounts of scientific and clinical data all combine to create challenges for achieving the interoperability, knowledge integration and accumulation necessary to efficiently harvest the full benefits of the existing knowledge base.

New models of health innovation and knowledge management are proving necessary for achieving a number of objectives:

- to support more radical innovation and to increase the efficacy of collaborative research
- to accelerate the scale-up of research from academia to industry and to the bedside of patients
- to increase the efficiency of therapeutic development and to tackle unmet medical needs and treatment paradigms, including high value-added diagnostics, antibiotics and neglected infectious diseases.

Changes in the pharmaceutical sector business models are driven by technological opportunities, medical needs and economic constraints. In order to bridge the innovation gap, the pharmaceutical industry is now engaging in collaborative research models to achieve a higher level of flexibility and to increase the output of therapies while remaining competitive. The overarching aim of these knowledge- and technology-sharing activities between pharmaceutical companies and academia is to increase the efficiency of preclinical and translational research through the use of global intellect, the avoidance of redundancy and ultimately, a faster translation of rapidly expanding biological data into drug discovery pipelines. However, an issue relates to the financial incentives for moving new biomedical technologies from academic research to industrial use. Measures need be taken to develop greater information sharing, increase flexibility and provide sufficient incentives for risk-taking investment. Meeting public health priorities will also require alternative approaches to forming capital for R&D, such as bond issues or better uses of philanthropic programmes, in order to finance health innovation programmes. Governments should be better aware of the options and their relative merits.

Reducing the cost of drug development by streamlining clinical trials to make them smaller and faster is a shared government-industry goal.[2] Work needs to be done to simplify, co-ordinate and process the permissions needed for clinical trials; to make advice consistent and standardised; to create model contracts; and to develop early-warning systems for problems. This can be achieved through early interaction between regulators and companies, which is critical to build support for the development of new clinically valid endpoints – biomarkers essential to personalised medicine, clinical practice and genetic databases – in order to improve clinical trials. Discussions between regulatory agencies and industry can help establish stable, predictable, transparent regulatory pathways; improve biomarker validation and pave the way for regulatory acceptance; take on the challenge of personalised medicines and targeted therapies; tackle methodologies for the design of next-generation clinical trials; and create safe havens for new approaches to knowledge sharing and risk sharing.

The move towards more effective and safer therapies tailored for individuals can be further facilitated by a broader use of biomarkers to make early go/no-go decisions in the development process and to better define diseases at the molecular and genetic levels. The development of specific and reliable diagnostic biomarkers is closely linked with the advancement in research technologies. Funders and payers (e.g. governments and insurers) need to better understand the advantages and disadvantages (i.e. cost-effectiveness) of using biomarkers as diagnostic tools when establishing their payment and reimbursement plans. The strengthening of evidence-based medicine and use of new diagnostic devices will require educating physicians and healthcare providers and giving them statistical training to understand tests and results. Information about the clinical utility of biomarkers will also be needed at point of care as it may affect the process of care and effective implementation.

New tools, frameworks and processes for evaluating new technologies in research and healthcare may need to be developed to capture aspects such as cost-effectiveness and return on investment. Pharmaco-economic assessment offers promise for achieving socially optimal outcomes, in terms of promoting the right level and type of R&D investment, through better signals to industry as to which innovations are most highly valued. It can also be used to establish market-based incentives for investment in treatments for rare and complex diseases with limited market size and high risk of failure and financial loss. Meeting global health challenges has spurred innovative approaches to inter-firm collaboration, access to and use of IP, and financing mechanisms. They complement traditional development programmes rather than compete with them. Because no individual tool (e.g. public-private partnerships for product development, advanced market commitment, prizes) can be used to address all problems, governments need to better understand the available mix of approaches in order to achieve different policy goals. Representative cases may yield lessons for a lower-cost approach to innovation in healthcare. Governments should identify the lessons contained in these innovative approaches to global health challenges and should try to apply them to health innovation more generally.

Integration and coherence of innovation policies in areas such as health, science, development, trade and industry would be of enormous help in addressing public health priorities. This is difficult to achieve, however, owing in part to institutional barriers to co-operation in policy making and implementation. Governments may also want to facilitate patients' and/or their organisations' more active role in innovation policy and policy making related to clinical trials and access to new products. Patients are important sources of innovations that remain underutilised. New modes of communication and networking among health systems, end users and innovators are emerging that may provide a better match between global health objectives and investment in R&D. These need to be better understood by governments.

Public acceptance and trust is a critical factor in uptake and diffusion. Direct-to-consumer tests and services are increasingly available. There is no consensus on whether and what oversight and governance are needed. This is a subject that calls for further consideration by governments. Establishing clear policies with regard to the privacy and security of personal data is fundamental for a wide range of health technologies (e.g. genetics and genomics, electronic health records). Similarly, the public is concerned with equity of access. Governments have a central role to play in finding a balance between individual rights and public health/research priorities.

From biomedical research to drug development

Drug discovery and development systems are rapidly evolving in order to respond to a diversification and globalisation of stakeholders, a need for cost-intensive resources, and a growing dependence on data-driven innovation. Traditional R&D systems have been suffering from declining productivity and escalating costs, even though innovation as measured by patenting has stayed relatively robust (Figure 7.7). The attrition rate in medicines' development has stayed high in complex therapeutic indications, despite the promise of new technology in drug discovery and preclinical research. The market perceives a pipeline problem in, for example, central nervous system disorders, coupled with a number of mergers and acquisitions in the pharmaceutical industry. In short, there is uncertainty over the future structure of the health biotechnology and pharmaceutical industry. Some drivers of change in business strategies for health innovation include:

- the productivity decline in innovative R&D and regulatory uncertainty
- technology opportunities and expansion of medical scientific knowledge, including the detection and potential therapy of genetic disorders
- recognition of the value of preventive medicines and companion diagnostics
- the entrance of new players as research capacity globalises and the number of competitive firms from emerging economies grows
- uncertainty over economic developments such as how precision medicines will be delivered, and the role of high-priced therapies in public health systems
- the changing nature of demand and outcome expectations, particularly the growing role played by payers and patient organisations.

Figure 7.7. **Health-related patents, 1999-2001 and 2009-11**

As a percentage of total Patent Cooperation Treaty patent applications

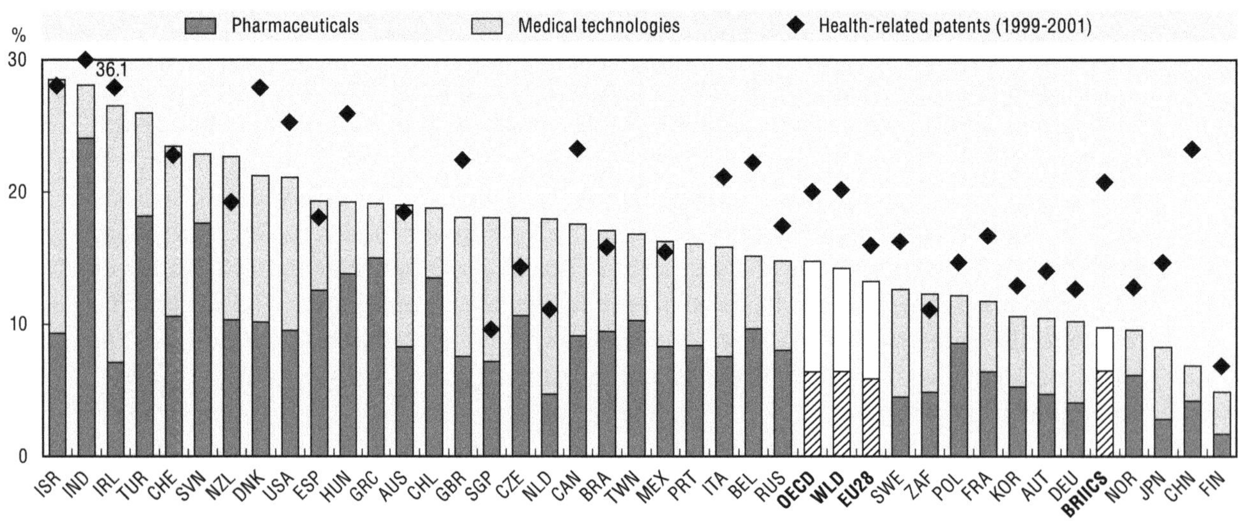

Source: OECD (2013h), *Science, Technology and Industry Scoreboard 2013*, *http://dx.doi.org/10.1787/sti_scoreboard-2013-en*, based on OECD Patent Database, *www.oecd.org/sti/inno/oecdpatentdatabases.htm*.

The future organisational structure of the biopharmaceutical industry is a hotly debated topic. Some see the rise of outsourcing and knowledge management services as part of the move away from fully integrated pharmaceutical companies to a more networked pharmaceutical industry ecology. Others, on the contrary, believe that the complexity and difficult capital investment decisions in the health sector portend well for a return to a more vertically integrated structure.

The search for funding in early stages of research and translation, the high costs and administrative burden of product development and regulatory approvals, the financial pressures due to valuation of biotechnology and pharmaceutical firms, and constraints on the public purse have been a constant in this sector. Particular issues are:

- escalating costs of long development programmes and regulatory complexity
- growing constraints on public health expenditures and return of investment
- funding gaps, e.g. the research, but also gaps in available funding for diseases/approaches
- efforts to improve valuation of assets and enterprises based on the knowledge they hold.

As an example of an effort to start addressing these challenges in a specific context, the recent OECD workshop in Lausanne was aimed at Alzheimer's disease and other dementias as an unmet medical need of ageing populations (Box 7.8). This workshop and the work on integrated development undertaken by the UK government demonstrate that progress in drug development, even for complex diseases such as Alzheimer's, can be made through well-designed public-private partnerships, involving all the relevant actors, including governments and regulators, scientific advisory groups, business, and patients (OECD, 2015c).

Towards better health data governance

The Rosetta stone for healthcare therapy innovators, clinicians, regulators, health system managers and governments is reliable data that tracks patients' progress over time and includes essential background characteristics such as genetic profile, the therapies and services provided throughout the healthcare and social care system, and patient-reported and experienced outcomes, including death. Clean, valid and comprehensive healthcare pathway and outcomes data simultaneously meet the need for discovery and evaluation including uncovering optimal care paths and optimal responders to treatment; cost-effective treatments and care pathways; adherence to and outcomes of clinical guidelines; and adverse events, medical errors and fraud. Calls for this type of data have been made recently in international conferences including a Toronto workshop in September 2014 on data development for innovation in dementia treatment and care organised by the OECD, Ontario Brain Institute, and Institute for Health Metrics and Evaluation (OECD, 2015e); an international meeting convened by the Royal Society of Physicians in the UK in October 2014 on better science for better health; and a European meeting on delivering better outcomes for sustainable healthcare systems in November 2014.

New opportunities are also emerging for public-private partnerships and collaborations toward the development and use of health data to support a wide range of innovations such as better, faster and less expensive approaches to R&D and clinical trials; comparative effectiveness of existing and new therapies for more comprehensive health technology assessment; deeper and comprehensive assessment of healthcare expenditures for planning and budgeting; clinical decision-support tools that enable patients to receive the right care, at the right time, the first time; and patient healthcare management tools to enable better experiences and outcomes.

Box 7.8. **Drug development for Alzheimer's and other dementias**

The development of new therapies for dementia relies on biomedical research, the translation of research into therapies and the development of new biomedical technologies. Translational neuroscience focuses on the linkages between basic neuroscience and the development of new diagnostic and therapeutic products that will improve the lives of patients or prevent the occurrence of brain disorders. Emerging biomedical technologies, such as synthetic biology, nanotechnology, pharmacogenomics and stem cell technology, offer the tools and techniques to shift the conventional drug development model to one that allows for more disruptive innovation. However, in order to realise opportunities in health, most of the emerging biomedical technologies still need to transition from the academic research environment to an industrial scale and to clinical use.

Over the past decades, the main driver of most drug discovery and development were large pharmaceutical companies. Partly due to more resource-intensive research approaches, shrinking returns on investment and persistent knowledge gaps in the biomedical underpinnings of diseases, traditional drug discovery and development processes have become unsustainable. To accelerate the development of new therapies for Alzheimer's disease and other dementias and to manage high financial risks, stakeholders are increasingly following cross-sectoral, collaborative strategies. Public-private partnerships, in particular, have the potential to facilitate a reform of traditional research and health innovation models towards more efficient innovation strategies (Galea and McKee, 2014). In providing a neutral environment, public-private partnerships can help to accelerate the development of effective therapies for Alzheimer's disease by supporting the mission of each stakeholder and bridging the innovation gap in neuroscience – ultimately, by reducing the attrition rate during clinical research and by reducing financial risks.

As the many failures in clinical research have shown, drug development for Alzheimer's disease remains a high-risk endeavour. Governments, in close collaboration with other stakeholders, are exploring new funding and risk-sharing mechanisms to support resource-intensive research (Feldman et al., 2014; Scott et al., 2014). To translate therapeutic efficacy into effective public health systems, cost-benefit measures are gaining importance for the regulatory assessment and marketing approval of therapies. However, when a new therapy is introduced into the market, there often remains uncertainty about its performance within the public health system – outside the strict controls of a clinical trial. Therefore, besides the assessment of efficacy and safety parameters, the successful development of innovative medicines should include the early consideration of future treatment costs (Foster et al., 2014). Striking an appropriate balance in the respective financial risks of producers (researchers, manufacturers) and purchasers (insurers, patients) could help to support innovative research, enhance access to medicines, and foster rational use and cost containment. Policies to support price transparency and the measurement of therapeutic effectiveness can help to assure value for money and the responsible use of limited resources.

At present, the probability of success for the development of central nervous system (CNS) drugs, Alzheimer's disease and other dementias in particular, is lower than in many other disease areas (Tufts Center for the Study of Drug Development, 2014). In general, CNS drugs are more challenging to develop than other medicines because the nervous system disorders that the drugs aim to treat are often chronic and complex, and outcomes of clinical trials are difficult to measure. In addition, the brain is inaccessible to study and to treat – making it difficult to develop accurate models and to reach therapeutic targets. The low success rate of clinical trials coupled with high investment costs has been a significant factor in the withdrawal of some pharmaceutical companies and funding organisations from neuroscience research. However, recent government-led initiatives, for example the commitments made by G7 governments following the 2013 Dementia Summit in London and the development of national dementia plans, have contributed to a range of actions by public and private stakeholders to increase investment.

A recent OECD workshop (OECD, 2015d) provided an international forum for all stakeholders to drive forward a change in the global paradigm in biomedical research and health innovation for Alzheimer's disease and other dementias. It demonstrated that therapeutic research needs to shift from established

Box 7.8. **Drug development for Alzheimer's and other dementias** *(cont.)*

dementia to preclinical stages, which will require adequate diagnostic tools, new trial designs and more flexible regulatory processes. It will be important to incorporate the learnings from clinical trials (positive and negative results) into regulatory science and agency qualification processes as rapidly as possible in order to enable policy development and regulatory decision making based on the best available science.

Sources: Galea, G. and M. McKee (2014), "Public-private partnerships with large corporations", *http://dx.doi.org/10.1016/j.healthpol.2014.02.003*; Feldman, H.H. et al. (2014), "Alzheimer's disease research and development", *http://dx.doi.org/10.1111/nyas.12424*; Scott, T.J. et al. (2014), "Economic analysis of opportunities to accelerate Alzheimer's disease research and development", *http://dx.doi.org/10.1111/nyas.12417*; Foster, N.L. et al. (2014), "Justifying reimbursement for Alzheimer's diagnostics and treatments", *http://dx.doi.org/10.1016/j.jalz.2014.05.003*; Tufts Center for the Study of Drug Development (2014), "CNS drugs take longer to develop and have lower success rates than other drugs"; OECD (2015c), *Addressing Dementia – the OECD Response*; OECD (2015d), "Enhancing Translational Research and Clinical Development for Alzheimer's Disease and other Dementias" *http://dx.doi.org/10.1787/5js1t57jts44-en*.

With the strength of the benefits of leveraging electronic patient data in this key sector, comes corresponding challenges that must be addressed before the benefits can be realised. Challenges include the need for data standards that increase the potential for analysable data to be shared within and between countries; investments in data infrastructure including electronic clinical record systems and data linkage services and improving data storage, processing and analytics; and investments in data quality including usability assessments, incentives, certification and auditing. Problems of the quality of data and missing data also need to be tackled; recent initiatives of major pharmaceutical firms to make the (negative) results of clinical trials available for research are important for further research, but also help inform policy makers and patients.

The challenges that tend to place a brake on leveraging health data for innovation, however, relate to how health data are governed. Governance frameworks are needed that maximise the benefits to society from health data use while at the same time minimising risks to individuals from health data misuse. Harms from data misuse include identity theft, discrimination in health insurance or employment, stigma and emotional harm, and loss of trust in healthcare providers and governments.

Health data that measure pathways and outcomes are usually both personal and sensitive, and include information from the confidential relationship between patients and their healthcare providers. While obtaining patient consent to use their data is feasible in a purpose-specific study, such as inviting patients to a clinical trial, it does not work when large-scale population-level health data are reused. This is because the consented data will be biased towards survivors and are often unaffordable to collect, and its collection can become intolerable to the public if a high volume of projects is anticipated. New approaches are needed to protect the privacy of patients while at the same time improving approaches to achieving patient consent.

Evidence of the role of health data governance in innovation

While all countries are investing in data infrastructure, there are significant cross-country differences in data availability and use, with some countries standing out with innovative practices enabling privacy-protective data use, and others falling behind with insufficient data and restrictions that limit access to and use of data, even by government itself (OECD, 2013j).

Among 22 countries surveyed in 2013, only one-half are regularly linking data across national data sets to understand pathways and outcomes. Leading countries include Canada, Denmark, Finland, Israel, Korea, New Zealand, Singapore, Sweden and the United Kingdom. Similarly, half of 25 countries surveyed in 2012 planned to analyse data collected within electronic health record systems for research purposes, and 10 were planning to allow the data to facilitate or contribute to clinical trials. Countries that had started using data extracted from electronic clinical records for research were Belgium, Finland, France, Iceland, Indonesia, Japan, Korea, Poland, Portugal, Singapore, Slovenia, Sweden and the United Kingdom. Four countries reported that data from electronic health record systems were already facilitating or contributing to clinical trials (France, Indonesia, Sweden and the United Kingdom). At the same time, Austria, Germany, Netherlands and Switzerland considered it very unlikely that the legislative and technical barriers preventing data use would be overcome within the next five years.

In 2013, countries were asked about a set of data accessibility factors that are directly linked to legislation for the protection of health information privacy and its interpretation in practice. These included whether or not identifiable personal health data are ever shared among data processors or government entities and whether personal health data, after they have been de-identified, can be approved for access by applicants from the academic, government and commercial sectors within the country and outside of the country. Data accessibility was highest in the United Kingdom, New Zealand, Sweden and the United States and was lowest in Turkey, Italy, Japan and Israel.

Good policy practices to strengthen health data governance

Effective collaboration among health ministries, justice ministries and data privacy regulators is essential if governments are to evolve toward a situation where societal benefits from data use are maximised and risks to society from data use are minimised. At the same time, government needs clear and open channels to engage with stakeholders in the development and use of data, so that data governance frameworks and practices reflect societal values and priorities. An overview is shown in Figure 7.8.

The OECD has brought together experts in privacy law and regulation, health statistics, health policy, research, IT, and civil society to review current data governance practices in OECD countries and to identify practices that support countries in strengthening health data governance. Leading countries provided good examples of practices that enable privacy-protective data use. Key elements of data governance frameworks include public consultation and information about the collection and use of personal health data and the safeguards applied; legislation that enables data sharing and use, subject to suitable safeguards; data processors that are held to high standards for data governance; data-use approval processes that are fair and transparent; and use of best practices in data de-identification and secure data access. The publication of these recommendations is expected in 2016.

Strengthening the role of information technology for innovation

The rapid pace of ageing creates an economic and social imperative for innovation – social, organisational and technological – that must be underpinned by policy reform and profound cultural change. Innovation has the potential to mitigate the impact of demographic change on society as well as create new sources of growth (OECD, 2013k). Solutions to the social and economic challenges posed by ageing populations require

Figure 7.8. **Data governance and health innovation**

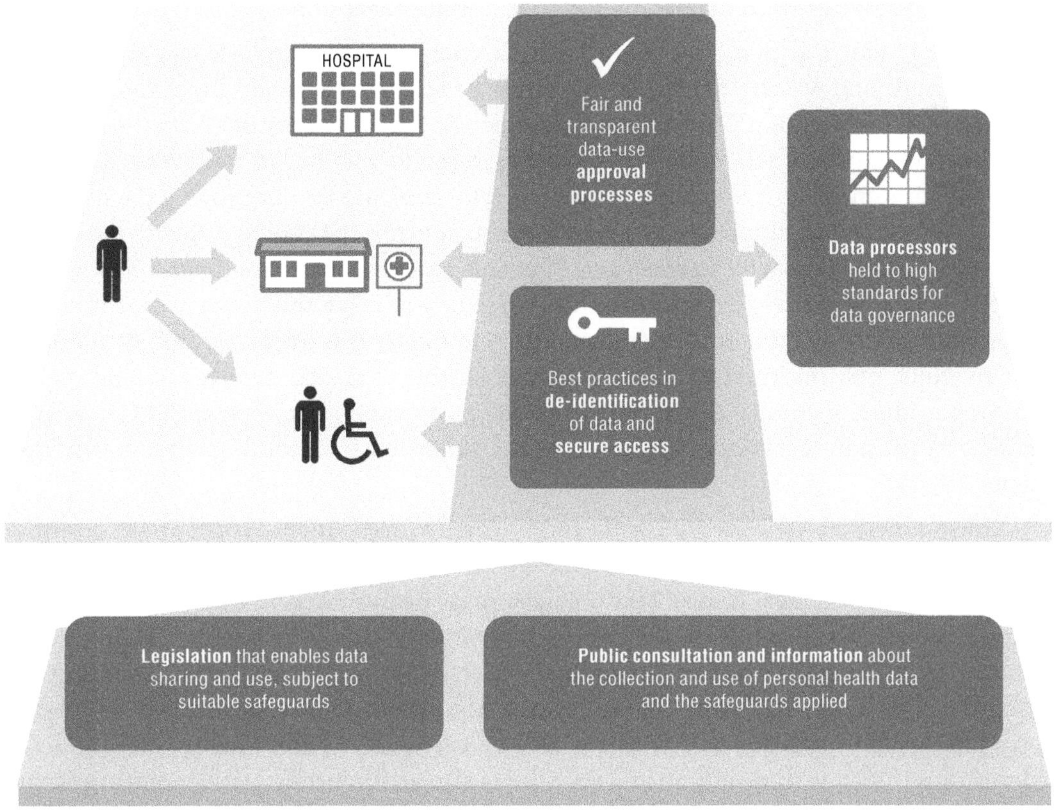

Source: OECD (2015e).

policies and measures to support and accelerate innovation, the development of new ideas, products, processes, organisational forms and services for healthy and active ageing. Innovation in health and social services, whether supplied by the public, private or third sector, will be important to meeting the challenges of an ageing society. At the same time it is essential to evolve current conceptions of middle and old age; older people should be considered a rich resource who can contribute to society whether as workers, consumers, volunteers or caregivers.

Moreover, with differential impacts on different sectors of the economy, population ageing leads to a changing demand structure in the economy. Information on older peoples' needs and consumption patterns can maximise opportunities for service sector transformation and innovation. Innovation has to be led by their desires, needs and abilities. Older populations are often excluded from market and social research. The rapidly growing elderly/silver market segment remains, therefore, one of the most unexplored and misunderstood of all markets. Government and industry support is needed to collect this information to maximise opportunities for service sector transformation and innovation to meet the needs of the older population. In particular, through this information, services providers can also seize upon unique opportunities, working alongside policy makers, to design innovative products and services that can aid the subset of older consumers most at risk of poverty. It is also important to recognise that older adults are the key informants, designers and customers of the technologies and services that are designed and developed on their behalf. Innovation has to be led by their desires, needs and abilities.

Rapid prototyping, responsiveness, and involvement of the user in the development of new products and services need to become the rule, not the exception.

ICTs are today a critical part of what enables new services innovation, particularly to address health and welfare needs. All countries need, however, to do more to unlock the potential of ICTs for ageing. Ensuring that older people are able to benefit from the advances in ICTs is an important policy objective. The Internet and web-based welfare technologies, especially for remote care, are going to be critical to servicing and improving the quality of life of older people, and their ability to continue to contribute to society. Ready availability of information; the capacity to access and receive services, including health and welfare services; new approaches to learning without needing to leave the home (e-learning); and the potential to connect with the community and family members are just some of the ways in which new information technology will be able to enrich the lives of older people and support their ongoing participation in society. For example, in several OECD countries, almost 70% of individuals now use the Internet to search for health-related information (Figure 7.9).

These potential benefits will, however, not be realised unless appropriate attention is paid to technological barriers such as high-speed broadband penetration as well as usability, ethical and legal issues. Technologies must be available and user-friendly, at the right place and time; poor user interfaces or connectivity can doom adoption. Older people will also need the skills, trust and confidence to use the new technologies. Ensuring that their privacy is respected and that they perceive the systems as secure, as well as creating opportunities for education and training, will be essential. Content and presentation for older people in online information and services is also an important issue requiring further attention if older people are going to embrace and benefit from these services.

Reimbursement mechanisms and models designed to encourage new welfare technologies and services that maintain functional independence and avoid hospitalisation are among the most important policy changes that could be made to promote a silver market. For this to happen, public and private payers need to better understand the returns on investments and address misalignment of incentives between buyers and beneficiaries. Many initiatives end up only as "successful technical pilots" and never achieve more widespread implementation and economies of scale, largely because of financial constraints. Given the upfront costs entailed, public or private payers are unlikely to increase funding or coverage for welfare technologies and home care services unless returns on investment are clear. In particular, public payers will want to know not only what may be the cost savings but also the benefits to the individual and the broader social opportunities generated. Public and private sectors must also co-operate to ensure that relevant new technologies and services respond to needs and priorities.

It is necessary to develop collaborative ways to promote innovation and shape the "silver/ageing society" of the future, using an integrated and whole-of-government approach to stimulate action on joint solutions. Success will require diverse and targeted innovative approaches and new metrics to benchmark progress and monitor impact. There are many obstacles for innovators today, whether they are business entrepreneurs or social entrepreneurs: access to capital, the characteristics of the silver market, the regulatory environment and lack of standards, the approval process including reimbursement practices, the sustainability of current business models. Alignment needs to happen among governance, finance, policy, business, the third sector and technology development to achieve appropriate innovative solutions. This requires collaboration among multiple actors and a different approach – getting out of the silos, encouraging broad-based co-operation at community level, public-private partnerships and using a whole-of-government approach. A good example is the American Medicare Readmissions Reduction Program, which stimulated innovations in predictive analysis as well as tele-health and home monitoring. Additional new metrics are needed, however, to monitor the impact of policy and benchmark progress.

Figure 7.9. **Individuals who searched for health-related information online, 2008 and 2013**

(as a percentage of individuals who used the Internet in the last three months)

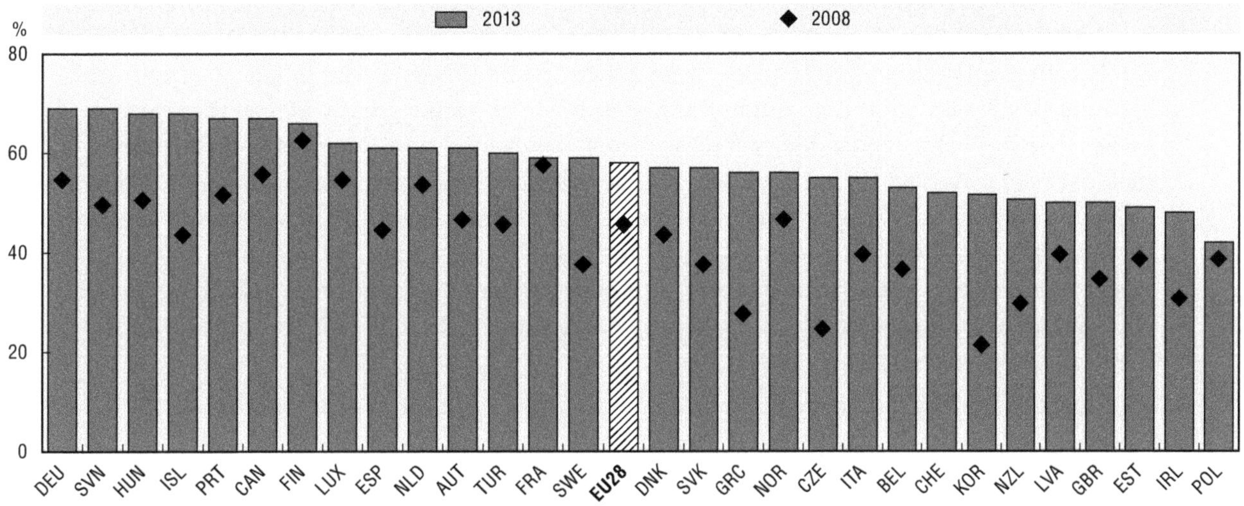

Source: OECD (2014q), *Measuring the Digital Economy, http://dx.doi.org/10.1787/9789264221796-en.*

Steering innovation by paying for value in the health sector

In many sectors of the economy, innovation is steered by consumers' willingness-to-pay for new products and new technologies. The health sector is very special: consumers (patients) generally trust doctors for the choice of therapeutic options, and most often, they do not pay the price of healthcare services or only a small share of it. Hence, what drives innovation is how much third-party payers – governments or individual health insurers – are ready to pay for the technology.

While payment schemes and prices paid for healthcare services and goods differ widely across countries, recent analyses suggest that they are not enough oriented towards the "value" they generate for patients (Porter, 2010). Changing payment methods to reflect their value to consumers requires first to measure patient outcomes. Such practices, however, are not yet commonly used in health systems. Another option to increase "value-based" payments is the use of economic evaluation at the time of coverage and pricing decisions for new technologies. Indeed, public payers have increasingly used health technology assessment (HTA) to make decisions on the reimbursement of new technologies, especially medicines.

The concept of value-based pricing (VBP) for pharmaceuticals has attracted a lot of attention. In the context of a publicly funded system, Claxton defined a value-based price as "(the price) that ensures that the expected health benefits [of a new technology] exceed the health predicted to be displaced elsewhere in the NHS, due to their additional cost. [...]

VBP is expected to ensure in the short run that technologies are accepted for use in health systems only if they are cost-effective, while providing in the longer run clear signals and incentives for manufacturers to invest in the development of technologies which are more likely to be cost-effective" (Claxton, 2007).

In a report published in 2013, the OECD examined how 13 countries refer to the "value of pharmaceuticals" when making decisions on reimbursement or price. Among these countries, some use formal economic assessment to make funding decisions; others evaluate the added therapeutic benefit of the new technology against existing therapeutic alternative and use this assessment to negotiate price with the manufacturer (Paris and Belloni, 2013). This study showed that payers are in principle ready to pay a price premium or additional costs for an innovative product. However, establishing a clear link between the level of innovativeness and the price premium seems impossible. The price countries are ready to pay for an additional quality-adjusted life year varies greatly across therapeutic areas. It is typically higher for severe and rare conditions and where there are unmet medical needs. Actually, countries do not base their decisions to cover a new technology only on its cost-effectiveness; they take into account other criteria, such as equity. Payers, however, are not always very explicit about the set of criteria they consider and how they weigh these criteria against one another. This is not an easy task, indeed, but making more explicit the rationales for choices could help provide the right signals to innovators about the innovations desired.

Some of the key messages of the OECD's work on health innovation are below.

Main policy messages on health innovation

- Health innovation is a complex and expensive undertaking that involves multiple actors from the public and private sectors. Government plays an important role in funding basic and clinical research; regulating safety and efficacy; determining availability, equity and access; and often in pricing and reimbursement of health services.
- Technology and innovation offer significant opportunities to improve the efficiency of global health systems and the quality of human health, e.g. through genomics and better use of ICTs.
- Better models of health innovation are needed to enable the translation of scientific discovery into effective therapies. Progress in drug development increasingly requires well-designed public-private partnerships, involving governments, regulators, scientific advisory groups, business and patients.
- The use and sharing of (big) data offer new opportunities for health innovation, but require sound frameworks for data governance that enable privacy-protected use of data.
- ICTs offer a strong potential to support healthy ageing. Realising these benefits requires attention to technological barriers, as well as usability and ethical and legal issues, including respect for the privacy of older people. Public-private partnerships are needed to ensure that the incentives between buyers and beneficiaries are well aligned, and to ensure that new technologies and services respond to the greatest needs.
- Payment systems and prices paid for healthcare products and services need to be more oriented towards the value they generate for patients. This requires better measurement of patient outcomes and evaluations.

7.4. Innovation and the green growth agenda

Green innovation today

Innovation is essential to achieve greener economic growth. However, inducing innovation that reduces the impacts of economic activity on the environment is a significant challenge to policy makers. There are two related market failures: 1) negative externalities that arise from the adverse impacts of economic activities on the natural environment; and 2) positive externalities associated with the generation and diffusion of knowledge that gives rise to innovation.

In both cases some agents (households, firms, public authorities) bear costs (enjoy benefits) without receiving (paying) compensation. In the absence of public policies designed to overcome the environmental and knowledge market failures, households and firms pollute too much, exploit natural resources too intensively and innovate too little compared with the social optimum. If the policy response is too slow, innovation efforts will have to be redirected more and more towards adaptation to the changing ecological conditions, almost certainly a more costly strategy.

The focus of this section is on those policy measures that seek to directly affect the **direction** of innovation by addressing the environmental market failure. The tools necessary to address the knowledge market failure – and thus increase the **rate** of innovation in the economy more generally – are discussed in other parts of this report. These tools include investment in basic research, IPRs, tax incentives and direct support for business R&D, public-private co-operation and networks, and policies to foster the growth of young firms.

However, it must be emphasised that the design of more general innovation policies may have indirect but important implications for "green" innovation. For example, R&D tax incentives that favour incumbents are likely to have particularly pernicious effects for new and fast-growing technology fields such as many of those related to green innovation (see Chapter 6). More directly, there have been discussions about differentiating IP regimes in order to meet specific policy objectives, including the realisation of environmental objectives. The benefits of doing so through measures such as expedited patent examinations for environmental technologies remain an open question. (See Maskus [2010] for a discussion.)

An overarching question is whether there is a trade-off between measures that seek to induce environmentally friendly innovation and the effects on the rate of innovation more generally. For example, efforts to induce innovation related to the environment may "crowd out" innovation in other domains, with possible negative implications for the social returns on innovation and downstream economic outcomes, such as productivity.

In general, little evidence has been found for such negative outcomes, although this can vary by sector.[3] Moreover, recent OECD evidence linking MFP growth to changes in environmental policy stringency (EPS)[4] suggests no harm to aggregate productivity growth from tighter environmental policies (Albrizio et al., 2014). Macro, industry and firm-level results show that the effects of incremental tightening of environmental policies on productivity are likely to be incurred through small and short-term adjustments.

At the firm level, impacts depend on technological advancement – more innovative firms tend to see a boost in productivity growth while less advanced ones are likely to see a temporary fall. Highly innovative firms may be best suited to profit rapidly from changing conditions – seizing new market opportunities, rapidly deploying new technologies or reaping previously overseen efficiency gains (Lanoie et al., 2011). Less technologically advanced firms may need higher investments to comply with new regulations, implying a temporary fall in productivity growth. Part of the adjustment, particularly for less technologically advanced firms, may take the form of firm exit.[5] Irrespective of the channel through which environmental policy affects productivity, the role of innovation is central.

Figure 7.10 shows that for a wide variety of domains, "green" innovation (as reflected in patented inventions) has been progressing more rapidly than innovation more generally. The exceptions are technologies related to waste management and water pollution treatment, which could be said to be mature technology domains and, not coincidentally, ones for which initial policy incentives date from many years or even decades. Analogously, those fields with relatively high rates of innovation in recent years are those that have become a greater focus of policy activity.

Figure 7.10. **Inventive activity in technologies included in the "green" patent index**

(World total, change on 1990, high-value inventions PF2)

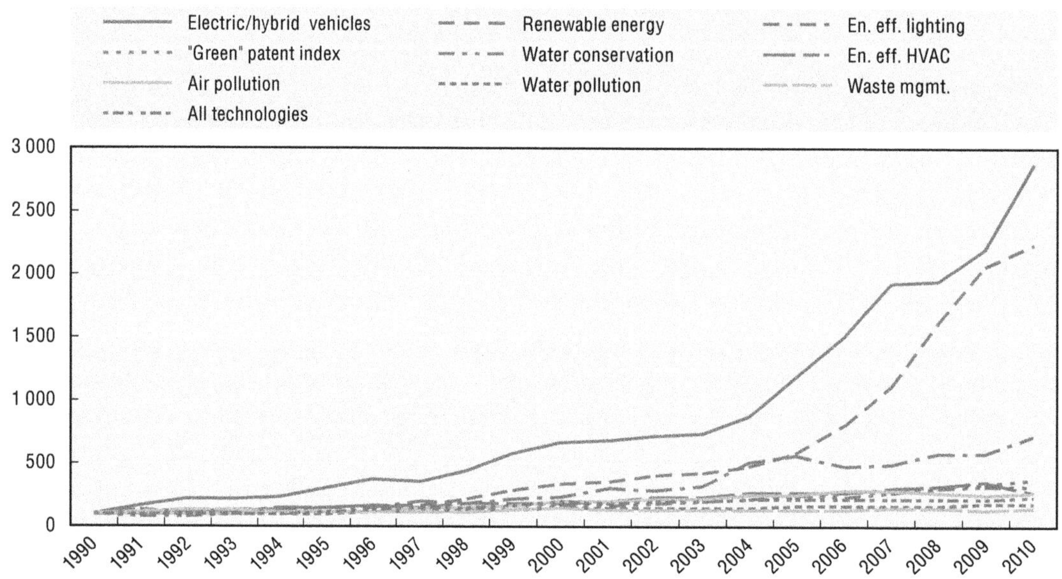

Notes: Only patents in which protection has been sought in at least two IP offices (PF2) are included. This has been found to be a good proxy measure for high-value patents. HVAC = heating, ventilation and air conditioning.

Source: Haščič and Migotto (2014), "Measuring environmental innovation using patent data", *http://dx.doi.org/10.1787/5js009kf48xw-en*.

However, it must be emphasised that developments in technological opportunities and the relative ambition of policy objectives can be mutually reinforcing. For example, with advances in fields such as biotechnology and nanotechnology, even some of the more "mature" fields may see resurgence in innovation, lowering the costs of meeting more stringent policy objectives, and perhaps launching a further wave of innovation.[6]

Policy lessons learned

Policy design is clearly important. Some environmental policies incorporate features that can hamper competition in a similar way to general product market regulation (Nicoletti and Scarpetta, 2003). Examples of environmental policies that can provide advantages to incumbent firms include: 1) high administrative burdens to entry; 2) vintage-differentiated regulations, where new firms are subject to stricter environmental limits; 3) subsidies or other benefits for a historical environmental record that new firms may not be able to demonstrate even if "cleaner" than older firms; 4) tax breaks for investments in improving environmental performance from which new firms that are not yet profitable will not be able to benefit; and 5) the grandfathering of licences and permits.[7]

While high barriers to entry and competition may be unintended by-products of some environmental policies, appropriate design can avoid these adverse effects. Indeed, there is some evidence to support the view that policy stringency generally does not go hand in hand with policy failures of this kind.[8] This is not surprising, since well-designed environmental policies allow policy makers to meet more stringent environmental objectives at lower cost. Conversely, badly designed policy constrains ambitions, perhaps below the level that would be optimal from a social perspective.

What is a "well-designed" policy in terms of incentives for innovation? The essential point is that potential inventors need to be given incentives to identify (search) for least-cost opportunities to realise given environmental objectives. And the policy should give the heterogeneous group of potential adopters the freedom to invest in those technologies that are least-cost given specific household and firm characteristics. It is for this reason that market-based incentives such as environmentally related taxes and charges and tradable permit systems are generally advocated since they affect relative prices, without identifying specific abatement options. More prescriptive forms of policy such as technology-based standards will induce agents to innovate for the regulation and not for the underlying objective (OECD, 2010c).[9]

The effect of relative prices on the nature and direction of innovation in the energy sector has been assessed in a recent OECD analysis (Lanzi, Haščič and N. Johnstone , 2012). Drawing upon data from 18 countries over three decades, the results shows that the price of oil has a positive effect on innovation in renewable energy generation. However, at a relatively low oil price, the effect of rising oil prices is first to encourage innovation targeted at improving the efficiency of electricity generation technologies that rely upon fossil fuels. It is only at an oil price in the region experienced during the most recent oil shock in the late 2000s that the pace of innovation in the alternative technology (renewables) comes to dominate innovation targeted at efficiency gains in the conventional technology (fossil fuels).

Of course, bringing about such changes in relative prices requires both the implementation of measures to internalise the externality and the removal of environmentally perverse subsidies. Recent work at the OECD and the International Energy Agency (IEA) has drawn attention to just how significant such support measures are in the case of fossil fuels (OECD, 2013l). Getting the prices right therefore requires action on both sides, and in order to ensure that the policy framework has the desired effect on innovation, it is arguably more important to ensure that perverse subsidies are first removed, or else the incentives induced by the environmental policy measure will be "swimming against the tide".

In addition to setting the right price in terms of EPS, in order to spur innovation it is important that market players have some certainty with respect to future developments in that price. And since prices related to the environment are strongly influenced by policy design, the predictability and credibility of the policy framework is essential for inducing innovation. Uncertain policy signals give investors incentives to postpone investments, including the risky, non-transferable and capital-intensive investments associated with technological invention and adoption.

For investors, and particularly those who are more risk-averse, there is an advantage to waiting until the policy dust settles.[10] By adding to the risk that investors face in the market, policy uncertainty can, therefore, serve as a brake on innovation. This implies that governments have an interest to behave in a predictable manner if they wish to induce innovations that achieve environmental objectives at lower cost (Kalamova, Haščič and Johnstone, 2013).

For at least some environmentally pressing concerns, meeting stated environmental objectives is likely to require radical breakthrough innovations, for which marginal adjustments in relative prices may be necessary, but insufficient. The case of climate change stands out as an example. Due to the long-lived nature of greenhouse gases (GHGs) as stock pollutants, concentrations are likely to continue to rise for many decades, even with the realisation of significant emissions reduction in the short to medium term. This can be compounded by the presence of path dependency that favours incumbent technologies and firms, slowing the transition to low-emission trajectories.[11]

Therefore, if the international community is to meet the declared objectives under the United Nations Framework Convention on Climate Change (UNFCCC) – global warming limited to 2°C above pre-industrial levels – more radical innovations need to be encouraged. In order to induce radical innovation, the introduction of "prices" through the use of market-based instruments, while the necessary backbone of any policy framework, may not be enough to spur the innovation required. More directed or targeted policies may be an important complement. For example, work undertaken at the OECD (Haščič and Johnstone, 2012) analysed the relative importance of prices and standards on environmental innovation in motor vehicles. While there is some overlap in the types of technologies associated with hybrid and electric vehicles, it is the development of batteries with the necessary quality attributes (i.e. capacity, reliability, weight) that presents the biggest challenge for electric vehicles, requiring more radical innovation.

More specifically, the analysis found that while changes in relative fuel prices induced innovation in hybrid propulsion, performance standards were necessary to induce development of technologies related to purely electric vehicles. More specifically, in order to induce a 1% increase in electric vehicle innovations, the stringency of California's Zero Emission Vehicle standard would have to be increased by 2%, while to induce a 1% increase in hybrid vehicle innovations, fuel prices would have to increase by 5% (see also Figure 7.11).

Figure 7.11. **The effect of prices, standards and R&D on alternative-fuel vehicle innovation**

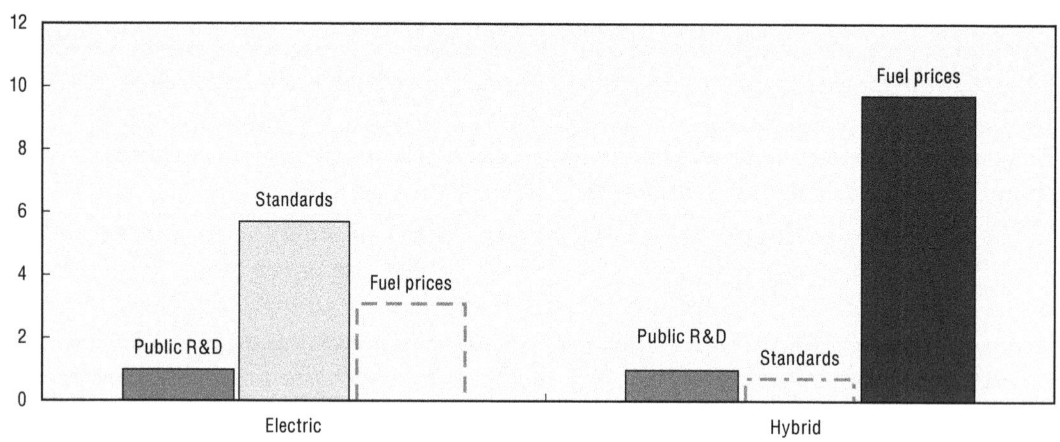

Note: The histogram shows empirical elasticities, evaluated at sample means and normalised in terms of the effect of "public R&D spending" (R&D = 1.0). Bars shown without fill represent estimates that are not statistically significant at the 5% level.

Source: Haščič and Johnstone (2012), "Innovation in electric and hybrid vehicle technologies: The role of prices, standards and R&D", *http://dx.doi.org/10.1787/9789264115620-5-en.*

This case also illustrates the importance of the complementarity between technological innovation and organisational innovation. Whereas electric vehicles have captured less than 1% of the global vehicle market, they make up more than 10% of the fast-growing car-sharing market. More generally, linking the storage potential of electric vehicles with efforts to develop a "greener" (but less predictable) fuel mix in the electricity sector will require organisational changes in households and firms, as well as a regulatory regime that can accommodate and encourage such changes.

Organisational factors are also important complements to technological adoption at the level of the firm. There is a link between firm-level organisational structures and practices on the one hand, and technology choices on the other hand. Environmental policy design can play an important mediating role, with market-based instruments helping to ensure that such choices are at the core of firm-level strategies. For instance, there is evidence that the use of flexible environmental policy instruments such as taxes and tradable permits correlates with: 1) firms allocating managerial responsibility for environmental matters to more senior managers with broader corporate responsibility; and 2) investment in technologies to address environmental concerns that are more integrated with the firm's more general production strategy.[12]

In addition to the effects of individual instruments, the manner in which policies interact is fundamentally important. In some cases this involves the use of measures that support green innovation at different stages of technological development and maturity. The case of alternative vehicle propulsion discussed above is an example, but is not exceptional. Different measures may be required to induce incremental and radical innovations. This shows that in green innovation – as in other areas of innovation – the functioning of the overall system of innovation matters.

The issue of policy mixes extends beyond the environmental policy toolbox. Those areas where the environmental benefits from innovation are greatest are often those areas in which there are a number of other market conditions that can have adverse impacts at all stages of the innovation cycle. Many environment-intensive sectors (e.g. transport, energy) are characterised by a number of market failures, above and beyond the environmental and knowledge market failures. This would include at least the following:

- natural monopoly conditions that can result in imperfectly competitive market conditions in the absence of effective market regulation
- consumption and network externalities that can result in slow take-up of "innovative" products whose characteristics have not been fully demonstrated in the market
- capital market failures associated with information gaps and uncertainty concerning the evaluation of high-risk/high-return technologies, resulting in suboptimal provision of finance for research.

In such cases a suite of policies will be required, ideally targeting each market failure as directly as possible. For example, public procurement can be an effective tool to address consumption/adoption externalities, and has often been used to "kick-start" a market (albeit with varying success).[13] Private-public partnerships in infrastructure development can be important for the investment in infrastructure that complements less environmentally damaging economic activities.

Addressing failures associated with financial markets may be the most important due to the preponderance of cases in the environmental domain that involve the co-existence of novel technologies with uncertain performance characteristics and young firms with little track record or collateral. Moreover, relative to other fields in the environmental domain, market and technology risk is compounded by policy risk, with important implications for access to finance. Considerable work has been undertaken on the central role of policy settings on the provision of finance for "clean" technologies in OECD economies.[14] However, more work is required to assess the effectiveness of targeted financial policy interventions (e.g. public investment in ''green" venture capital funds) relative to more general policies associated with the functioning of capital markets.[15]

The use of multiple instruments raises the question of co-ordination. Biotechnology is a case in point with a number of policy developments emerging in this area (see OECD, 2013m), with overlapping strategies ranging from very specific domains (e.g. bio-refineries), to broader sectors or applications (e.g. industrial biotechnology, synthetic biology), all the way up to general strategies for the "bio-economy" as a whole. In order to ensure that such measures are mutually reinforcing and not redundant, governments need to ensure co-ordination (see OECD, 2013m; Schieb and Philp, 2014).

More generally, recent work at the OECD on system innovation[16] underscores the importance of policy co-ordination. In recognition of the complementary nature of many of the investments (technological and other) that need to be made, a number of OECD countries are implementing system innovations in areas such as "smart cities". For example, the benefits associated with innovation in one area (i.e. transport), will be only partially realised in the absence of innovation in other complementary areas (i.e. housing). However, case studies undertaken in these areas have shown that policy success in fostering transitions is hampered by a reliance on single market failure rationales, short-term political processes, and fragmented governance structures and processes (OECD, forthcoming).

Targeted support for green innovation

The use of more discretionary policy interventions raises the question of selection. How to identify those fields of knowledge and technologies that are likely to bring about significant environmental benefits at least cost? This is, of course, a hazardous exercise with dangers of "lock-in" (technological and institutional). The 2010 OECD Innovation Strategy drew attention to the complexity of this task in the environmental sphere by illustrating the wide breadth of sources of knowledge upon which environmental innovations are dependent. The original impetus for the development of technologies that prove to have positive environmental outcomes can be far removed from the environmental domain. For example, carbon capture technologies originally arose to serve high-value chemical uses, before their potential applications in power plant pollution control became evident (see IEA, 2015, and also Box 7.9).

The breadth of domains that can generate environmentally friendly innovations strengthens the case for greater public investment in basic research, a point made in Chapter 5 of this report, particularly with respect to transformative technologies. In the face of fundamental uncertainty concerning which specific technologies are likely to yield significant environmental benefits in the future, an essential component of a risk-averse, portfolio approach is to support basic research even if the immediate commercial applications are less evident.

Moreover, the breadth of domains that can yield environmentally friendly innovation also underscores the importance of supporting technologies that can have wide applications in the economy. At a narrow level, work at the OECD has shown that innovation in "local general-purpose technologies" such as energy storage and grid quality may yield greater benefits than placing bets on specific renewable energy-generating technologies that are

Box 7.9. **Innovation in the energy sector**

The transformation of the energy sector is one of the key challenges for green innovation. The IEA's *Energy Technology Perspectives (ETP)* series highlights the urgent need to deploy a range of low-carbon technologies in order to achieve the goal of limiting long-term temperature increases to 2°C. The *ETP* 2°C scenario (2DS) outlines a pathway to decarbonising the energy sector and delivering secure, reliable and affordable energy needed to support wider economic and social development.

The need for new energy technology development is undisputed, and all major economies have recognised the importance of transforming energy systems. However, only 4% of public research funds in IEA countries are devoted to energy. IEA analysis shows that the gap between estimated needs and actual investments is too wide, and that government support for energy research, development and demonstration (RD&D) should at least triple (IEA, 2013) if a transition to a low-carbon energy system is to be achieved.

To ensure that new technologies are widely adopted, governments must supplement public funding schemes for RD&D (e.g. grants, loans and tax credits) with non-RD&D support of business innovation (e.g. support for venture capital, public-private partnerships and nascent entrepreneurial activities) and targeted policies that foster demand and markets for clean energy (e.g. pricing mechanisms, public procurement, labelling schemes, minimum energy performance standards and mandatory targets).

RD&D and innovation efforts should focus on a portfolio of technologies, taking into account existing domestic resources, skills and knowledge. The development and deployment of this "portfolio" should be supported by policy frameworks and market mechanisms that encourage the development and deployment of the targeted technologies. Such a process should also help in identifying priority partnerships for international co-operation, and improve efficiency of domestic efforts. *Energy Technology Perspectives 2015* focuses on innovation and explores the potential for energy technology innovation in achieving ambitious climate change targets (IEA, 2015). The objective is to help decision makers identify tools to spur innovation and evaluate their economic efficiency.

Technological change and development will significantly enhance the portfolio of options available to bring about a low-carbon economy, and lower the cost of achieving green growth and global climate change goals. Governments can help bring about this transformation by creating an attractive environment for energy RD&D. Well-designed energy technology policies targeted on both the supply and demand sides are a fundamental ingredient in a strategy to accelerate energy innovation. While the specific combination of policy measures will depend on country circumstances, in all cases it will be important to set the appropriate framework to allow breakthroughs to happen.

The IEA's Accelerating Energy Innovation project proposed six recommendations for good practice in the development of an energy RD&D policy framework (see figure below). Establishing a vision for what the government aims to achieve in the energy sector and through RD&D is a crucial first step. Achieving a vision requires a comprehensive strategy that integrates a portfolio of policy tools adapted to both national systems and individual technologies. Once a strategy is in place, governments need to demonstrate their own commitment to it through adequate and stable government RD&D funding and policy support.

The management (governance) of funding is as critical as the level of funding. Improving the structure and co-ordination of the various institutions that have a role in funding is important in the development of low-carbon technologies. Collaboration and networking are vital to using effectively the limited RD&D resources. This includes collaboration between private and public bodies, at the national and international level. Effective and well-designed monitoring and evaluation of the performance of energy RD&D policies and programmes are key to ensuring effective and efficient implementation of a country's energy RD&D strategy. Finally, strategic international collaboration should be implemented to enable governments to conduct more RD&D at a lower cost and with less duplication. This collaboration should involve all major economies to help support wider adoption of low-carbon technologies globally.

Box 7.9. **Innovation in the energy sector** *(cont.)*

An energy innovation policy framework based on good practices

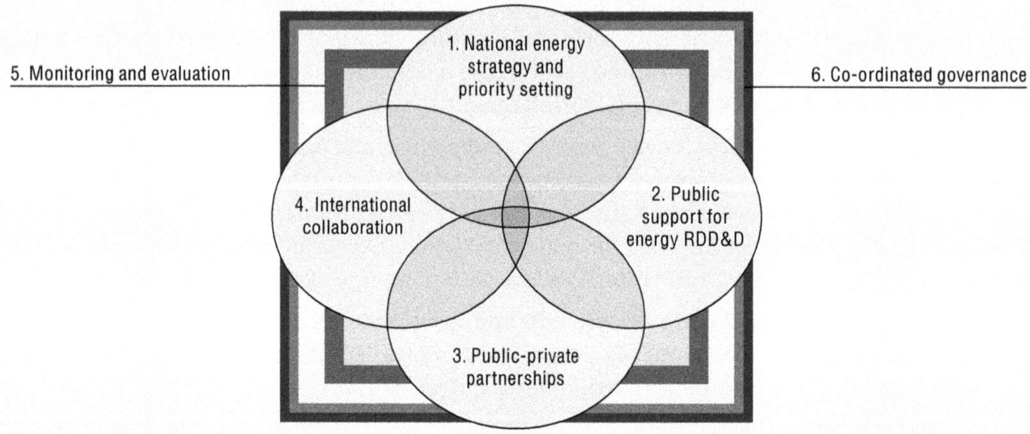

The IEA's *Energy Technology Roadmap* series identifies priority actions for governments, industry, financial partners and civil society that will advance technology development and uptake of the most important energy technologies needed to achieve climate change goals. Each roadmap represents international consensus on milestones for technology development, legal/regulatory needs, investment requirements, public engagement and international collaboration. The IEA has a long history of facilitating international energy RD&D co-operation through its technology network made up of 40 Implementing Agreements, which involve co-ordinated research, joint projects, information exchange, modelling, databases and capacity building. The IEA Implementing Agreements provide a flexible mechanism for governments, industries, businesses and NGOs to leverage resources and improve results of research in energy technologies and related issues. Through regular in-depth reviews of both member and non-member country energy policies, the IEA also undertakes evaluation of energy RD&D policies and makes recommendations to support accelerated energy innovation.

Note: RDD&D = research, development, demonstration and deployment.

Source: Chiavari and Tam (2011), *Good Practice Policy Framework for Energy Technology RD&D*, IEA, 2012.

thought to be promising. Innovation in the former will be of value irrespective of differences in the future cost trajectories of the latter (Johnstone and Haščič, 2013).

Nanotechnology is one interesting case of a technology with broad applications (see OECD, 2013k). While the initial spur for research on nanotechnology was not related to environmental considerations, "green" nanotechnology is increasingly being applied in fields such as green chemistry, sustainable manufacturing, and monitoring and control applications (e.g. nano-sensors). ICTs provide another example. Decentralised production of energy, which has seen a massive surge in the last few decades, has become possible because of ICTs. Other applications of ICT that can yield environmental benefits include transport logistics and environmental monitoring.

Given the difficulties involved in identifying emerging technologies that could prove to have significant environmental applications, recent work at the OECD has sought to develop a set of "early" or "leading" indicators of breakthrough environmental technologies that can help provide some guidance in the allocation of support measures. Figure 7.12 compares climate mitigation technologies with a group of counterfactual non-environmental technologies, in terms of two of these indicators: 1) originality – an indication of the "breadth" of the technology fields on which a patent relies;[17] and 2) generality – an indication of the range of technology fields in which a patent is subsequently cited.[18]

Applying these measures, it is found, for example, that solar photovoltaic (PV) technologies are very general, with innovations in this area subsequently providing knowledge of value to a wide variety of fields – a "giant upon whose shoulders many others build". On the other hand, transport-related environmental technologies appear to be original, implying they are "building on the shoulders of many different giants". More generally, it can be seen that climate mitigation technologies are on average more "original" and "general" than the counterfactual technology representative of inventions in general.[19] Subsequent work has shown that at least some of these indicators, and particularly the one that relates to generality, are robust indicators of future technological and market developments.

Figure 7.12. **Key characteristics of climate mitigation technologies**

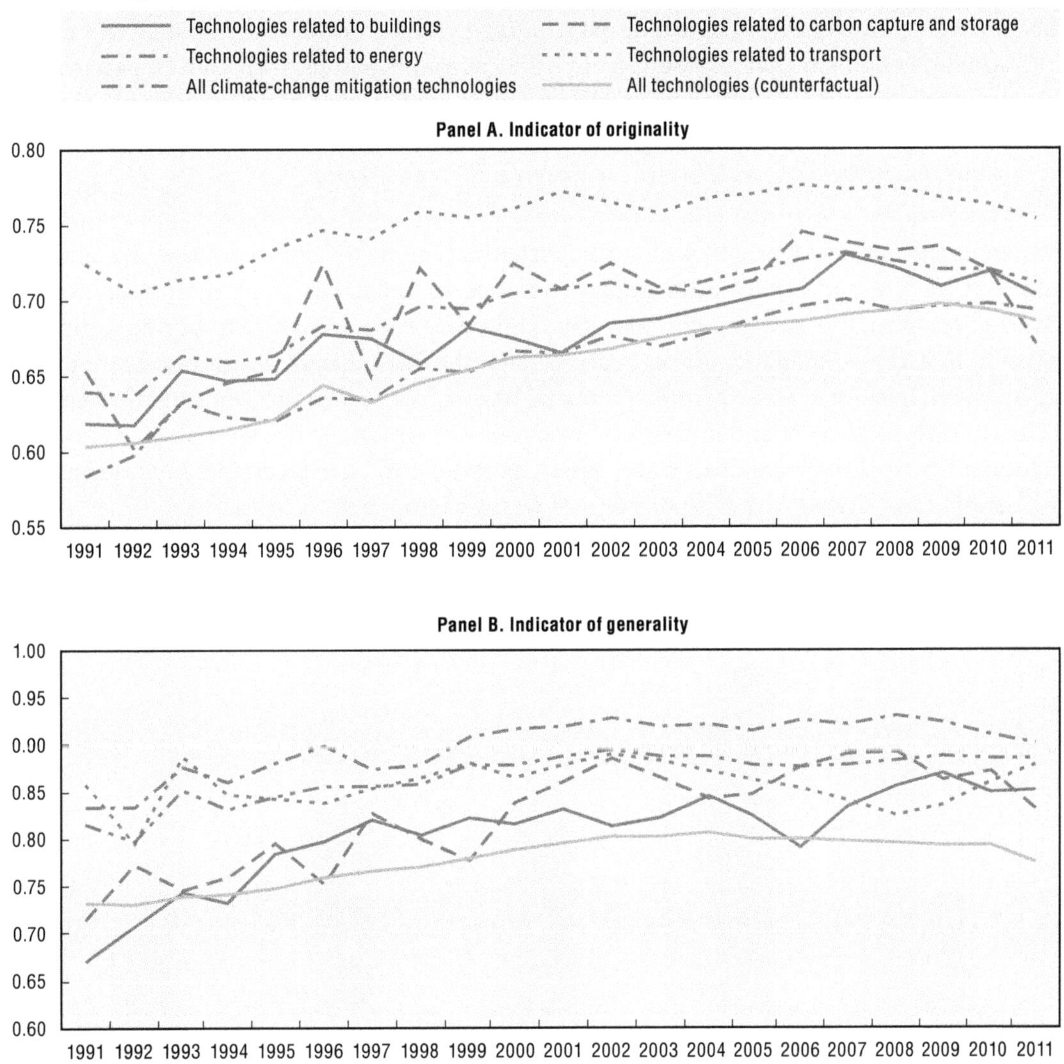

Note: For each technological class, the index corresponds to the HHI measuring the distribution of patent shares across the three-digit NACE industry sectors of the patent applicant:

$IGI_k = 1 - \Sigma_q \left(\frac{applications_q}{applications_k} \right)^2$, where k indexes the patent technological class and q the applicant's NACE three-digit sector.

Source: Egli, Menon and Johnstone (2014), "Identifying and inducing breakthrough inventions".

It has to be re-emphasised that providing discretionary support is a hazardous exercise, and so even if further work allows for the identification of indicators that appear to be reliable leading indicators to future technology and market developments, great uncertainty will always remain. It is important to bear in mind the suggestions of Rodrik (2004), who emphasises the importance of transparent and unambiguous exit mechanisms in the context of government support for specific technologies. In fact, the government needs to exit not only if a technology does not prove to be promising, but also if a technology proves to be successful enough to be driven by private incentives.

Technology transfer and the need for international co-operation in innovation

The international dimension of public policy incentives in the area of green innovation differ from those in most other areas since the benefits associated with the development and adoption of technologies related to environmental concerns cross borders, with impacts that can be regional or even global. The externality is trans-frontier,[20] and as such there is an additional mechanism that is absent from other domains through which a "downstream" country gains from the adoption of environmental technologies in an "upstream" country (OECD, 2011c). For global pollutants (GHGs, ozone-depleting substances) and resources with global benefits (biodiversity), this is even more pronounced.[21]

Moreover, even if the benefits of technology development and adoption are exclusively domestic, the areas of world in which the potential benefits of their adoption are greatest may not be the sources of innovation. The case of pollution from mobile sources is illustrative, with the pioneer country (the United States) initially benefiting primarily from innovations arising in another country (Japan) (OECD, 2008d). In terms of innovation related to the use of scarce natural resources, Figure 7.13 shows the relationship between different regions of the world in terms of innovation in water-saving technologies and their vulnerability to climate-induced water loss. The latter can be interpreted as a measure of the benefits associated with the adoption of water-saving technologies.

Figure 7.13. **Relative technological advantage (RTA) and water vulnerability**

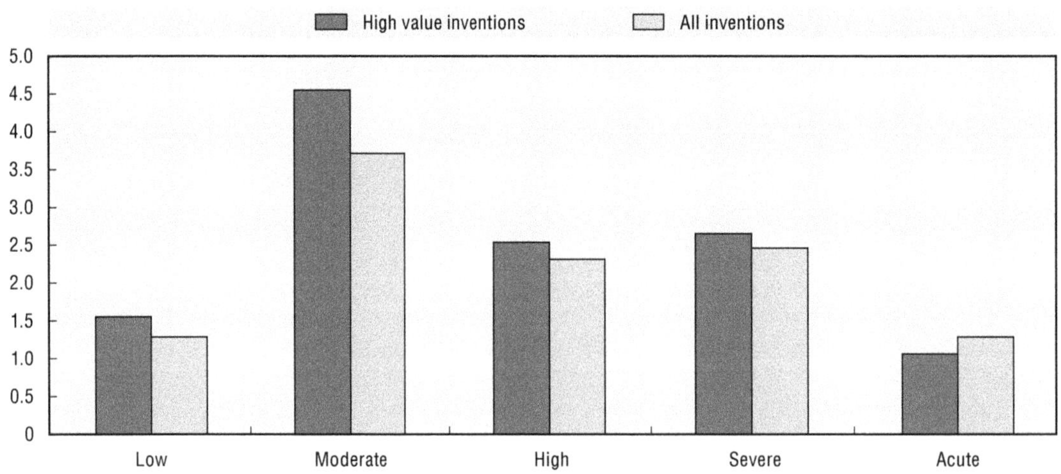

Note: Only countries with more than 100 patents are included. The RTAs for all regions are greater than 1 due to the skewed nature of the distribution, with minimum values of 0 and maximum of almost 12 for this sample.

Source: Dechezleprêtre et al. (2015).

Given that many of the countries that are particularly subject to water vulnerability are less developed economically, it is not surprising to find that innovation capacity in the field in absolute terms is limited. However, the figure goes one step further, showing that such regions are less specialised in the development of water-saving technologies than in other technology fields more generally. Encouraging technology transfer to countries where potential environmental benefits are high is, therefore, a key priority.

While technology transfer is essential, development of local innovation capacity is even more important for at least two reasons: 1) local innovation capacity allows countries to fully exploit the benefits associated with innovations arising elsewhere in the world economy; and 2) many innovations need to be adapted to reflect local conditions. While the first point is of general relevance for all technology fields, the second is likely to be particularly important for environmental technologies since it is not just market conditions (i.e. local preferences) that matter, but also ecological conditions. In the absence of domestic innovation capacity, it may be difficult to develop technologies that are appropriate for local ecological conditions.[22]

Box 7.10. **Innovation in nuclear energy**

Nuclear energy is a mature technology, powering over 430 reactors worldwide that generated 11% of the world's electricity in 2013. Within the OECD region, nuclear power represents the first source of low-carbon electricity in OECD countries, and the second at global level behind hydropower. Nuclear energy will continue to play an essential role in the necessary decarbonisation of the world's energy system. The *ETP* 2DS foresees the need to deploy more than 900 gigawatts (GW) of capacity by 2050, more than double the current capacity. Innovation will play a major role in ensuring this build-up rate, not only by addressing the need to improve the performance and cost-effectiveness of the current technology or to develop new applications, but also by alleviating public concern about safety. Addressing concerns about hazardous waste management and the safe decommissioning of plants will be essential for nuclear becoming a key component of a low-carbon energy strategy.

The update of the *Nuclear Energy Technology Roadmap* from the IEA and the Nuclear Energy Agency (NEA) (IEA and NEA, 2015) provides an extensive overview of current trends in the area of reactor technology and nuclear fuel cycle, identifying some possible developments up to 2050 and policy actions to be undertaken to facilitate the deployment rate of nuclear power plants. Innovations in nuclear technology over the next decades will cover areas such as enhanced safety features, design simplification and modularisation to improve the constructability of the large Generation III reactor units (typical size 1 GW or more), the development of small modular reactors (SMRs) with innovative passive systems, and in the area of decommissioning, advanced robotics to facilitate the dismantling of components of shut-down reactors. Future advances in the nuclear fuel cycle are targeted at the development of so-called accident-tolerant fuels and higher performance fuels. Innovations are also needed to develop the next generation of reactor systems such as Generation IV systems as well as advanced fuel cycles, for instance materials that can withstand higher temperatures, technologies to allow multi-recycling of spent fuel and further reduce the volume and radio-toxicity of the final waste, in-service inspection and repair technologies especially for liquid metal coolant systems, and advanced power conversion systems to replace the current steam turbine technologies.

The update of the *GIF (Generation IV International Forum) Technology Roadmap*, published in 2014 by the NEA, provides a comprehensive overview of these R&D needs. Innovations are also needed to develop non-electric applications of nuclear energy, such as desalination or large-scale hydrogen production, which could displace fossil-fuel usages and reduce the associated carbon emissions. Finally, innovations are needed to address the long-term challenge of adaptation to climate change: future nuclear power plants, just like other thermoelectric plants, will need to be more resilient against extreme weather events and increasing air and water temperatures. Work is already ongoing to develop strategies to reduce water dependency of cooling systems and improve their thermal efficiency to compensate for higher cooling water temperatures.

Finally, the further development of nuclear technology, including in newcomer countries interested in developing this source of low-carbon base-load power, requires an educated and trained workforce. Putting in place innovation strategies and comprehensive R&D programmes can help attract young talents to the industry and build this necessary competence base.

As noted in Chapter 5, general education, skills and specific innovation policies are important contributors to the development of this capacity. However, international research co-operation is also important, particularly so in the environmental context since as noted above the positive spillovers of innovation in this area are not just economic, but also ecological.

Inventors from different countries are already actively co-operating in the development of environmental technologies. Indeed, the rate of international co-invention in some areas such as carbon capture and storage (CCS) is almost twice the rate in other fields (see Kahrobaie, Haščič and Johnstone, 2012). Table 7.2 presents data on top co-inventing country pairs in a number of environmental technologies. Moreover, it is important to note that this co-operation at the "frontier" is not exclusively composed of inventors from OECD countries, and cases of collaboration involving inventors from emerging economies are highlighted in bold.

Table 7.2. **Top ten co-inventor country pairs, levels, 2000-08**

Co-invention pairs including at least one non-OECD country in bold

Sector rank	Biofuels	CCS	Fuel cells	Solar PV	Hydrogen	Energy storage	Wind	All technologies
1	DK-US	CA-US	JP-US	JP-US	DE-US	GB-US	DK-GB	GB-US
2	NL-US	NL-US	CA-US	DE-US	JP-NZ	CA-US	DE-US	DE-US
3	CA-US	GB-US	DE-US	GB-US	CH-DE	DE-US	CA-US	CA-US
4	DE-US	FR-US	GB-US	CH-DE	IT-US	JP-US	DE-NL	CH-DE
5	**CN-DK**	DE-US	**CN-US**	AT-DE	CA-US	JP-KR	NL-US	JP-US
6	DE-GB	AU-NL	KR-US	CA-US	CH-US	FR-US	DE-DK	FR-US
7	GB-US	DE-GB	FR-US	**CN-US**	FI-SE	CH-DE	**IN-US**	NL-US
8	CH-DE	GB-NL	CH-DE	DE-FR	DE-FR	CA-FR	**BE-ZA**	DE-FR
9	GB-NL	NO-US	CA-FR	DE-NL	DE-GB	**CN-US**	**RU-US**	CH-FR
10	JP-US	**CN-US**	CA-DE	GB-IT	**IN-US**	KR-US	DK-ES	CH-US

Note: The two-letter standard international country codes refer to: Austria (AT), Australia (AU), Belgium (BE), Canada (CA), Switzerland (CH), China (CN), Germany (DE), Denmark (DK), Spain (ES), Finland (FI), France (FR), United Kingdom (GB), India (IN), Italy (IT), Japan (JP), Korea (KR), the Netherlands (NL), Norway (NO), New Zealand (NZ), the Russian Federation (RU), Sweden (SE), the United States (US) and South Africa (ZA). See also Haščič and Migotto (2014).

Source: Kahrobaie, Haščič and Johnstone (2012), "International technology agreements for climate change".

On the one hand, the benefits arising out of international research co-operation strengthen the case for the use of more flexible policy instruments, such as market-based measures. As was noted above, more prescriptive policies encourage innovation efforts that are narrowly targeted according to the characteristics of the regulation. If such regulations differ – even subtly – across countries, the potential market for any individual innovation is constrained, with firms facing a set of fragmented national markets. Domestic firms will innovate to domestic regulations, and potential economies of scale will be forsaken. Conversely, with more flexible policy instruments, co-operative research efforts go hand in hand with the realisation of international market opportunities.[23]

On the other hand, given the central role of research co-operation and technology transfer, many commentators have argued that an important complement to any future emissions reduction agreement in the area of climate change will be an international technology-oriented agreement (Popp, 2011; Ockwell et al., 2010).[24] Ongoing examples include the IEA's Implementing Agreements in areas related to climate change (e.g. CCS, fuel cells, solar PV).[25] Such agreements potentially contribute to achieving faster technological innovation at lower cost through knowledge-, cost- and task-sharing of the participating countries. Collaborations bring the benefit of scale, and permit research in instances where the scale or scope is too large for a national project.

Indeed, recent work at the OECD has found that participation in such agreements can have a significant impact on co-inventive activities between inventors residing in different member countries (see Kahrobaie, Haščič and Johnstone, 2012). Moreover, other work has found that international co-operation at the level of knowledge creation (as reflected in co-authorship of scientific articles) leads to the generation of higher-value technologies in the area of wind power relative to the case where the authors are from the same country. Interestingly, this is particularly true if countries at different stages of economic development collaborate (Poirier et al., 2014).

However, even with the development of innovation capacity, technologies are unlikely to be adopted in the absence of available finance. As pointed out above, the policy context is central to ensuring that this is the case, and as such this issue is even more pertinent for emerging and developing countries where the policy incentives are less often present. For example, based on data from approximately 14 000 financial deals in the renewable energy sector, Haščič et al. (2015) find that approximately 10-20 times the private finance could have been mobilised in developing and emerging economies during the 2000-11 period if a policy framework had been implemented similar to that which is in place in OECD countries. (See also OECD, 2015f.)

A final point that warrants re-emphasis is that in many cases the policy response has lagged, resulting in disruptions to ecosystems that can affect the economy in a host of different ways. The case of climate change is an obvious example, but this has also been true of other areas (e.g. emissions of ozone-depleting substances, over-exploitation of natural resource stocks). Economic agents will, of course, adjust to changing conditions, in terms of both their behaviour and their technological choices. However, in many cases targeted support will be required, particularly when ecological conditions change in a sudden and unpredictable manner (Mullan et al., 2013). In such cases, innovation for adaptation is a necessary complement to innovation for mitigation. Some of the main policy messages emerging from the OECD work on green innovation are below.

Main policy messages on green innovation

- The co-existence of knowledge and environmental market failures means that in absence of appropriate policies, households and firms pollute too much, exploit natural resources too intensively and innovate too little compared with the social optimum.
- In order to induce innovation, the environmental policy framework must be such that the social cost of polluting the environment and using natural resources is reflected in the market prices faced by different economic agents (households, firms, public authorities).
- Ensuring that this opportunity cost is fully reflected in market prices requires both the introduction of efficient environmental policy measures and the removal of distortions such as subsidies that undermine the effects of such measures.
- In order to encourage innovation to meet public environmental objectives, the design of policy measures must allow for flexibility in the response of potential inventors and adopters of environmental innovations. In a world of imperfect information, research must be encouraged.

Main policy messages on green innovation *(cont.)*

- Due to the sunk and irreversible nature of the investments necessary to bring about such innovation, policy predictability and credibility are essential, encouraging investors to take the risks associated with development and adoption of environmental technologies.
- Many of the sectors that have potentially important environmental implications are subject to other market imperfections (natural monopoly, network externalities, etc.), placing a premium on policy co-ordination. The systemic nature of some areas, notably energy, reinforces the need for a comprehensive approach to innovation.
- Some environmental concerns are pressing (e.g. climate change), for which marginal changes in relative prices may not suffice, requiring more targeted innovation support measures.
- The issue of targeted support measures raises the issue of selection. Given the presence of fundamental uncertainty and information asymmetry as well as the diverse sources from which environmental innovation can arise, designing targeted support measures that support green innovation remains an open policy research question.
- Essential elements will include support for basic research, transformative technologies with wide applications, and identification of a portfolio of more specific and promising breakthrough technologies. However, any policy of targeted support requires a clear and unambiguous strategy for policy exit.
- Given the mutual, and in some cases global, nature of the benefits of environmental innovation, there are incentives for co-operation in the development, transfer and adoption of environmental technologies.

7.5. Fostering public-sector innovation

The public sector faces both challenges and opportunities that make innovation in the public sector a salient issue for governments across the OECD and beyond. Social challenges, demographic change and weak economic growth have increased demands on the public sector, while fiscal restraints constrain the capacity of government to respond. Innovation offers a lever to help governments improve public services so that they are better able to meet society's needs. This is the case across all the diverse activities undertaken in the public sector, including health, education, justice, defence, police and social policy. More innovative and efficient public services are also important in supporting innovation in other parts of the economy and throughout society, implying that public-sector innovation needs to be an important part of the overall innovation agenda.

Surveys of the public sector indicate that public agencies are innovating. The European Union's Innobarometer 2010 found, for example, that across the 27 member countries, 66% of organisations on average reported having implemented a service innovation. Likewise, Nesta's 2011 pilot survey found that over 90% of local government authorities and National Health Service Trusts in the United Kingdom reported a service, process or management technique innovation.[26] Case evidence of innovations in the public sector at local, national and international levels also stand testament to the fact that the public sector is innovating. Many countries operate innovation awards schemes that collect, judge and award the best innovations based on comparative review. Work undertaken by the OECD Centre for Educational Research and Innovation (CERI) shows that, contrary to common belief, there is also a fair level of innovation in the education sector (see Box 7.11).

Box 7.11. **Towards innovation strategies in education**

As in other sectors, innovation in education could be a major driver of productivity and welfare gains. On average, countries spend 6% of their national income on educational institutions, and in spite of progress in some countries, it is not always clear that education systems have reaped all the opportunities that could have made them improve learning outcomes, enhance equity and equality, improve efficiency, and adapt to societal needs. In many countries, the awareness that a co-ordinated educational innovation policy could be beneficial is just emerging.

Recent work shows that, contrary to common belief, there is a fair level of innovation in education, both relative to other sectors of society and in absolute terms. While education is below average in terms of the speed of adoption of innovations, 58% of tertiary-educated professionals in education hold a highly innovative job, that is, a job contributing to the innovation process in an organisation at the forefront of absorbing innovations, slightly above the 55% average in the economy. Within education, higher education is the most innovative sub-sector, but examples of innovation exist at all levels (OECD, 2014r, 2013o).

Having a fair level of innovation does not necessarily imply that the education sector has a strong innovation ecosystem. Some innovations can be the outcome of policy reforms that mandate change. However, a good co-ordination of the different levers of innovation policy is often lacking, and the knowledge generated by past innovative pilots or experiments is not always shared and used in a cumulative way.

The education regulatory framework may or may not be conducive to innovation. For example, curriculum and assessment policies have an impact on the scope for innovation, and most countries have checks and balances to ensure that grass-roots innovation is possible, but controlled (Kärkkäinen, 2012). As in other sectors, quasi-markets have been used with the objective to foster innovation, and they have been found to help diffuse a variety of alternative models of schooling; however, they do not seem to lead to the emergence of new ones (Lubienski, 2009). Access to finance for innovation, dissemination strategies and staff development policies are also key elements of this regulatory framework.

A second pillar of innovation policy lies in investment in R&D, even though this typically varies by sectors. Given the significance of the sector, public spending on educational research is likely to be below what is needed. In 2012, research on education was the least funded of all socio-economic objectives for which information is available. There are also often few incentives for companies producing educational resources and devices to invest in development, even though there has been a rise in specialised innovative educational companies over the past decade (Foray and Raffo, 2012).

A third pillar of innovation policy lies in work organisation and the extent to which individual, organisational and sectoral learning occurs in the education sector. The role of professional learning communities is often highlighted, as is the importance of leadership (OECD, 2013p).

A fourth pillar is the application of general-purpose technology, and notably ICT, to the education sector. Some countries such as Italy have tried to change their systems by developing digital plans for education (Avvisati et al., 2013). Technology is also used as a means to transform and enhance pedagogy (Kärkkäinen and Vincent-Lancrin, 2013) or to modify the business model of education, for example with the support of open educational resources or of massive open online courses (MOOCs). Technology also transforms education through data-driven innovation, which is increasingly facilitated by the establishment of administrative longitudinal information systems that follow students throughout their school and university years.

A few countries already have an innovation policy for education. For example, France has a chapter on innovation in its education law. The United States has several programmes to support innovation and improvement efforts at the federal level to support innovation, notably the Invest in Innovation (i3) Fund of the Department of Education. What is still missing in most if not all countries is a holistic and explicit strategy that takes into account the different dimensions and drivers of innovation to create an innovation-friendly ecosystem for the education sector. The OECD Centre for Educational Research and Innovation is currently working with countries on the development and monitoring of such strategies.

Box 7.11. **Towards innovation strategies in education** *(cont.)*

Sources: OECD (2014r), *Measuring Innovation in Education, http://dx.doi.org/10.1787/9789264215696-en*; OECD (2013o), *Innovative Learning Environments, http://dx.doi.org/10.1787/9789264203488-en*; Kärkkäinen, K. (2012), "Bringing about curriculum innovations", *http://dx.doi.org/10.1787/5k95qw8xzl8s-en*; Lubienski, C. (2009), "Do quasi-markets foster innovation in education?", *http://dx.doi.org/10.1787/221583463325*; Foray, D. and J. Raffo (2012), "Business-driven innovation", *http://dx.doi.org/10.1787/5k91dl7pc835-en*; OECD (2013p), *Leadership for 21st Century Learning, http://dx.doi.org/10.1787/9789264205406-en*; Avvisati, F. et al. (2013), "Review of the Italian strategy for digital schools", *www.oecd.org/edu/ceri/Innovation%20Strategy%20Working%20Paper%2090.pdf*; Kärkkäinen, K. and S. Vincent-Lancrin (2013), "Sparking innovation in STEM education with technology and collaboration", *http://dx.doi.org/10.1787/5k480sj9k442-en*.

Despite evidence that the public sector is innovating, it is clear that its innovative performance remains below potential. The OECD's Observatory of Public Sector Innovation (OPSI) was established to support countries in strengthening their public sector capacity to innovate by collecting, sharing and analysing innovations from across the public sector.[27] Evidence from the OPSI database reveals that countries encounter important challenges in generating adequate support and finding the right resources to foster innovation. The operating framework – or the "ecosystem for innovation" – is also not optimal. Public agencies lack the flexibility, culture and guidance that could help innovation to flourish.

Metrics that can shed light on innovation drivers and processes in public organisations would help leverage the public sector as a key innovation actor. Despite the existence of a framework to measure innovation in firms, challenges related to the measurement of public-sector innovation are multiple and non-trivial. Recent OECD work has led to the development of a set of working definitions and proposals for international surveys of innovation in public-sector organisations (see Box 7.12), which could complement work on innovation in specific sectors, such as education.

The components of public-sector innovation

Drawing on research and countries' learnings reported through OPSI, the OECD has started to set out some of the key components for an operating framework of innovation in the public sector and has identified four areas that government policies need to address to strengthen public-sector organisation capacity to innovate.[28] Within each of these components, there are policies and tools that public organisations can use to foster innovation. These areas are:

- **People:** How civil servants are motivated within an organisational setting to explore new ideas and experiment with new approaches, and how this affects their propensity to innovate. Leadership and the way staff are selected, rewarded, socialised and managed also shapes an organisation's innovative capacity.
- **Data, information and knowledge** are essential to innovation, and the way they are managed can support or hinder innovation. The challenge is to build the capacity to pool available knowledge to improve public decisions about innovative solutions and to share knowledge to encourage social innovation.[29]
- **Organisational design:** The way work is structured within and across organisations may have an impact on innovation in the public sector. This includes the development of spaces and innovative methods to structure teams, break down silos and work in partnerships across organisations and even sectors.

- **Rules and processes:** Formal and informal rules, processes and procedures that guide the daily operations of public agencies may provide an architecture that creates a flexible environment for innovation, or results in a web of complexity, hindering the capacity to innovate.

Box 7.12. Defining and measuring innovation in the public sector

Innovation has become a ubiquitous concept. Its relevance as key driver in the transformation of economies and societies has spurred interest in obtaining statistical measures of this concept for organisations, sectors and countries, for uses that range from benchmarking to the analysis of factors and policies that underpin innovation and their impacts. While the OECD framework for defining and measuring innovation (the *Oslo Manual*, OECD/Eurostat, 2005) in the context of business activities has been implemented in national statistical surveys since the early 1990s in almost 100 countries, there is limited statistical evidence on the extent of innovation in other sectors. A recent OECD project has set out to consider how the experience with business innovation surveys as well as the lessons from various other fields could help in defining and measuring innovation in the public sector.

Defining innovation for statistical purposes must take into account how users intend to use the data, as well as the feasibility of collecting the information in a consistent and accurate fashion. The OECD measurement guidelines for the business sector emphasise the role played by firms as actors responsible for innovating, for which it is possible to define a population that can be surveyed and compare the reported innovation performance of firms to other features such as productivity or employment. This subject-based approach can be complemented by data on innovations introduced by firms and their characteristics, i.e. object-based approaches. Subject approaches have been recently applied to the public sector in a number of countries, targeting its organisational units and their heads for information on new services and methods introduced over the reference period. While these surveys have been useful to test the applicability of private-sector framework to the public sector, they underscored the difficulty of accounting for the specificities of innovation within public-sector organisations (e.g. the cross-boundary nature of some innovations).

A working definition for surveys of public-sector organisations identifies public-sector innovation as the implementation by the unit of a new or substantially changed service or good, operational process, or organisational or communication method aimed at improving a public-sector unit's operations or outcomes. In this context, implementation refers to the provision of (or, making available to potential users) products, or the use of methods, as part of the unit's functions. Innovations must be new to an organisation, although they may have been developed by others, in the public sector or beyond. This definition has been cognitively tested in a series of co-ordinated interviews with public-sector managers across several OECD countries and provides a starting point for measurement initiatives. Public-sector managers typically understand this definition, but often find it difficult to ascertain when a change is sufficiently large to count as an innovation, requiring further guidance and examples. This definition is expected to be refined building on the evidence emerging from the joint work of OPSI and the OECD's Working Party of National Experts for Science and Technology Indicators (NESTI), which will also focus on operationalising some concepts and classifications for collecting statistical data on innovation in the public sector.

Innovation in the public sector has some specific features that bear on the appropriate measurement approaches to be used by countries that wish to introduce these measurements. The specific combination of multiple objectives under political oversight, the absence of external (market) data validation and the limited availability of performance data present important challenges to measurement. Furthermore, an innovation can be easily attributed to more than one organisation, depending on the allocation of various decision-making, development and implementation roles.

Thus, measurement of innovation at the level of individual public-sector units can be highly revealing but can offer limited insights if no measures are available on the outcomes of the activities of such units and the resources they can draw upon. The measurement of productivity in non-market public services, for

Box 7.12. **Defining and measuring innovation in the public sector** *(cont.)*

example, is still in its infancy, which complicates the scope for using innovation indicators to assess the link between innovation and productivity in the public sector – a major concern for policy makers and a key rationale for measuring innovation in the public sector. The measurement of public-sector innovation needs to go hand in hand with improvements in the measurement of productivity and knowledge-based assets in public services, at both the micro and macro levels. For example, the OECD Government at a Glance work is developing both public-sector efficiency and effectiveness indicators, including a "serving citizens framework" that considers four dimensions of public service quality: satisfaction, access, responsiveness and reliability. Since very different performance measures can apply in different parts of the public sector, this justifies the use of more targeted approaches aimed at specific activities within the public sector. For example, innovation data from schools could in principle be linked to data on learning outcomes and on other educational objectives.

Given the generalised lack of a market/competitive test for innovation in the public sector and the diversified interest and demands from policy makers in terms of developing a sound evidence base in this area, it is advisable to aim to collect information on public-sector innovations from different perspectives, using whenever possible already-available data collection instruments. For example, surveys of public-sector employees typically intended to gauge staff satisfaction can add some valuable insights to complement and also validate the perspectives on innovation reported by the heads invited to report on behalf of their organisations, following the so-called employer-employee model. These "personnel" surveys are becoming commonplace in several OECD countries' public administrations and can provide a more sustainable vehicle for statistical enquiries on public-sector innovation. Likewise, the perspectives of public-service users and citizens at large can provide a further perspective on the process and outcomes of innovations introduced by public-sector organisations. Dedicated surveys of public-sector organisations should therefore be complemented with various other sources to develop indicators of public-sector innovation.

Non-statistical approaches focused on collection of innovation cases can be particularly informative about the common aspects across innovations and promote learning through knowledge sharing. These databases can be subject to quantitative analysis, but they may not be entirely representative of the innovative activity of public-sector organisations, for example, if there is an incentive to report successes instead of failures (e.g. entries to innovation awards). Data collected for performance management purposes may not always be suitable as a source of statistical information, if those reporting have incentives to overstate their innovativeness. For subject- and object-based approaches alike, carefully designed policy and documented experiments can be a fruitful source of information on the effectiveness of innovation activities and measures to promote public-sector innovation more generally.

Facilitating a more comprehensive measurement of innovation across different sectors is a key objective of the forthcoming revision of the *Oslo Manual*. The innovation examples and survey initiatives undertaken by individual countries and organisations provide the main source of evidence for testing and recommending approaches that allow statisticians to capture reliable, internationally comparable data on the role of the public sector on driving innovation.

People

Civil servants are at the centre of public-sector innovation, and their commitment and determination drives every stage of the innovation process. Research shows that innovations are born from ideas that come from staff at all levels of an organisation. Front-line staff and middle managers who interact with clients and operationalise policies often have the best understanding of the need to innovate. This is reflected in various OPSI cases, where employee involvement has been cited as key.

Recognising that the human factor is core to innovation leads to questions regarding human resources policies and practices and their role in strengthening the ability, motivation and opportunity to innovate. The following human resources management levers are particularly important:

- **Work organisation:** Research suggests that the organisation of work can have a significant impact on innovative capacity. Innovation is positively correlated with teamwork, autonomy, employee engagement and commitment, and flexibility. Working in teams enables a greater convergence of perspectives, knowledge and skills than individual workers would have access to on their own. Engaging staff members in the decisions of the team/unit/organisation also likely boosts their motivation to innovate. Some OECD countries are establishing networks of change agents to enable more innovation by stimulating interaction among diverse actors.
- **Recruitment and selection:** Research suggests that diversity of ideas is positively correlated with innovative problem solving. Policies to foster recruitment and diversity can include mobility programmes that move staff across ministries or levels of government or between sectors. A diverse set of competencies related to innovation, such as problem solving, creativity, communication, analysis and synthesis, and foresight, among others, suggest the need to fine-tune selection processes that measure and value more than education and hard skills.
- **Performance management:** Fostering innovation relies on establishing approaches that incentivise safe risk-taking and embrace early failure as part of the learning process. Developing performance objectives that focus on taking safe risks, and setting up creative safe spaces to try and fail and learn, can reduce risk aversion. Some governments are beginning to introduce innovation-oriented behavioural competencies into performance management systems.
- **Training and development:** Learning is at the heart of innovation. Rotating staff members through a wider variety of tasks can help build their competencies and learning. Taking advantage of external conferences, partnerships and opportunities to share ideas with others is another option. Performance management and assessment process can be used as an opportunity to discuss learning needs and potential. The use of networks and cross-functional teamwork is also being increasingly recognised as a valuable source of learning that can spark innovation in public-sector organisations.
- **Compensation:** Using the financial compensation mechanism to motivate innovation can prove challenging and should be approached with caution to ensure that financial rewards incentivise the right kind of behaviour, such as teamwork, open information sharing and co-operation. Taking a broader perspective to include awards and recognition may prove more effective in motivating innovation. Many countries are developing various innovation awards schemes to recognise innovators and build a culture that values innovation.
- **Leadership:** Leaders set the tone of an organisation; play a large role in establishing, reinforcing and changing the culture of the organisation; and act as role models. Senior management support for innovative projects is one of the key factors for their success, but it can often be challenging for leaders to connect with front-line service delivery staff who may have innovative ideas to bring forward, particularly in large hierarchies. This places additional focus on the communication mechanisms and strategies internal to organisations, both upward and downward.

Data, information and knowledge

Developing capacity for innovation requires that an organisation change and adapt by learning from its past experiences while anticipating future challenges through organisational foresight. Data, information and knowledge have important roles to play as the building blocks for creating a learning organisation that displays these attributes. They support daily operations, help an organisation understand its evolving context and support evidence-based decisions. When used strategically they can help an organisation adapt and compete through learning to promote and sustain employee and organisational learning.

Data, information and knowledge about an organisation exist in many different forms and locations; developing as a learning organisation means being able to harness each of them to support continuous learning. This means identifying their different sources, exploiting what they say about an organisation by regularly and systematically integrating them into the decision-making process, and sharing them openly across the relevant actors both within and beyond government. As highlighted already in the main policy messages of green innovation, developing new knowledge may also require a systematic effort through R&D, which may be insufficient in several parts of the public sector.

Using information to improve the innovative capacity of public organisations gives rise to three interrelated issues:

- **Sourcing:** The identification of the different types and sources of data, information and knowledge that are relevant. This may also involve explicit efforts to generate new knowledge.
- **Exploiting:** Organisations need to channel data, information and knowledge into a usable form so that it can be fully exploited to support evidence-based decision making and organisational renewal (to support the development of "learning organisations").
- **Sharing:** Organisations need to share information collected with wider sets of actors including other public-sector organisations and members of the public to support decision making, accountability and co-innovation and facilitate value creation elsewhere in the economy.

Measuring how governments are using data, information and knowledge to support innovation remains challenging. While individual cases demonstrate how the use of data, information and knowledge can support the public sector to identify needs and opportunities and generate new insights that foster innovation, further work is required to understand how they need to be used to support innovation systematically.

One indication of how far governments are sharing information beyond organisational borders, and with society as a whole, is the availability of information in the public domain. Sharing public information widely, such as making public data sets available and accessible in open government data portals (Figure 7.14), can support other actors to co-innovate and create value elsewhere in the economy.

Rules and processes

People need the opportunity and the resources to innovate, and this suggests the need to consider how the rules, laws and bureaucratic processes that regulate the public sector can be designed to encourage public-sector innovation to flourish.

Public-sector organisations are regulated by a complex web of laws, rules and procedures. These include budgeting, resource management, reporting obligations, project management and approval processes, communication protocol, and legal frameworks that regulate public-sector organisations' activities in areas such as privacy, security or procurement.

Figure 7.14. **Number of data sets in centralised "one-stop shop" open government data portals, 2013**

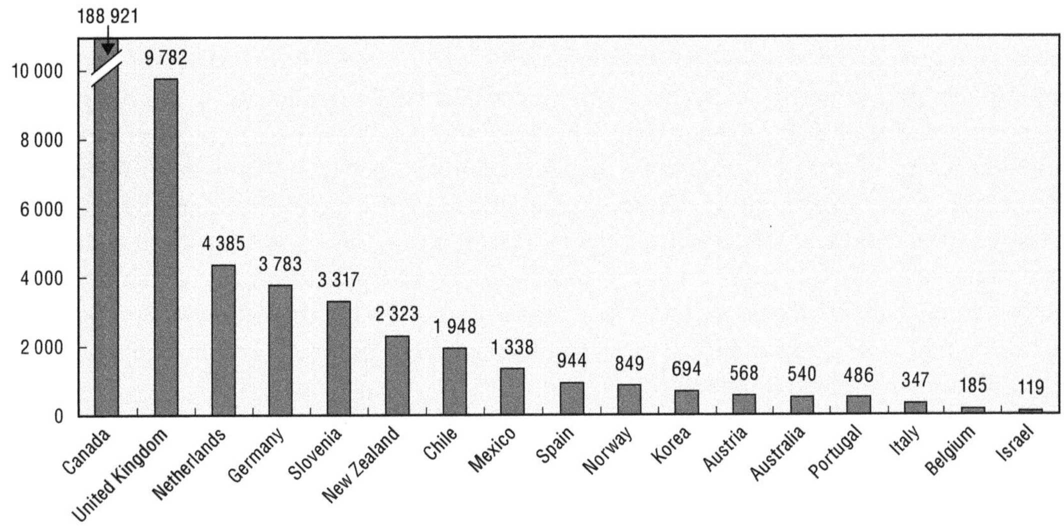

Source: 2013 OECD Survey on Open Government Data.

While these rules are established for good reasons (protecting the public interest, ensuring ethical use of resources, promoting accountability, establishing common operating procedures for consistency and efficiency), their design may have unintended effects that can inhibit individual and organisational capacity to innovate. For example, regulation may constrain programmatic changes or inhibit co-operation across ministries or in partnership with other sectors.

There are three discrete but interrelated areas that can impact the public sector's capacity to innovate in terms of rules and processes:

- clarifying and, in some cases, simplifying the legal and regulatory context to encourage public sector innovation
- looking at the relationship among resource flexibility, budget agility and innovation in a public-sector setting
- the innovation processes – how innovation requires more flexible and experimental approaches to project management and public service design.

The relationship between rules and processes, and innovation is complex. In some cases rules and processes may act to support innovation, providing the guiding architecture that sets the course of action for public organisations. In other contexts, however, they may create a complex web that is difficult to navigate and inhibits civil servants to act.

Further work is required to understand more fully what policies should be put in place so that rules and processes foster innovation, but one aspect of the relationship between rules and processes and innovation relates to their degree of flexibility. In particular, when it comes to resources, it is important that budgets and people can be swiftly moved to where they are most needed. One way to measure this is the degree of executive budget flexibility. For example, in only 6 out of 33 OECD countries for which data are available are line ministries able to borrow against future appropriations to fund investment expenditure (OECD, 2012b). On the other hand, in 25 countries such ministries can carry over unused funds from one year to the next to finance investment expenditure.

Organisational design

Organisational design looks at the work that needs to be achieved, how its component parts can be divided, how these parts connect and how organisational activities will be managed. Moving to a government context this means taking an issue such as youth unemployment; thinking about the components that shape it, such as economic growth, labour market policies and education; thinking about where the knowledge, resources and execution of these components lie, for example, policy making in different line ministries, policy execution in schools, universities and businesses, or expert knowledge from academic experts; considering how work can be managed so that the resources of each actor are channelled effectively to provide a joined-up solution; and establishing an organisational design that supports this. Organisational design therefore has implications for how individuals, organisations, the public administration as a whole and society in general interact and work together.

Rethinking some elements of organisational, system-wide and institutional designs may offer one lever to improve collaboration within and across entities. The public sector is often charged with having a weak capacity for integrated, collaborative ways of working. There is "silo" working both within and across organisations, with managers arguably devoting more energy to defending their own fiefdoms than working corporately. At the same time, an inward focus can lead to policy conversations being limited to government. Rethinking organisational design, from the individual jobs, teams and upward, could be one effective approach to overcoming some of these tendencies, and better support information and data sharing.

Collaboration beyond government with society is perhaps more important than ever. Citizens interact with public services in a variety of ways whether in relation to their health, their family, their employment, their education and their income to name a few. Increasingly, such interaction is based on ICTs (Figure 7.15), which offer further opportunities for closer interaction and citizen engagement. Furthermore, issues such as

Figure 7.15. **Individuals using e-government services, 2010 and 2013**

Percentage of individuals obtaining information and sending completed forms on government websites in the last 12 months

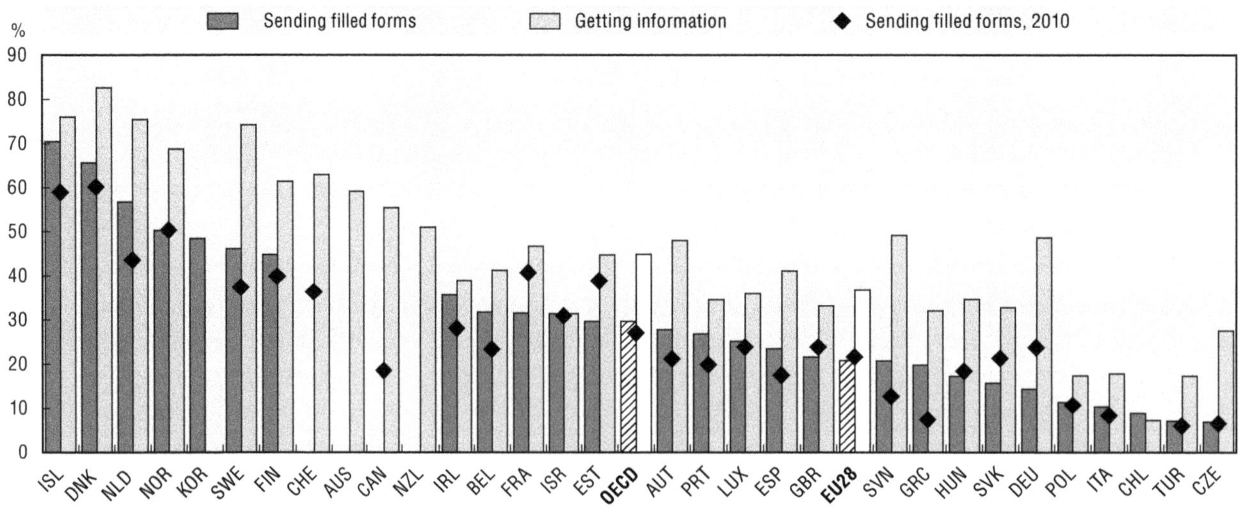

Source: OECD (2014q), *Measuring the Digital Economy, http://dx.doi.org/10.1787/9789264221796-en.*

population ageing or inequality are not the domain of any single organisation but touch different bodies across different parts of the public, private and civil society sectors.

If public organisations and public institutions are going to have the capacity to develop innovative solutions that tackle these issues, they need to be able to collaborate and partner effectively with actors from across society. Some countries are experimenting with new structures and units that facilitate the public sector's ability to engage the broad range of actors who are relevant to any single issue. Such innovation teams, units or organisations "i-teams" are operating in different ways. Nesta has recently started looking at these teams, how they work, their structures and how they are governed (Puttick, Baeck and Colligan, 2014). Strengthening the public sector's ability to innovate may also benefit from specific initiatives to encourage innovation, e.g. challenge prizes. Some preliminary policy messages emerging from the OECD's work on public-sector innovation are highlighted below.

Main policy messages on public sector innovation

Meeting the innovation challenge requires taking a systemic approach – focusing on the people involved, the information they are using, the ways in which they are working together, and the rules and processes that govern their work. Governments are called upon to urgently address these four areas to promote and enable public-sector innovation. They are also encouraged to strengthen measurement and the evidence base for policy making.

1. People matter: Governments must invest in the capacity and capabilities of civil servants as the catalysts of public-sector innovation. This includes building the culture, incentives and norms to facilitate new ways of working. No government can build a strong and secure country without a professional, capable and innovative civil service. As demands on the public sector and corresponding resources continue to move in opposing directions, an innovative public service will be increasingly vital to ensuring both domestic and global success. Given the far-reaching impact of the public sector, all governments have a stake in ensuring that their public services are equipped with the skills, incentives and scope for smart risk-taking and problem solving to spur innovation and drive better outcomes for citizens.

2. Knowledge is power: Governments must facilitate the free flow of information, data and knowledge across the public sector and use it to respond creatively to new challenges and opportunities. The information, data and knowledge that inform strategic and operational decision making are critical to fuelling public-sector innovation. Harnessing the innovative potential of information requires that information be open and available, and that organisations carefully consider what information is required and how best to systematically integrate it into the decision-making process to support continuous learning. Organisations that fail to learn risk incurring higher costs and repeating their errors, while failing to realise new possibilities.

3. Working together solves problems: Governments must advance new organisational structures and leverage partnerships to enhance approaches and tools, share risk, and harness available information and resources for innovation. The complex problems faced by governments today require new ways of working. These include approaches based on collaboration and partnership that integrate the vital perspectives of citizens, civil society, academia and business, as well as exchange within the public sector. They also require the creation of more open, networked and horizontal organisations, adept at collaboration inside and outside government. So too are more flexible approaches to working, including pooling talent and creating multidisciplinary management teams to strengthen collaboration. Temporary teams, pilot projects and short-term assignments are all ways for governments to experiment and better align talent and resources to encourage dialogue, experimentation, risk-taking, problem solving and innovation.

Main policy messages on public sector innovation *(cont.)*

4. Rules and processes to support, not hinder: Government must ensure that internal rules and processes are balanced in their capacity to mitigate risks while protecting resources and enabling innovation. While internal controls, rules and processes are required to ensure sound stewardship and accountability, they can inadvertently stifle innovation. To guard against this, governments must ensure that their public services are marked by reasonable rules and lean processes. Such efforts should ensure that the accountability regime is both robust and easy to navigate while also more systematically focused on enabling innovation. New outcome-focused approaches to project management are one step in moving away from the rigidities of a command-and-control model.

5. Strengthen measurement of public-sector innovation. Better measurement is key to strengthening the evidence base for public-sector innovation, benchmarking performance and enabling the evaluation of specific actions.

Notes

1. See the 2008 *OECD Review of Innovation Policy* for Hungary (OECD, 2008c).
2. At the same time, trials need to be better co-ordinated so that they can draw on larger population cohorts.
3. One of the few studies to examine this question is Popp and Newell (2012).
4. EPS is constructed as a composite index of environmental market-based policies and command-and-control policies. See Albrizio et al. (2014) for further details.
5. The OECD results contained in Albrizio et al. (2014) shed light on the effects on MFP growth. Work on other components of GDP, such as capital and labour, is currently under way to provide a fuller picture of the economic impact.
6. Note that under certain market conditions this "ratcheting effect" may actually lower the rate of innovation relative to the optimum. See Milliman and Prince (1989) for an analysis of the manner in which strategic behaviour can affect the relationship between policy stringency and incentives for innovation.
7. See Heyes (2009) for a discussion of the competition effects of environmental policy design. While competition-inhibiting environmental policies have little direct impact on large swathes of the economy since the most burdensome procedures and designs are concentrated in high-polluting sectors and industries, studies of regulatory impact show that regulations impacting on competition and entry in a specific sector tend to trickle through the entire economy (Barone and Cingano, 2011; Bourlès et al., 2013).
8. See Koźluk (2014).
9. See also Greene and Braathen (2014).
10. See Baker, Bloom and Davis (2013) for a general discussion.
11. See Aghion et al. (2014).
12. For example, the introduction of market-based instruments correlates with the adoption of technologies that involve integrated changes in production processes rather than "stand-alone" end-of-pipe abatement technologies. This allows for the realisation of economies of scope. At the same time, the adoption of such strategies is dependent upon overarching responsibility for environmental matters being allocated to managers with broad oversight of firm strategy, and not in a dedicated environmental department. See Johnstone, Labonne and Thevenot (2008).
13. See OECD (2003).
14. See Cárdenas Rodríguez et al., (2014), Criscuolo and Menon (2014), and Criscuolo et al., (2014). See also OECD (2015b).
15. See Lerner (2011) for a discussion.

16. "System innovation" is a concept used to illustrate a horizontal policy approach that mobilises technology, market mechanisms, regulations and social innovations to solve complex societal problems in a set of interacting or interdependent components.
17. More specifically, the indicator measures the degree to which citations to different technological classes are dispersed in different fields.
18. More specifically, for each technological class, the index corresponds to the Hirschman-Herfindahl Index (HHI) measuring the distribution of patent shares across the three-digit NACE industry sectors of the patent applicant.
19. This could also imply that such green innovations are special, and may have larger impacts than other technologies. However, this question still needs further research.
20. This is analogous to other cases such as health innovation to address international pandemics.
21. See Barrett (2010).
22. See, for example, Beattie et al. (2012) for a discussion of the case of solar PV.
23. For evidence of this see Johnstone and Haščič (2011).
24. De Coninck et al. (2008) define technology-oriented agreements as "including those international agreements that are aimed at advancing research, development, demonstration, and/or deployment of technologies" rather than on emissions reduction per se.
25. See discussion in IEA (2015).
26. OECD, Working Party of National Experts on Science and Technology Indicators, "Measuring Public Sector Innovation: A Review and Assessment of Recent Studies", 15 June 2011, DSTI/EAS/STP/NESTI(2011)8.
27. *www.oecd.org/governance/observatory-public-sector-innovation/home/*.
28. Daglio, M., .D Gerson and H. Kitchen (2014), "Building organisational capacity for public sector innovation", OECD Working Paper, OECD Publishing, Paris; and OECD (2015g), *Promoting Public Sector Innovation*, OECD Publishing, Paris.
29. Social innovation refers to innovation that occurs beyond government among actors in civil society and citizens, targeting social problems with the benefits accruing to society as a whole.

References

Acs, Z.J., D.B. Audretsch and M.P. Feldman (1994), "R&D spillovers and recipient firm size", *The Review of Economics and Statistics*, 76/2, pp. 336-40.

Aghion, P. et al. (2014). "Path dependence, innovation and the economics of climate change", *Policy Paper*, Centre for Climate Change Economics and Policy, Grantham Research Institute on Climate Change and the Environment.

Albrizio, S. et al. (2014), "Do environmental policies matter for productivity growth? Insights from new cross-country measures of environmental policies", *OECD Economics Department Working Papers*, No. 1176, OECD Publishing, Paris, *http://dx.doi.org/10.1787/5jxrjncjrcxp-en*.

Alston, J. (2010), "The benefits from agricultural research and development, innovation, and productivity growth", *OECD Food, Agriculture and Fisheries Working Papers*, No. 31, OECD Publishing, Paris, *http://dx.doi.org/10.1787/5km91nfsnkwg-en*.

Arrow, K. (1962), *The Rate and Direction of Inventive Activity: Economic and Social Factors*, National Bureau of Economic Research, Princeton University Press, Pri

Arthur, W.B. (2007), "The structure of invention", *Research Policy*, 36/2, pp. 274-287.

Audretsch, D. and M. Feldman (1996), "R&D spillovers and the geography of innovation and production", *The American Economic Review*, pp. 630-640.

Avvisati, F. et al. (2013), "Review of the Italian strategy for digital schools", *OECD Education Working Papers*, No. 90, OECD Publishing, Paris, *www.oecd.org/edu/ceri/Innovation%20Strategy%20Working%20Paper%2090.pdf*.

Baker, S., N. Bloom and S. Davis (2013), "Measuring economic policy uncertainty", *Chicago Booth Research Paper*, No. 2.

Barone, G. and F. Cingano (2011), "Service regulation and growth: Evidence from OECD countries", *Economic Journal*, Vol. 121/555.

Barrett, S. (2010), *Why Cooperate? The Incentive to Supply Global Public Goods*, Oxford University Press.

Beattie, N.S. et al. (2012) "Understanding the effects of sand and dust accumulation on photovoltaic modules", *Renewable Energy*, Vol. 48, December, pp. 448-452.

Bloom, N., M. Schankerman and J. Van Reenen (2013), "Identifying Technology Spillovers and Product Market Rivalry", *Econometrica*, Vol. 81(4), pp. 1347-1393.

Bourlès, R. et al. (2013), "Do product market regulations in upstream sectors curb productivity growth? Panel data evidence for OECD countries", *Review of Economics and Statistics*, Vol. 95/5, pp. 1750-1768.

Braconier, H., G. Nicoletti and B. Westmore (2014), "Policy *Challenges for the Next 50 Years*", *OECD Economics Department Policy Papers*, No. 9, OECD Publishing, Paris, *http://dx.doi.org/10.1787/5jz18gs5fckf-en*.

Cárdenas Rodríguez, M. et al. (2014), "Inducing private finance for renewable energy projects: Evidence from micro-data", *OECD Environment Working Papers*, No. 67, OECD Publishing, Paris, *http://dx.doi.org/10.1787/5jxvg0k6thr1-en*.

CASTED (Chinese Academy of Science and Technology for Development) (2014), *China's Report on Inclusive Innovation*, CASTED, Beijing.

Chiavari, J. and C. Tam (2011), *Good Practice Policy Framework for Energy Technology RD&D*, OECD/IEA, Paris.

Claxton, K. (2007), "OFT, VBP: QED?", *Health Economics*, Vol. 16/6, pp. 545-558.

Crescenzi, R., A. Rodríguez-Pose and M. Storper (2012), "The territorial dynamics of innovation in China and India", *Journal of Economic Geography*, Vol. 12, pp. 1055-1085.

Criscuolo, C. and C. Menon (2014), "Environmental policies and risk finance in the green sector: Cross-country evidence", *OECD Science, Technology and Industry Working Papers*, No. 2014/01, OECD Publishing, Paris, *http://dx.doi.org/10.1787/5jz6wn918j37-en*.

Criscuolo, C. et al. (2014), "Renewable energy policies and cross-border investment: Evidence from mergers and acquisitions in solar and wind energy", *OECD Science, Technology and Industry Working Papers*, No. 2014/03, OECD Publishing, Paris, *http://dx.doi.org/10.1787/5jxv9f3r9623-en*.

Daglio, M., D. Gerson and H. Kitchen (2014), "Building organisational capacity for public sector innovation", background paper prepared for the OECD conference *Innovating the Public Sector: from Ideas to Impact*, Paris, 12-13 November 2014.

De Coninck, H. et al. (2008), "International technology-oriented agreements to address climate change", *Energy Policy*, Vol. 36/2008, pp. 335-356.

Dechezleprêtre, A., N. Johnstone and I. Haščič (2015), "Invention and international diffusion of water conservation and availability technologies", *OECD Environment Working Papers* No. 82, OECD Publishing, Paris, *http://dx.doi.org/10.1787/5js679fvllhg-en*.

Ding, W.W. et al. (2010), "The impact of information technology on academic scientists' productivity and collaboration patterns", *Management Science*, Vol. 56/9, pp. 1439-1461.

Egli, F., C. Menon and N. Johnstone (2015), "Identifying and inducing breakthrough inventions: An application related to climate change mitigation", *OECD Science, Technology and Industry Working Papers*, 2015/04, OECD Publishing, Paris, *http://dx.doi.org/10.1787/5js03zd40n37-en*.

European Commission, *2013 EU Industrial R&D Investment Scoreboard*, Publications Office of the European Union, Luxembourg.

Fernandes, A. and C. Paunov (forthcoming), "The Risks of Innovation: Are Innovating Firms Less Likely to Die?", *The Review of Economics and Statistics*, forthcoming.

Görg, H. and D. Greenaway (2004), "Much ado about nothing? Do domestic firms really benefit from foreign direct investment?", *The World Bank Research Observer*, Vol. 19(2), pp. 171-197.

Gupta, R., S. Sankhe, R. Dobbs, J. Woetzel, A. Madgavkar and A. Hasyagar (2014), *From poverty to empowerment: India's imperative for jobs, growth, and effective basic services*, McKinsey Global Institute, McKinsey & Company, New York.

Feldman, H.H. et al. (2014), "Alzheimer's disease research and development: A call for a new research roadmap", *Annals of the New York Academy of Science*, Vol. 1313, *http://dx.doi.org/10.1111/nyas.12424*.

Foray, D. and J. Raffo (2012), "Business-driven innovation: Is it making a difference in education? An analysis of educational patents", *OECD Education Working Papers*, No. 84, OECD Publishing, Paris, *http://dx.doi.org/10.1787/5k91dl7pc835-en*.

Forbes Africa (2012), "Ranking of top 20 African startups", *Forbes Africa*, February, Forbes Magazine Press.

Foster, N.L. et al. (2014), "Justifying reimbursement for Alzheimer's diagnostics and treatments: Seeking alignment on evidence", *Alzheimer's & Dementia*, Vol. 10/4, pp. 503-508, *http://dx.doi.org/10.1016/j.jalz.2014.05.003*.

Frey, C.B. and M. Osborne (2013), "The future of employment: How susceptible are jobs to computerisation?", *Oxford Martin School Working Paper*, University of Oxford.

Galea, G. and M. McKee (2014), "Public-private partnerships with large corporations: Setting the ground rules for better health", *Health Policy*, Vol. 115/2-3, *http://dx.doi.org/10.1016/j.healthpol.2014.02.003*.

Grameen Bank (2013), *Annual Report 2012*, Grameen Bank.

Greene, J. and N.A. Braathen (2014), "Tax preferences for environmental goals: Use, limitations and preferred practices", *OECD Environment Working Papers*, No. 71, OECD Publishing, Paris, *http://dx.doi.org/10.1787/5jxwrr4hkd6l-en*.

Haščič, I. and N. Johnstone (2012), "Innovation in electric and hybrid vehicle technologies: The role of prices, standards and R&D", in OECD, *Invention and Transfer of Environmental Technologies*, OECD Publishing, Paris, *http://dx.doi.org/10.1787/9789264115620-5-en*.

Haščič, I. and M. Migotto (2014), "Measuring environmental innovation using patent data", *OECD Environment Working Papers*, No. 89, OECD Publishing, Paris, *http://dx.doi.org/10.1787/5js009kf48xw-en*.

Haščič, I. et al. (2015), "Public Interventions and Private Climate Finance Flows: Empirical Evidence from Renewable Energy Financing", *OECD Environment Working Papers*, No. 80, OECD Publishing, Paris, *http://dx.doi.org/10.1787/5js6b1r9lfd4-en*.

Heyes, A. (2009), "Is environmental regulation bad for competition? A Survey," *Journal of Regulatory Economics*, Vol. 36/1, pp. 1-28.

Hsieh, C.-T. and P. Klenow (2009), "Misallocation and Manufacturing TFP in China and India", *Quarterly Journal of Economics*, Vol. 124, pp. 1403-1448.

Hsieh, C.-T. and B.A. Olken (2014), "The Missing "Missing Middle'", *Journal of Economic Perspectives*, American Economic Association, Vol. 28(3), pp. 89-108, summer.

Hu, A.G., G.H. Jefferson and Q. Jinchang (2005), R&D and technology transfer: Firm-level evidence from Chinese industry, *Review of Economics and Statistics*, 87(4), pp. 780-786.

IEA (International Energy Agency) (2015), *Energy Technology Perspectives 2015*, OECD Publishing, Paris.

IEA (2013), *Tracking Clean Energy Progress*, OECD Publishing, Paris.

IEA and NEA (Nuclear Energy Agency) (2015), *Nuclear Energy 2015*, IEA Technology Roadmaps, OECD Publishing, Paris, *http://dx.doi.org/10.1787/9789264229938-en*.

IFC (2013), *IFC and Microfinance*, fact sheet, IFC, October 2013, *www.ifc.org/wps/wcm/connect/0cf7a70042429b19845aac0dc33b630b/Fact+Sheet+Microfinance+_October+2013.pdf?MOD=AJPERES*.

ITU (International Telecommunication Union) (2014), data on telecommunication statistics, *www.itu.int/en/ITU-D/Statistics/Pages/stat/default.aspx*.

Isen, A., M. Rossin-Slater and W.R. Walker (2014), "Every breath you take – every dollar you'll make: The long-term consequences of the Clean Air Act of 1970", *NBER Working Paper*, No. 19858, National Bureau of Economic Research, Cambridge.

Jaffe, A.B., M. Trajtenberg and R. Henderson (1993), "Geographic knowledge spillovers as evidenced by patent citations", *Quarterly Journal of Economics*, 108(3), pp. 577-598.

Johnstone, N., J. Labonne and C. Thevenot (2008), "Environmental policy and economies of scope in facility-level environmental practices", *Environmental Economics and Policy Studies*, Vol. 9/3, pp. 145-166.

Johnstone, N. and I. Haščič (2013), "Increasing the penetration of intermittent renewable energy: Innovation in energy storage and grid management", in R. Fouquet (ed.), *Handbook on Energy and Climate Change*, Edward Elgar.

Johnstone, N. and I. Haščič (2011), "Environmental policy design and the fragmentation of international markets for innovation", in V. Ghosal (ed.), *Reforming Rules and Regulations*, MIT Press, Cambridge, Massachusetts.

Kahrobaie, N., I. Haščič and N. Johnstone (2012), "International technology agreements for climate change: Analysis based on co-invention data", in OECD, *Energy and Climate Policy: Bending the Technological Trajectory*, OECD Publishing, Paris, *http://dx.doi.org/10.1787/9789264174573-8-en*.

Kalamova, M., I. Haščič and N. Johnstone (2013), "Implications of policy uncertainty for innovation in environmental technologies: The case of public R&D budgets", in V. Constantini and M. Mazzanti (eds.), *The Dynamics of Environmental and Economic Systems*, Springer, Berlin.

Kärkkäinen, K. (2012), "Bringing about curriculum innovations: Implicit approaches in the OECD area", *OECD Education Working Papers*, No. 82, OECD Publishing, Paris, *http://dx.doi.org/10.1787/5k95qw8xzl8s-en*.

Kärkkäinen, K. and S. Vincent-Lancrin (2013), "Sparking innovation in STEM education with technology and collaboration: A case study of the HP Catalyst Initiative", *OECD Education Working Papers*, No. 91, OECD Publishing, Paris, *http://dx.doi.org/10.1787/5k480sj9k442-en*.

Koh, H., A. Karamchandani and R. Katz (2012), *From Blueprint to Scale: The Case for Philantropy in Impact Investing*, The Monitor Group.

Kokko, A., R. Tansini and M.C. Zejan (1996), "Local technological capability and productivity spilloversfrom FDI in the Uruguayan manufacturing sector", *The Journal of Development Studies*, Vol. 32(4), pp. 602-611.

Kothandaraman, P., and S. Mookerjee (2008), "Healthcare for all: Narayana Hrudayalaya, Bangalore". *GIM Case Study*, No. A031, United Nations Development Programme, New York.

Koźluk, T. (2014), "The indicators of the economic burdens of environmental policy design: Results from the OECD questionnaire", *OECD Economics Department Working Papers*, No. 1178, OECD Publishing, Paris.

Krugman, P. (1991), "Increasing Returns and Economic Geography", *The Journal of Political Economy*, Vol. 99(3), pp. 483-499.

Lanjouw, J. and M. Schankerman (2004), "Protecting intellectual property rights: Are small firms handicapped?", *The Journal of Law and Economics*, Vol. 47(1), pp. 45-74.

Lanoie, P. et al. (2011), "Environmental policy, innovation and performance: New insights on the Porter hypothesis", *Journal of Economic Management and Strategy*, Vol. 20/3, pp. 803-842.

Lanzi, E., I. Haščič and N. Johnstone (2012), "The determinants of invention in electricity generation technologies: A patent data analysis", *OECD Environment Working Papers*, No. 45, OECD Publishing, Paris, *http://dx.doi.org/10.1787/5k92v111shjc-en*.

Lee, K. (2014), "Innovation and Upgrading for Inclusive Growth: Implications for LICs/MICs", presentation given at the Growth Dialogue Forum in Malaysia.

Lembcke, A., R. Ahrend and K. Maguire (2015), "Spatial Aspects of Inclusive Growth: The Distribution of Regional Benefits from Innovation", *OECD Regional Development Working Papers*, forthcoming.

Lerner, Josh (2011), "Venture capital and innovation in energy" in R. Henderson and R. Newell (eds.), *Accelerating Innovation in Energy*, University of Chicago Press.

Lubienski, C. (2009), "Do quasi-markets foster innovation in education? A comparative perspective", *OECD Education Working Papers*, No. 25, OECD Publishing, Paris, *http://dx.doi.org/10.1787/221583463325*.

Maes, J.P. and L.R. Reed (2012), *State of the Microcredit Summit Campaign Report 2012*, Microcredit Summit Campaign (MCS), Washington, DC.

Maskus, K. (2010), "Differentiated intellectual property regimes for environmental and climate technologies", *OECD Environment Working Papers*, No. 17, OECD Publishing, Paris.

Mendoza, R.U. (2011), "Why do the poor pay more? Exploring the poverty penalty concept", *Journal of International Development*, Vol. 23/1, pp. 1-28.

Milliman, S.R. and R. Prince (1989), "Firm incentives to promote technological change in pollution control", *Journal of Environmental Economics and Management*, Vol. 17, pp. 247-265.

Mullan, M. et al. (2013), "National adaptation planning: Lessons from OECD countries", *OECD Environment Working Papers*, No. 54, OECD Publishing, Paris, *http://dx.doi.org/10.1787/5k483jpfpsq1-en*.

Muto, M., and T. Yamano (2009), "The impact of mobile phone coverage expansion on market participation: Panel data evidence from Uganda", *World Development*, Vol. 37/12, pp. 1887-1896.

NInC (National Innovation Council, Government of India) (2013), "Financing innovation: The India Inclusive Innovation Fund", *Report to the People: Third Year*, National Innovation Council, pp. 12-14.

Nicoletti, G. and S. Scarpetta (2003), "Regulation, productivity and growth: OECD evidence," *OECD Economics Department Working Papers*, No. 347, OECD Publishing, Paris, *http://dx.doi.org/10.1787/5k483jpfpsq1-en*.

Ockwell, D. et al. (2010), "Enhancing developing country access to eco-innovation: The case of technology transfer and climate change in a post-2012 policy framework", *OECD Environment Working Papers*, No. 12, OECD Publishing, Paris, *http://dx.doi.org/10.1787/5kmfplm8xxf5-en*.

OECD (forthcoming), *Systems Innovation and Green Growth*, OECD Publishing, Paris.

OECD (2015a), *Innovation Policies for Inclusive Growth*, OECD Publishing, Paris, *http://dx.doi.org/10.1787/9789264229488-en*.

OECD (2015b), *The Future of Productivity*, OECD Publishing, Paris, *www.oecd.org/eco/growth/OECD-2015-The-future-of-productivity-book.pdf*.

OECD (2015c), *Addressing Dementia – the OECD Response*, OECD Publishing, Paris, *http://dx.doi.org/10.1787/9789264231726-en*.

OECD (2015d), "Enhancing Translational Research and Clinical Development for Alzheimer's Disease and other Dementias", *OECD Science, Technology and Industry Policy Papers*, No. 22, OECD Publishing, Paris. *http://dx.doi.org/10.1787/5js1t57jts44-en*.

OECD (2015e), *Dementia Research and Care: Can Big Data Help?*, OECD Publishing, Paris, *http://dx.doi.org/10.1787/9789264228429-en*.

OECD (2015f), *Mapping Channels to Mobilise Institutional Investment in Sustainable Energy*, OECD Publishing, Paris, *http://dx.doi.org/10.1787/9789264224582-en*.

OECD (2015g), *The Innovation Imperative in the Public Sector - Setting an Agenda for Action*, OECD Publishing, Paris, *http://dx.doi.org/10.1787/9789264236561-en*.

OECD (2014a), *Science, Technology and Industry Outlook 2014*, OECD Publishing, Paris, *http://dx.doi.org/10.1787/sti_outlook-2014-en*.

OECD (2014b), *Going for Growth Interim Report*, OECD Publishing, Paris, *http://dx.doi.org/10.1787/growth-2014-en*.

OECD (2014c), *OECD Reviews of Innovation Policy – France*, OECD Publishing, Paris, *http://dx.doi.org/10.1787/9789264214026-en*.

OECD (2014d), *OECD Reviews of Innovation Policy – Netherlands*, OECD Publishing, Paris, *http://dx.doi.org/10.1787/9789264213159-en*.

OECD (2014e), *Perspectives on Global Development 2014: Boosting Productivity to Meet the Middle-Income Challenge*, OECD Publishing, Paris, *http://dx.doi.org/10.1787/persp_glob_dev-2014-en*.

OECD (2014f), *OECD Reviews of Innovation Policy – Industry and Technology Policies in Korea*, OECD Publishing, Paris, *http://dx.doi.org/10.1787/9789264213227-en*.

OECD (2014g), *OECD Reviews of Innovation Policy – Colombia*, OECD Publishing, Paris, *http://dx.doi.org/10.1787/9789264204638-en*.

OECD (2014h), *OECD Reviews of Innovation Policy – Science, Technology and Innovation in Viet Nam*, OECD Publishing, Paris, *http://dx.doi.org/10.1787/9789264213500-en*.

OECD (2014i), *Dynamics of Dairy Farm Productivity Growth: Cross-Country Comparison*, OECD, Paris [TAD/CA/APM/WP(2014)37].

OECD (2014j), *Innovation for Agriculture Productivity and Sustainability: Review of Australian Policies*, OECD, Paris. [TAD/CA/APM/WP(2014)22/FINAL]

OECD (2014k), *Innovation for Agriculture Productivity and Sustainability: Review of Brazilian Policies*, OECD, Paris. [TAD/CA/APM/WP(2014)23/FINAL]

OECD (2014l), *Innovation for Agriculture Productivity and Sustainability: Review of Canadian Policies*, OECD, Paris. [TAD/CA/APM/WP(2014)24/FINAL]

OECD (2014m), *Agricultural Policy Monitoring and Evaluation 2014: OECD Countries*, OECD, Paris, *http://dx.doi.org/10.1787/agr_pol-2014-en*.

OECD (2014n), *All on Board: Making Inclusive Growth Happen*, OECD Publishing, Paris, *http://dx.doi.org/10.1787/9789264218512-en*.

OECD (2014o), "Time to terminate termination charges?", *OECD Insights*, 13 June, *http://oecdinsights.org/2014/06/13/time-to-terminate-termination-charges/*.

OECD (2014p), "International cables, gateways, backhaul and international exchange points", *OECD Digital Economy Papers*, No. 232, OECD Publishing, Paris, *http://dx.doi.org/10.1787/5jz8m9jf3wkl-en*.

OECD (2014q), *Measuring the Digital Economy*, OECD Publishing, Paris, *http://dx.doi.org/10.1787/9789264221796-en*.

OECD (2014r), *Measuring Innovation in Education: A New Perspective*, OECD Publishing, Paris, *http://dx.doi.org/10.1787/9789264215696-en*.

OECD (2013a), *OECD Reviews of Innovation Policy – Sweden*, OECD Publishing, Paris, *http://dx.doi.org/10.1787/9789264184893-en*.

OECD (2013b), *OECD Reviews of Innovation Policy – Knowledge-Based Start-Ups in Mexico*, OECD Publishing, Paris, *http://dx.doi.org/10.1787/9789264193796-en*.

OECD (2013c), *OECD Reviews of Innovation Policy – Peru*, OECD Publishing, Paris, *http://dx.doi.org/10.1787/9789264128392-en*.

OECD (2013d), *Perspectives on Global Development 2013: Industrial Policies in a Changing World*, OECD Publishing, Paris, *http://dx.doi.org/10.1787/persp_glob_dev-2013-en*.

OECD (2013e), *OECD Compendium of Agri-environmental Indicators*, OECD Publishing, Paris, *http://dx.doi.org/10.1787/9789264186217-en*.

OECD (2013f), *Agricultural Innovation Systems: A Framework for Analysing the Role of the Government*, OECD Publishing, Paris, *http://dx.doi.org/10.1787/9789264200593-en*.

OECD (2013g), *Start-up Latin America: Promoting Innovation in the Region*, OECD Development Centre Studies, OECD, Paris, *http://dx.doi.org/10.1787/9789264202306-en*.

OECD (2013h), *OECD Science, Technology and Industry Scoreboard 2013*, OECD Publishing, Paris, *http://dx.doi.org/10.1787/sti_scoreboard-2013-en*.

OECD (2013i), *OECD Regional Statistics* (database), OECD, *http://dx.doi.org/10.1787/region-data-en*.

OECD (2013j), *Strengthening Health Information Infrastructure for Health Care Quality Governance: Good Practices, New Opportunities and Data Privacy Protection Challenges*, OECD Health Policy Studies, OECD Publishing, Paris, *http://dx.doi.org/10.1787/9789264193505-en*.

OECD (2013k), *ICTs and the Health Sector: Towards Smarter Health and Wellness Models*, OECD Publishing, Paris, *http://dx.doi.org/10.1787/9789264202863-en*.

OECD (2013l), *Inventory of Estimated Budgetary Support and Tax Expenditures for Fossil Fuels*, OECD Publishing, Paris, *http://dx.doi.org/10.1787/9789264187610-en*.

OECD (2013m), "Biotechnology for the environment in the future: Science, technology and policy", *OECD Science, Technology and Industry Policy Papers*, No. 3, OECD Publishing, Paris, *http://dx.doi.org/10.1787/5k4840hqhp7j-en*.

OECD (2013n), "Nanotechnology for green innovation", *OECD Science, Technology and Industry Policy Papers*, No. 5, OECD Publishing, Paris, *http://dx.doi.org/10.1787/5k450q9j8p8q-en*.

OECD (2013o), *Innovative Learning Environments*, OECD Publishing, Paris, *http://dx.doi.org/10.1787/9789264203488-en*.

OECD (2013p), *Leadership for 21st Century Learning*, OECD Publishing, Paris, *http://dx.doi.org/10.1787/9789264205406-en*.

OECD (2012a), *Agricultural Policy Monitoring and Evaluation 2012: OECD Countries*, OECD Publishing, Paris, *http://dx.doi.org/10.1787/agr_pol-2012-en*.

OECD (2012b), *Survey on Budgeting Practices and Procedures*, OECD Publishing, Paris, *www.oecd.org/gov/budgeting/2012-Budgeting-survey.pdf*.

OECD (2011a), *OECD Reviews of Innovation Policy – Russian Federation*, OECD Publishing, Paris, *http://dx.doi.org/10.1787/9789264113138-en*.

OECD (2011b), *Regions and Innovation Policy*, OECD Publishing, Paris, *http://dx.doi.org/10.1787/9789264205307-en*.

OECD (2011c), *The Invention and Transfer of Environmental Technologies*, OECD Publishing, Paris, *http://dx.doi.org/10.1787/9789264115620-en*.

OECD (2010a), *Science, Technology and Industry Outlook 2010*, OECD Publishing, Paris, *http://dx.doi.org/10.1787/sti_outlook-2014-en*.

OECD (2010b), *SMEs, Entrepreneurship and Innovation*, OECD Publishing, Paris, *http://dx.doi.org/10.1787/9789264080355-en*.

OECD (2010c), *Taxation, Innovation and the Environment*, OECD Publishing, Paris, *http://dx.doi.org/10.1787/9789264087637-en*.

OECD (2009a), *OECD Reviews of Innovation Policy – Mexico*, OECD Publishing, Paris, *http://dx.doi.org/10.1787/9789264075993-en*.

OECD (2009b), *OECD Reviews of Innovation Policy – Korea*, OECD Publishing, Paris, *http://dx.doi.org/10.1787/9789264067233-en*.

OECD (2008a), *OECD Reviews of Innovation Policy – China*, OECD Publishing, Paris, *http://dx.doi.org/10.1787/9789264039827-en*.

OECD (2008b), *The Internationalisation of Business R&D*, OECD Publishing, Paris, *http://dx.doi.org/10.1787/9789264044050-en*.

OECD (2008c), *OECD Reviews of Innovation Policy – Hungary*, OECD Publishing, Paris, *http://dx.doi.org/10.1787/9789264054059-en*.

OECD (2008d), *Environmental Policy, Technological Innovation and Patents*, OECD Publishing, Paris, *http://dx.doi.org/10.1787/9789264046825-en*.

OECD (2003), *The Environmental Performance of Public Procurement*, OECD Publishing, Paris, *http://dx.doi.org/10.1787/9789264101562-en*.

OECD-Eurostat (2005), *Oslo Manual – Guidelines for Collecting and Interpreting Innovation Data, 3rd Edition*, OECD Publishing, Paris, *http://dx.doi.org/10.1787/9789264013100-en*.

Paris, V. and A. Belloni (2013), "Value in pharmaceutical pricing", *OECD Health Working Papers*, No. 63, OECD Publishing, Paris, *http://dx.doi.org/10.1787/5k43jc9v6knx-en*.

Paunov, C. (2014), "Democratizing Intellectual Property Systems: How Corruption Hinders Equal Opportunities for Firms", unpublished manuscript.

Paunov, C. and C. Lavison (forthcoming), "How to scale-up inclusive innovation? Policy lessons from a cross-country perspective", *OECD STI Working Paper*, forthcoming.

Poirier, J. et al. (2014), "The benefits of international co-authorship in scientific papers: The case of wind energy technologies", *OECD Environment Working Papers*, No. 81, OECD Publishing, Paris, *http://dx.doi.org/10.1787/5js69ld9w9nv-en*.

Popp, D. (2011) "International technology transfer, climate change, and the clean development mechanism", *Review of Environmental Economics and Policy*, Vol. 5, pp. 131-152.

Popp, D. and R.C. Newell (2012), "Where does energy R&D come from?", *Energy Economics*, Vol. 34/4, pp. 980-991.

Porter M. (2010), "What is value in health care?", *New England Journal of Medicine*, Vol. 363/26, pp. 2477-2481.

Prahalad, C.K., and S. Hart (2002), "The fortune at the bottom of the pyramid", *Strategy + Business*, No. 26.

Puttick, R, P. Baeck and P. Colligan (2014), *I-teams: The Teams and Funds Making Innovation Happen Around the World*, Nesta and Bloomberg Philanthropies, London.

Radjou, N., J. Prabhu and S. Ahuja (2013), *L'Innovation jugaad. Redevenons ingénieux!*, Éditions Diateino, Paris.

Rodrik, D. (2004), "Industrial policy for the twenty-first century", paper prepared for UNIDO, Harvard University, *https://myweb.rollins.edu/tlairson/pek/rodrikindpolicy.pdf*.

Salmi, J. (2013), "The race for excellence – A marathon not a sprint", *University World News* 254, 13 January, *www.universityworldnews.com/article.php?story=20130108161422529* (accessed on 7 March 2013).

Schieb, P.-A. and J.C. Philp (2014),"Biorefinery policy needs to come of age", *Trends in Biotechnology*, Vol. 32, pp. 496-500.

Scott, T.J. et al. (2014), "Economic analysis of opportunities to accelerate Alzheimer's disease research and development", *Annals of The New York Academy of Sciences*, *http://dx.doi.org/10.1111/nyas.12417*.

Siemens (2011), *Annual Report 2011: Creating Sustainable Cities*, Siemens, *www.siemens.com/annual/11/_pdf/Siemens_AR_2011.pdf*.

Stiglitz, J.E., A. Sen and J.-P. Fitoussi (2009), *Report of the Commission on Measurement of Economic Performance and Social Progress*, *www.stiglitz-sen-titoussi.fr/en/index.htm*.

Suárez Franco, C.F. (2010), "EPM: Antioquia Iluminada", *Growing Inclusive Markets Case Study*, No. C109, United Nations Development Programme.

Syverson, C. (2004), "Product Substitutability and Productivity Dispersion", *Review of Economics and Statistics*, MIT Press Journals, Cambridge, MA.

Tufts Center for the Study of Drug Development (2014), "CNS drugs take longer to develop and have lower success rates than other drugs", *Impact Report*, Vol. 16/6.

Tybout, J. (2000), "Manufacturing Firms in developing countries: How well do they do, and why?", *Journal of Economic Literature*, Vol. 38/1, American Economic Association, pp. 11-44.

Von Hippel, E. (2005), *Democratizing Innovation*, MIT Press, Cambridge, Massachusetts.

The Innovation Imperative
Contributing to Productivity, Growth and Well-Being

Chapter 8

Governance and implementation of innovation policies

This chapter explores a number of issues that are key to the governance and implementation of innovation policies. The first section of the chapter focuses on how to improve multi-level governance, i.e. in aligning policy actions across different ministries, agencies and stakeholders, but also across different levels of governments and across borders. The second section focuses on questions of trust in policies related to innovation, including the role of public engagement and risk management related to innovation. The final section focuses on specific questions related to the implementation and evaluation of innovation policies, including how governments can overcome challenges related to the implementation of reforms.

The statistical data for Israel are supplied by and under the responsibility of the relevant Israeli authorities. The use of such data by the OECD is without prejudice to the status of the Golan Heights, East Jerusalem and Israeli settlements in the West Bank under the terms of international law.

8.1. Governance and policy coherence

Why governance matters for innovation

Innovation is an area of policy making that is characterised by multiple levels of governance, which implies that the public authorities in charge of innovation belong to various levels of authority and policy competences, and budgetary resources are distributed across these levels of government. Moreover, even within a single level of governance, effective innovation policies need to be joined up across a broad set of policy and delivery areas, including tax, science, education, immigration, enterprise, foreign direct investment (FDI) and even health policy. This increases the number of actors, organisations, agendas and policies to be co-ordinated in order to achieve coherent policies. Two important developments have contributed to an increase in the importance of various levels of government:

- The first is **globalisation**, which is characterised by the emergence of new powerful internationalised actors – multinational enterprises (MNEs) – and by the geographical extension of the scope of innovation partnerships and competitive pressures. Intergovernmental organisations and international instruments also increasingly shape governance regimes, notably in Europe where the European Commission plays a prominent role in supporting research and innovation agendas (OECD, 2010a).
- Secondly, **regionalisation and decentralisation** have made local and regional governments more powerful and increased their capacity to design and operate their own development policies (OECD, 2005a, 2010b, 2014a).

Research and development (R&D) and innovation activities are increasingly global, thanks to the shifting international organisation of functions within MNEs, which are internationalising their R&D at faster pace and on a larger scale than before (OECD, 2010a). The increasing complexity of innovation and pervasiveness of new technologies generate a drive towards much wider partnerships extending over national borders. As a response to this trend, national innovation policy is increasingly framed in global terms, reflecting a growing sense of global identity, the global nature of many problems and issues, and the globalisation of markets and production. National innovation policies aim at fostering domestic innovation potential, strengthening capabilities and developing local skills, and may also have an explicit goal of raising the attractiveness of the country for global innovation performers and mobile talents (OECD, 2014b).

Policies, both at national and at supranational levels (the European Union [EU] R&D framework programmes are an emblematic case), foster the participation of national actors in international R&D and innovation networks. Indivisibilities, notably in very large research infrastructures, create the need for international policy co-operation in funding and exploitation of such resources. Education systems are also increasingly open, through alignment of curricula and permeability of education trajectories across national borders as well as incentives to foster the international mobility of students, teachers and researchers.

The importance of regions in innovation policy is linked to the following two trends (OECD, 2011). First, the growing role of regions in national innovation policies, which reflects the importance of place and proximity for innovation. Second, a shift in regional development policy paradigms, with an increasing focus on the mobilisation of knowledge assets and the promotion of innovation for endogenous growth, while moving away from a redistribution policy paradigm.

Subnational authorities – mainly regions, but also cities in some countries – have therefore become important actors in shaping and implementing innovation policies with a territorial dimension. Exploiting benefits of spatial and cultural proximity between firms and other organisations, such as universities and public research institutes (PRIs), is at the root of these efforts, as proximity facilitates the flow of tacit knowledge that is important for developing new combinations of ideas, technologies and skills as well as for spurring co-operation and interactive learning processes (see Chapter 6).

As a result of these two major developments, the governance of innovation policy is characterised by a shift from the national level as the unique locus for the design and implementation of innovation policy, to both the international and subnational levels. Those two levels have experienced a rise in relevance and an increase in policy activity in innovation, while often targeting the same innovation actors as national policies. To achieve policy coherence, multi-level governance of innovation calls for a clear division of labour and strong complementarities among these various levels, and it seeks to avoid wasteful forms of policy overlap or policy gaps.

Multi-level governance is complex, however; it requires that authorities at various levels possess not only the right capabilities and resources to effectively run their own competencies, but also the capacity and means to enter into negotiations, align their policies and conclude agreements with authorities from other levels. Failures in capabilities and resources can have important effects on the processes and outcomes of multi-level governance efforts.[1] The recently adopted *OECD Recommendation of the Council on Effective Public Investment across Levels of Government* (OECD, 2014c) offers a framework for addressing public investment in a multi-level governance context, confronting challenges in co-ordination, capacity (particularly at the subnational level) and framework conditions.

Co-ordination across levels of government is just one dimension in the complexity of patterns for innovation governance. Another complementary dimension is that of co-ordination between policy domains at the same level of government, i.e. horizontal governance. Sound innovation policy governance incorporates both multi-level (or vertical) and horizontal considerations to achieve policy coherence.

In addition, the emergence of public-private partnerships in policy making result in more diversified patterns of stakeholders and further increase the complexity of multi-level governance. Public actors responsible for developing and implementing innovation strategies and policy instruments therefore need a skill set that increasingly involves a networking function with stakeholders outside of government.

Multi-level governance in innovation policy faces several challenges. First, articulating goals and means among various levels of government is a time- and resource-consuming process, involving high transaction costs. Moreover, achieving good co-ordination and alignment of policies is hampered by imperfect information. Also, weak horizontal governance at one level of government makes the negotiations and search for better articulation of policies with another level of government more cumbersome. And lack of resources and public budgets for innovation at one level acts as a barrier to co-ordination efforts with other levels. Finally, the need for flexibility in multi-level arrangements often comes in conflict with a need for stability and long-term perspective in policy.

Improving horizontal governance

Innovation, like growth/competitiveness, climate change mitigation, managing demographic change and crisis management, is an issue requiring policy responses that do not fit neatly with the competencies of any single governmental department or agency. Citizens and businesses increasingly expect public policies and services to be seamless and responsive, not defined by administrative structures. Good governance for innovation is therefore about joint action, where administrations work in a co-ordinated and collaborative manner across boundaries and take a user-based perspective, e.g. in providing a single point of contact for firms or academic institutions. Poor co-ordination can increase the risk of duplication, inefficient spending, a lower quality of service, and contradictory objectives and targets. In a context of budgetary pressure, improved co-ordination has become an imperative.

Yet government structures and policy toolkits struggle to keep pace with this rapidly evolving operating environment. Renewal of the strategic capacity, organisational design and management structures of the public sector can certainly be helped by new technologies, but better governance starts from more basic foundations, notably effective leadership and better mobilisation of resources across government. Some of these foundations are explored in Box 8.1.

Box 8.1. **Improving governance: The role of the Centre of Government**

In the 2013 OECD survey of Centres of Government (OECD, 2014d), a majority of countries (some 59%) confirmed that the number of cross-ministerial policy initiatives has increased over the past few years and almost all reported that leading policy co-ordination has now become one of the priority tasks of the Centre of Government. The Centre plays this leadership role in different ways: 1) integrating cross-disciplinary perspectives into policy advice for the head of government and/or cabinet; 2) leading policy co-ordination via both traditional committee architectures and more innovative and informal channels; 3) facilitating resource sharing through a closer partnership with ministries of finance; and 4) supporting experimentation and testing of new delivery systems, many of which are based on shared service models.

There is no doubt that across large organisations, co-ordination is always difficult to achieve, for many reasons including inertia and strong silo cultures; gaming and other incentive problems; and inflexible financial resource systems. The OECD survey of Centres of Government indicates that 62% of Centre of Government officials consider that they exert a moderate degree of influence over ministries to promote co-ordination. Maintaining this influence depends on avoiding certain risks; in particular, interaction between the Centre and departments must be consistent and structured, not an ad hoc approach determined by time and resource availability.

Co-ordination has been traditionally addressed by means of inter-ministerial bodies, usually chaired by a senior official from the presidency or prime minister's office. More than three-quarters of countries have co-ordination groups at the state secretary level, with

Box 8.1. **Improving governance: The role of the Centre of Government** *(cont.)*

additional co-ordination groups at the director level and at head-of-unit or expert level. The new emphasis on strategic management of cross-sectoral policy initiatives is testing these traditional methods of co-ordination, which were designed to manage more "routine" co-ordination issues. The Centre of Government is now playing more of a leadership role with respect to specific strategic priorities, designing action plans in co-operation with relevant departments and leading on project management. In some cases, these are issue-specific or one-off reform initiatives, though often wide-ranging and sometimes controversial. A prominent example is that of austerity-era public-sector reform, but others include economic recovery strategies and transition to a low-carbon economy.

Exercising influence and guiding reform across several ministries has to be done in a way that does not create resistance or suggest over-centralisation. Several countries' Centres now provide technical and advisory support to line ministries, e.g. project and programme management skills, to help them adjust and meet the extra demands of horizontal projects. This makes it easier to integrate horizontal working or participation in horizontal projects into performance management systems at either the organisational or the individual level. Such systems represent the main incentive to participate in cross-departmental initiatives (accounting for almost 60% of such incentives), with financial incentives less common (accounting for some 30%).

The horizontal co-ordination processes managed by the Centre of Government are increasingly broad in scope and participation. For example, multilateral action on economic, social and environmental issues now touches the core of domestic public policy in individual countries. Effective working at international level is an increasingly important aspect of good governance at the domestic level and comes increasingly into the sphere of the Centre of Government (even though, as the survey indicates, formal responsibilities for international aid and foreign policy usually lie outside the Centre itself). International regulatory co-operation as a way to unblock trade negotiations and co-ordination of complex climate change responses are two examples of global debates that are influencing domestic policy making.

Centres of Governments also understand the need to innovate to promote innovation. For example, having space to experiment is commonly mentioned as an important factor in innovation. The Centre of Government can play a crucial role in several of the enablers of innovation – for example, getting senior government leaders on board, making innovation an accepted "part of the day job", experimenting and promoting the adoption of new ideas across the whole of government. For some countries, this is very clearly part of the remit of the Centre, either by using its own resources or by working closely with other governmental and non-governmental organisations to develop "prototypes". The Centre can also play a role in working with departments individually and collectively to organise the transition from existing systems and infrastructure to new ones. For departments, and for the civil servants working directly with existing systems, it can be difficult to envisage adopting something entirely new.

Source: OECD (2014d), *Centre Stage – Driving Better Policies from the Centre of Government*, *www.oecd.org/gov/Centre-Stage-Report.pdf*.

One specific example of the need for better governance related to innovation concerns systems innovation. For example, there is a growing understanding that system-wide changes are necessary to make economies socially, economically and environmentally sustainable. Although many national governments have put sustainability and green growth objectives at the centre of their economic development strategies, achieving this goal will require wide-ranging changes in their underlying economic, technological and social systems, from transport, water and energy systems to modes of consumption and waste management. Ensuring that socio-technical systems move towards greater sustainability is a major challenge for governments, but also for civil society.

At the core of such transitions is a shift in governance structures that not only allows change to occur but also directs and orchestrates some of the changes. The "smart city" initiatives that mobilise technological and social innovations to make the production and consumption of a city's goods and services more sustainable illustrate this point (e.g. Santiago Smart City in Chile). At the national level also, improved governance mechanisms and better means of engaging a range of stakeholders are needed to facilitate system innovation. Finland and the Netherlands, for example, have developed public-private partnerships to foster co-ordination and alignment (Strategic Centres for Science, Technology and Innovation [SHOKs] in Finland and the Top Sectors approach in the Netherlands)]. Another example is Denmark's approach to Growth Teams (Box 8.2).

Box 8.2. **Denmark's approach to smart specialisation**

In order to exploit the opportunities of globalisation to the fullest, the Danish government introduced in 2011 a new approach to business and growth policy addressing areas where the Danish private sector had gained a strong position internationally. Eight "Growth Teams" were set up with business leaders and other experts as members tasked with evaluating how changes to regulation, public-private partnerships, etc. could promote further growth.

Based on their recommendations, the Danish government has published specific plans for growth for each of the following eight areas: Blue Denmark (maritime transport, etc.); Creative Industries and Design; Water, Bio and Environmental Solutions; Health and Care Solutions; Energy and Climate; Food Sector; Tourism and Experience Economy; and ICT and Digital Growth.

The rationale for this approach was not to "pick the winners", but rather to acknowledge the revealed comparative advantages of the eight areas and their significance for the Danish economy, their differing interactions with government and regulations, and their potential to contribute to solving grand societal challenges such as climate change, public health and a growing global population – an approach also reflected in the Danish Innovation Strategy. This approach did not mean focusing on subsidies for specific sectors but rather on complementary policy measures such as reduced or smarter regulation, a sufficient supply of labour with relevant competencies, and improving public-private partnerships, e.g. in R&D.

Thus the plans for growth address the sector-specific barriers to investment and focus on areas where new markets can be developed. A specific example is better government regulation of the waste water sector, where efficiency improvements will help to develop cost-effective technology that can also underpin further exports of Danish solutions. It could also be in relation to corporate R&D activities, where the creation of a single, transparent access point to Danish health data will strengthen medical research in Denmark and attract new R&D activities.

Improving the very general framework conditions remains a key pillar of Danish growth policies, but the work of the Growth Teams has shown that there are a range of more detailed policy issues that are addressed well only through a sector-specific approach.

Making the most out of regional and local science, technology and innovation efforts

Cities and regions are increasingly taking explicit policy measures to support innovation in their jurisdictions. Cities are focused more on urban development projects that seek to co-locate firms (often high-tech or creative industries) with universities, research centres, etc. This takes the form of a city district or a more contained science and technology (S&T) park or small business incubator, in some cases around selected "clusters". Regional governments have a wider range of tools at their disposal. In terms of high-level strategic bodies and technology foresight exercises, regions are almost as active as national governments. Regions are also financing R&D in public entities, but to a somewhat lesser extent in private entities. Technology transfer activities and innovation advisory services to existing and start-up firms are promoted by regions in most member countries. Programmes to support clusters and excellence hubs are frequently used, but more so at regional than national level. This is also the case with incubators and science and technology parks (OECD, 2011).

It is common for both national and regional levels in the same country to use similar policy instruments, which may be complementary or duplicative actions. There is not a strict division of labour across levels of government in terms of science, technology and innovation (STI) instruments. Country structure (federal, unitary elected regions, unitary administrative regions) does not appear to determine the number of instruments at regional level or the share in common with the national level (Figure 8.1). One explanation for the large number of instruments reportedly used in common at both levels is the diversity of ways an instrument may be used. The same type of instrument may be complementary by having a different configuration, target group, territorial scope or operating approach. In some cases the same instrument may be used by both levels with active co-financing and thus be aligned. Some redundancies across levels are difficult to avoid and may actually reinforce system stability.

However, excessive redundancy can be attributed to a lack of awareness of the instruments developed at another level of government or a failure to distinguish between target groups or topics in the instruments offered. The result can be wasted resources, insufficient scale of the smaller programmes and excessive complexity for target groups (i.e. firms, research institutions). Streamlining the range of policy instruments is therefore an important task for government, although one that often proves hard to achieve.

Furthermore, regional governments account for a notable and growing share of public spending on innovation-related matters. In countries such as Belgium, Germany and China, subnational government shares of different aspects of STI spending (typically public expenditure on R&D) can be 50% or greater. These counts exclude many business development programmes that also support innovation. In other countries, that share may be less than 10%, such as for Austria, a federal country, or Denmark, a unitary country (Figure 8.1). Countries almost uniformly reported an increase in the regional share over the last five years.[2] Increases in such STI spending by subnational levels result from broader decentralisation trends, including financing and policy responsibilities.

Regions and cities can also be a better scale for the development of certain policy innovations that can be generalised nationwide. Cities, in particular, are critical sources of national growth and play disproportionately large roles in countries' economies, knowledge generation and environmental performance. Compared with higher levels of government, cities offer more easily identifiable policy synergies and complementarities. Urban policy makers are more likely to identify and combine complementary climate policies within and across sectors, given the interconnectedness of urban systems such as transport, land-use planning and economic development (OECD, 2010c). For example, cities are responsible for a significant share of green infrastructure investments (Figure 8.2).

Figure 8.1. **Many STI instruments are used by both national and regional governments**

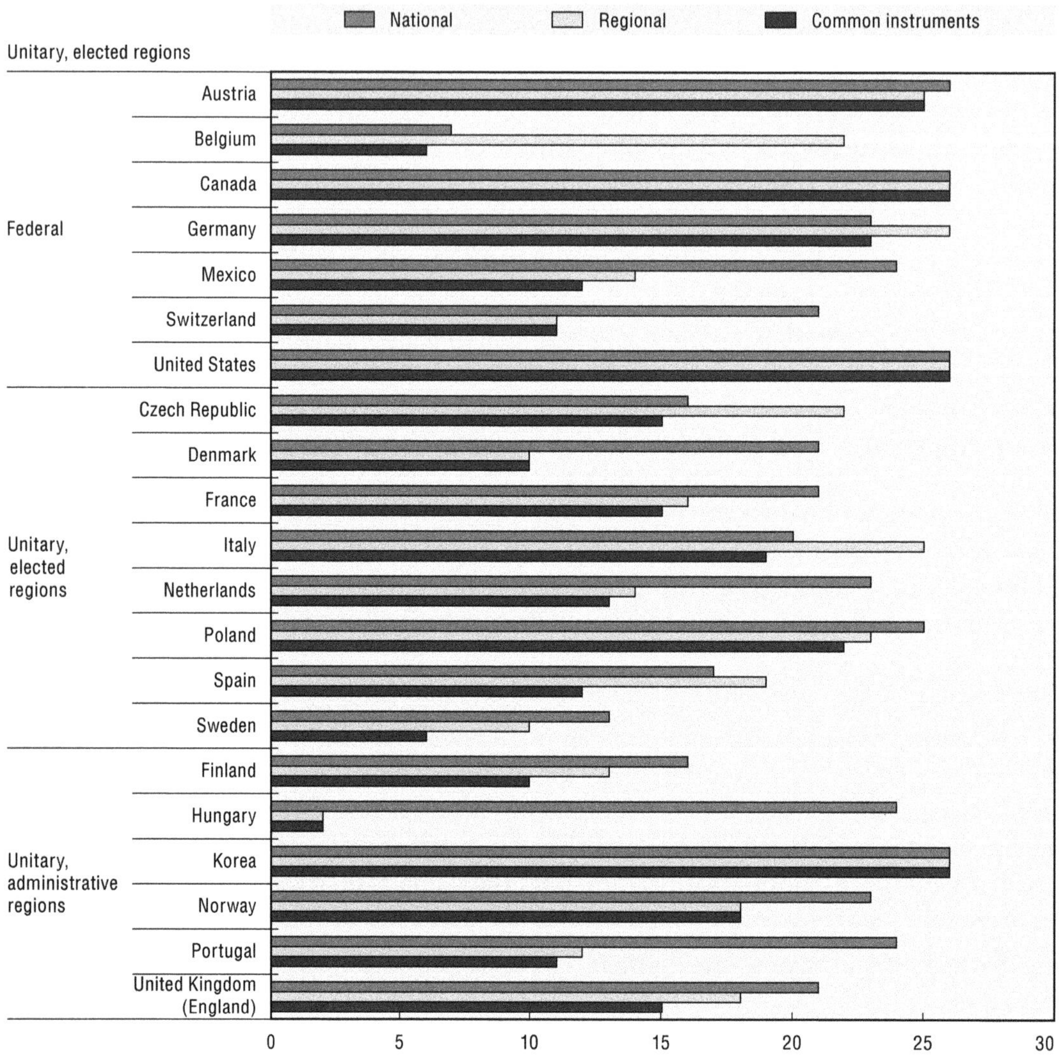

Notes: National refers to the number of instruments used at national level, regardless of whether they are used at other levels. Regional refers to instruments reported at regional level, regardless of whether they are used at other levels. The "common instruments" category refers to the number of instruments reported at both national and regional level, which includes those instruments reported in the tally of national and regional instruments.

Source: OECD (2011), *Regions and Innovation Policy, OECD Publishing, Paris, http://dx.doi.org/10.1787/9789264097803-en* using data from OECD (2009a), OECD-GOV Survey on the Multi-level Governance of Science, Technology and Innovation Policy.

Cities are also the places in which smart innovative approaches, driven by information and communications technologies (ICTs), analysis of (big) data and machine-to-machine communication, naturally arise. Smart cities often target different aspects of urban development, such as transport, electricity grids, buildings, or the delivery of public services in fields such as healthcare or education. Beyond governance challenges raised by the many levels of government and of stakeholders involved, smart cities are likely to improve citizens' well-being and increase the efficiency of the urban system as a whole.

Figure 8.2. **Gross capital formation in environmental protection by level of government, 2012**

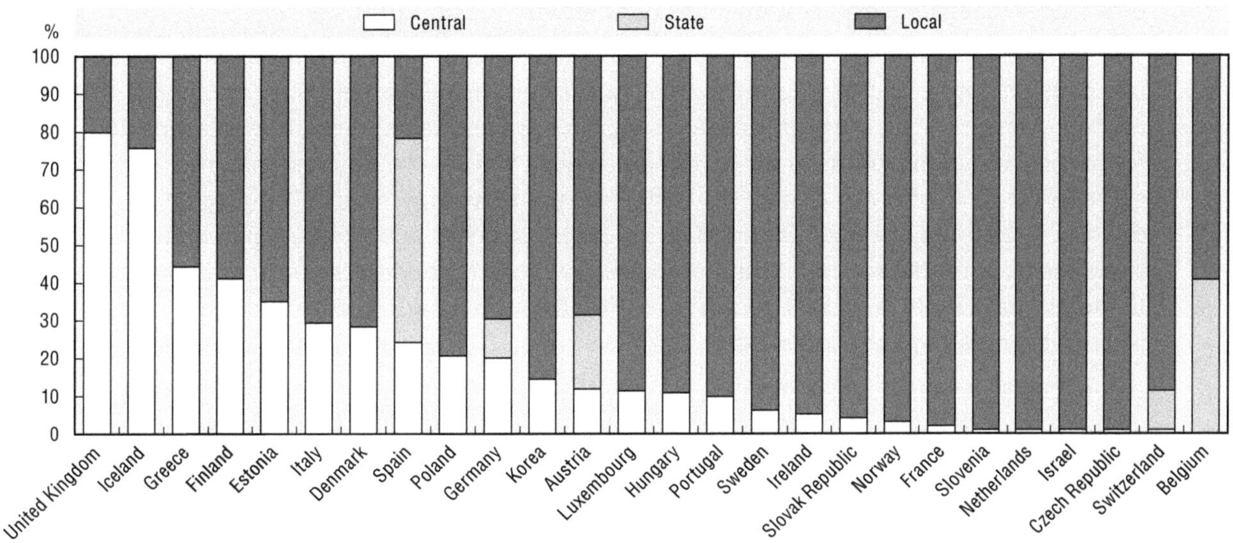

Note: State government data only for Austria, Belgium, Germany and Spain.

Source: OECD, *National Accounts Database*, April 2014, *http://data.oecd.org/*, based on OECD (2013a), *Regions at a Glance*, OECD Publishing, Paris, *http://dx.doi.org/10.1787/reg_glance-2013-en*.

The number and diversity of situations across regions and cities lends itself to policy experimentation in country contexts where they have the flexibility to so. A national government can then see what works and what doesn't before introducing a suitable adjusted national programme (see also the discussion on experimentation in Section 8.3 of this chapter). Better aligned national and regional strategies and policy actions are therefore essential, and can improve the policy effectiveness of both. Given the reality of increasing subnational policy action, the imperative for greater coherence is stronger than in the past. In addition, that coherence can provide greater impacts from respective policy action. For example, a national government may finance a public research entity, while a regional initiative to link up firms in the area with the research facility could help extract more economic benefits from that national investment. National policy makers may have the data or analyses to inform regional-level policy making while regional public agencies may actually have better information on the needs of local firms that could inform national policy. In both of these examples, sharing information across levels of government becomes critical for better intergovernmental co-ordination of STI policy.

Clarity with respect to the role of regions for STI policy in a given country context sets the stage for better co-ordination. Regions may be active in different parts of the national policy process: 1) setting the overall strategy and framework; 2) developing policy; 3) financing policy; 4) implementing programmes and instruments; and 5) assessment/ evaluation (of strategies, programmes and instruments). Some national STI strategies consider regions as relevant given the spatial dimension to the innovation process and the role of certain innovation hubs. In some countries with stark variations in regional capacity, national plans have highlighted that addressing this imbalance is a way to improve national performance. The role of regions for STI policy in federal countries may be set by the constitution, a specific S&T law that gives a role for regions or an administrative act outside of STI policy, such as in the regional development field. Generally, but not always, these formal roles are the same for all regions in a country. Even if regions do have similar formal powers, there may be de facto asymmetric decentralisation of STI policy due to differences in regional capacity, financial or otherwise (OECD, 2011).

Co-ordinating national and regional efforts takes multiple channels in OECD countries, but consultation and dialogue are ranked most common and effective (Figure 8.3). In addition, some national governments have agencies or national representatives responsible for particular regions that help to co-ordinate actions across levels of government. In other cases, some form of contract may be used to ensure financing for an STI policy objective, or co-financing may be used for specific STI projects. There is no right approach *a priori*, since the use of different co-ordination tools may prove more or less effective in practice. However, ultimately any tool to improve information sharing will underpin greater intergovernmental policy coherence.

Regional strategy development is another important tool for clarifying and communicating regional priorities, but such strategies need to be based on a valid assessment of regional strengths and needs. Generally OECD regions have some form of business development or explicit innovation strategy, of varying degrees of quality. The use of a smart specialisation strategy for European regions, a condition for receiving European Structural and Investment Funds, was implemented to help regions clarify their relative strengths to ensure a better use of funds for more realistic and less duplicative investments (see Chapter 6).

Figure 8.3. **Aligning STI actions between national and regional governments**

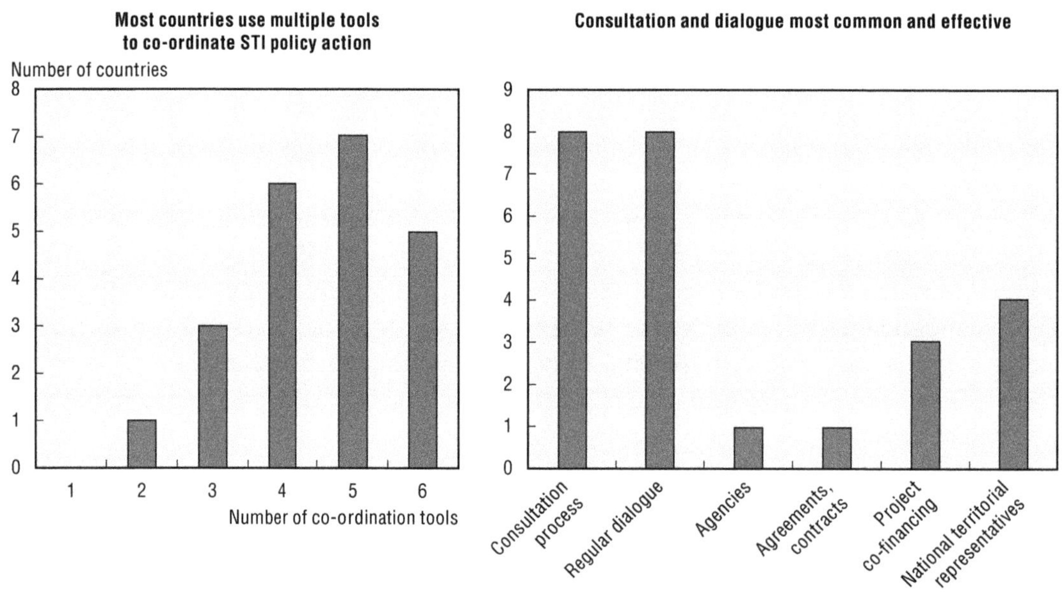

Source: OECD (2011), *Regions and Innovation Policy, OECD Publishing, Paris, http://dx.doi.org/10.1787/9789264097803-en* using data from OECD (2009a), *OECD-GOV Survey on the Multi-level Governance of Science, Technology and Innovation Policy.*

National strategies and labelling mechanisms can serve as strong alignment tools, but could be more effective if they involve regions in the development and implementation phases. It is very common for regions in a country to align their priority sectors with national priorities. They do so in part to better capture national STI funds in their regions. The quality and relevance of a particular region's role towards this national picture requires some clarifications, as not all regions can be strong in all priorities.[3] The strategy development process in some OECD countries involves a consultation process with regions of varying degrees (from participating in a workshop to a more formal role). A national government designation of a cluster, research centre or science park as "world class" or of national significance also serves to gather funding from different levels around commonly recognised innovation assets.

In some cases, the interests of national and regional innovation policy efforts may differ for good reasons, so national policies need to allow for some flexibility and promote capacity building. Ultimately a region needs to consider what is most relevant for its industrial structure and assets. In some regions, research centres and universities may simply be less competitive in national calls or specialised in research areas that are not top priorities at a national scale. However, these regions still need to support innovation-driven economic development, which may require some flexibility in certain national programmes. In cases where regions simply do not have assets that are considered of national priority, the national government may wish to support capacity building by actors in the regional innovation system, from firms and universities to the public sector itself.

Cross-border governance

Cross-border governance of STI involves at a minimum co-ordination, if not the partial or total delegation of policy making from the national to the international level. It implies, among other things, international co-ordination of national policy initiatives, removal of obstacles to the movement of resources, setting of international standards and regulations, and transfer of authority to intergovernmental organisations and supranational authorities. It is part of a wider dual delegation process that gives a greater say in STI matters to the international, but also to the subnational, level of governance (see Chapter 5).

In terms of international STI policy goals, most countries seek efficiency and/or effectiveness gains from complementarities in orientation, planning, regulation and resource pooling. However, national governments focus on domestic challenges and can be reluctant to take a global or even a collective view. The economic and financial crisis has increased that reluctance, as has the emergence of STI as a focus of industrial policy. Countries also have concerns about the appropriation of the benefits of public investments in education, research and innovation, given the increasing international competition for scarce talent and investment. As a result, narrower objectives often determine the nature and extent of national involvement in cross-border STI initiatives. These range from foreign policy and economic diplomacy, to access to funding for the development of national STI capabilities and access to international scientific networks. Commitment to cross-border STI policy is therefore often shaped by contingency and tends to vary over time. Ultimately, the reluctance to internationalise aspects of STI governance reflects the limitations of existing arrangements to provide credible assurances about the distribution of the resulting costs and benefits.

Extensive international policy co-operation frameworks for R&D have developed in Europe, but international frameworks in other regions and STI areas are still in their infancy. For instance, much can still be achieved by developing technological standards related to the environment or improving international co-ordination on cybersecurity.

Cross-border governance of STI can be achieved through arm's-length policies, such as bilateral or multilateral agreements of limited duration or co-ordination of national policies, without delegation to a supranational body. This seems to be the preferred approach to cross-border governance of STI outside of Europe. Even within Europe, international STI governance frameworks – by far the most developed of their kind globally – have historically been designed as complements, rather than substitutes, to national frameworks.

However, a number of STI policy areas can benefit from delegation of decision making and deeper integration. These are areas characterised not only by high fixed costs but also by high international transaction costs owing to the need for access on equal terms to highly specialised, single-purpose assets (e.g. the European Organization for Nuclear Research [CERN] and the International Thermonuclear Experimental Reactor [ITER]) as well as a high frequency of interaction and high uncertainty. A recent example of the latter is the high-risk, high-potential research funded by the European Research Council (ERC), which can maximise success by drawing from the largest possible pool of excellent scientists. Outside of Europe, the Consultative Group of International Agricultural Research (CGIAR) is an example of a long-term strategic arrangement with common R&D programming and performance functions (OECD, 2012b).

OECD analysis of the challenges in establishing international research infrastructures, including shared financing, governance and legal frameworks, suggest there is no one-size-fits-all model for such facilities. However, lessons can be learned, for example from different cost-sharing arrangements. Increasingly, and partly related to the move towards more open science and innovation, international distributed research infrastructures have become more prevalent. These are located across several countries, share a common purpose and are co-ordinated in some way but otherwise can be of very variable geometry.[4]

There have recently been ambitious initiatives to promote cross-border governance of STI in several regions, including Southeast Asia and Latin America, e.g. the Association of Southeast Asian Nations (ASEAN) Committee on Science and Technology. However, unlike Europe, they have a short history and limited continuity to date. The case of Europe is unique, in that its progress in cross-border governance of STI is part of wider economic integration.

Cross-border innovation policies are also relevant at the regional scale. When functional regions for innovation do not conform to national administrative boundaries, a need arises for alignment of policies across the borders involving, in most cases, both subnational and national authorities active in the cross-border area (OECD, 2013b). In Europe, policies dedicated to territorial co-operation across national borders add a supranational level that needs to also be aligned with strategies and policies at national and local levels.

In response to the *OECD STI Outlook 2014* policy questionnaire, many national authorities reaffirmed their commitment to cross-border governance of STI, specifying some of their reasons, but also listing important barriers and policy initiatives to lift them (OECD, 2014b). Mutual policy learning and the transfer of good practices appear to be important motivations for engaging in international STI forums. Global "grand challenges" such as climate change and threats to health and resource sufficiency are also strong motivators for international co-operation. Other countries see unexploited scale economies as the major challenge, though with some countries citing the lack of dedicated funding for large-scale and longer-term co-operation as a key obstacle. Fragmentation of funding agencies – and of the rules and procedures for research funding – is considered an important obstacle. A number of countries also mentioned barriers to cross-border governance of STI, including the absence of comprehensive national policies or mechanisms for domestic co-ordination of cross-border governance arrangements as an important constraint. Progress on these and other areas can potentially be made and could also help to strengthen international frameworks that affect innovation across the globe.

Below are some key policy learnings from the OECD's work on governance and innovation.

Main policy messages on governance of policies for innovation

- The governance and implementation of national innovation strategies are critical to their success. The process of developing a national strategy requires early and adequate involvement of stakeholders, including business, academia, social partners and key actors.
- Given the wide range of policies that affect innovation, it is important to ensure that the full set of government policies that affect innovation are well aligned, not only at the level of central government, but also between the central government and regional and local authorities, many of which are actively involved in innovation activities.
- The development and implementation of innovation policies requires strong capabilities within the public sector, including in ensuring the support of stakeholders for policy actions.
- The growing importance of governance also reflects a new approach to policies for innovation in many countries, where governments are increasingly acting as a facilitator in the face of complexity and uncertainty, enabling closer co-ordination between individual economic agents as well as fostering greater experimentation in the economy. This includes greater emphasis on building networks, improving co-ordination and regulation, promoting awareness, and less reliance on government funding.
- Cross-border governance is of growing importance for innovation, to help address common challenges, share costs and benefit from mutual learning. But governance mechanisms are underdeveloped and are affected by several barriers, including fragmentation of funding and lack of dedicated funding for long-term co-operation.

8.2. Trust, public engagement and risk governance

As discussed in the previous section, the success of innovation strategies is heavily influenced by the engagement of stakeholders in the development of policies and strategies. It is also influenced by a number of broader factors, notably the degree of trust in government and the degree of public engagement in science and innovation policies. Moreover, how governments manage and govern risks related to innovation is a major factor for innovation. This section discusses these three aspects.

Ensuring trust in government

On average, only 40% of OECD citizens today report that they trust their government, while 57% feel that corruption is widespread in business (Gallup, 2013). Low levels of trust can reduce compliance with laws and regulations, diminish investor confidence, and increase risk aversion, which is likely to have an impact on innovation. Levels of trust were already low, but were particularly hard hit by the crisis, which stemmed from numerous regulatory failures and mismanagement not only by government but also by business. As a result, trust in public institutions and in certain private sector institutions – banks and financial institutions as well as major corporations – suffered. Strengthening trust will help restore the predictability in the economic environment that is necessary for long-term and risk-laden investment decisions linked to innovation.

Restoring trust involves action across a set of inter-related drivers that encapsulate what citizens expect from government, for example:

- **Reliability:** The ability of governments to minimise uncertainty in the economic, social and political environment of their country and to act in a consistent and predictable manner; this has fairly clear implications for the propensity of firms to invest in innovation in a particular country. Previous sections of this report have pointed to the importance of stable and predictable policies related to innovation, given the long-term nature of investment decisions in this area.
- **Responsiveness:** The provision of accessible, efficient and citizen-oriented public services that effectively address the needs and expectations of taxpayers; this relates to the ability of government to identify the needs of businesses and provide appropriate support, including support for innovation activities.
- **Openness and inclusiveness:** Institutionalising a two-way communication with stakeholders to improve transparency, accountability and engagement; this helps to promote collaborative behaviour between the government and non-government sectors, including with respect to joint innovation-related activities.
- **Integrity:** The alignment of government and public institutions with broader principles and standards of conduct that contribute to safeguarding the public interest while preventing corruption; this dimension has a direct impact on innovation as firms are less likely to conduct innovation activities in a country in which respect for the rule of law and integrity are low.
- **Risk management:** Trust in the innovation area also depends on sound public policies to manage the risks associated with innovation, as discussed further below.

While many of these policies fall outside the focus of this report, they have an important impact on the effectiveness of government efforts to strengthen innovation and therefore need to be kept in mind. The forthcoming OECD Trust Strategy will look at measurement of trust, fairness in decision making and the impact of trust in specific policy domains, including tax policy, corporate governance, education and regulation.

One policy domain area that is particularly important for innovation concerns trust in science. Trust in scientists, the science system and the way science is used to inform policies are major issues that affect the ability of governments to promote change. Increasingly governments are turning to the scientific community to provide advice and evidence that can inform decisions and policies across a range of issues, from short-term public health emergencies to longer-term challenges, such as energy security. Such advice can be a valuable, or even essential, input to policy making, but its usefulness depends on how it is formulated and communicated as well as how it is perceived by its target policy audience and by other interested parties. It is rare that scientific evidence is the only consideration in a policy decision and, particularly for complex issues, many interests have to be balanced in situations where the science itself may be uncertain. The rapid evolution of ICTs and moves towards more participative democratic decision making have put additional pressure on science to help provide answers and solutions, while opening up the academic world to surveillance and criticism. While science advice used to be most often formulated behind closed doors, the new norms for science advisory systems are openness, transparency and accountability. Ongoing OECD work is exploring the appropriate mechanisms to provide scientific advice to policy makers, including the responsibilities and legal liabilities of institutions and individuals.

Public engagement and public perception

Closely related to the discussion on trust are questions related to public engagement and public perception. The public debate on the impact of science and technology on people's lives has been ongoing for centuries and has evolved in response to crises and new developments. For example, rapid advances in automation and robotics are raising concerns about technical change and its implications on future employment levels and the distribution of wealth within and across countries. Many societal concerns about the applications of science and technology appear to arise in conditions where the scientific evidence appears to be persuasive, but not complete, yet the socio-economic implications and the appropriate policy solutions from a broad societal perspective are much less certain.

It is a critical issue for governments to find effective ways of consensus building, especially for difficult S&T issues, involving the relevant parties, including citizens and scientists, but also media, as well as fostering public engagement in science. Gauging the public perception regarding these processes and actual degree of participation will inform the design of policies. Another important issue is how scientists view and engage with the public. Policy makers are increasingly interested in encouraging scientists to demonstrate or at least articulate the beneficial impact of their activities on the economy and society (OECD, 2014b).

While the importance of innovation for sustaining economic growth and driving improvements in living standards is generally acknowledged, there is also widespread evidence of significant attitudinal and knowledge "gaps". Public perception surveys in a large number of countries indicate that, although most people have a positive view of the impact of S&T on their personal well-being, a significant proportion have mixed or negative opinions about the effects of scientific research (Figure 8.4; OECD, 2013c). From the perspective of the adoption of new goods and services, a European poll found that nearly half of the EU25 population was significantly hostile to new innovations or very reluctant to try new products or services or pay a premium for them (European Commission, 2005). Public engagement and better risk management (see below) are some of the key responses to this challenge.

Policy makers will need to identify and monitor systematically skills and attitudes of relevance to science and innovation in order to improve them. Individual and collective attitudes are complex and constantly evolving phenomena, although some changes occur only over generations. At the same time, some social and environmental challenges require more immediate action in terms of consumption behaviour and social habits, for instance. Efforts to promote a science and innovation culture can be undermined not only by high-profile incidents and crises of confidence (e.g. Fukushima), but also by a less apparent erosion of trust in the decision-making process and in its use of science and evidence. This has triggered some serious rethinking about the impacts of S&T on the economy and society and a reassessment of the appropriate policy responses, including in the area of scientific advice (OECD, 2015).

Figure 8.4. **Public perception of scientific research benefits, 2010**

Responses to the question: "Have the benefits of scientific research outweighed the harmful results?"

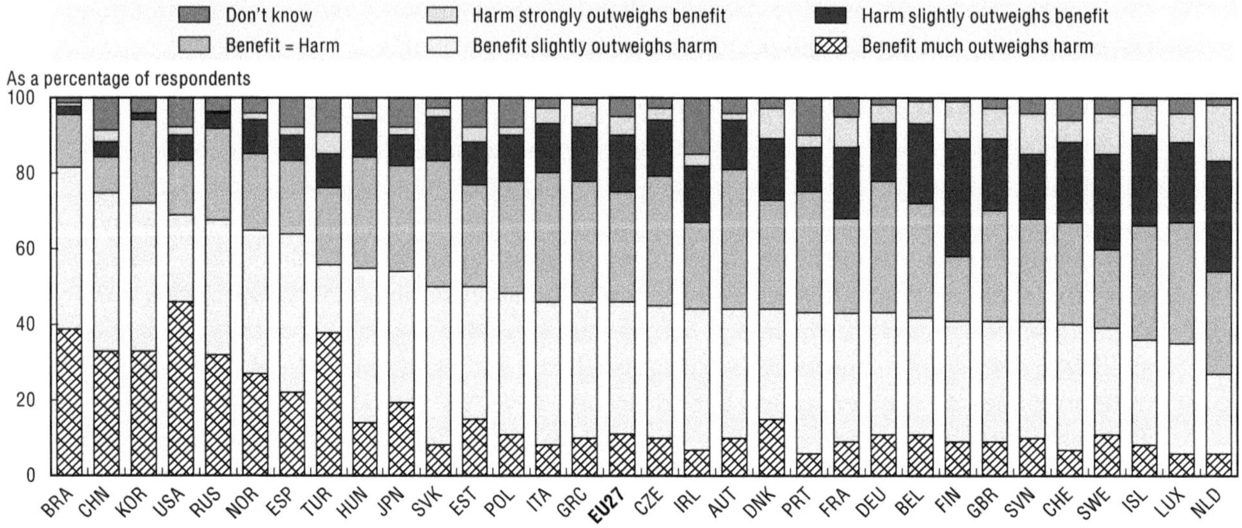

Note: International comparability may be limited. Results are based on surveys conducted by means of face-to-face interviews. For Japan, Mexico and the Russian Federation, data refer to 2011. For Korea, data refer to 2006. See source for further detail.

Source: OECD (2013c), *OECD Science, Technology and Industry Scoreboard 2013*, *http://dx.doi.org/10.1787/888932890542*, based on EU and national sources.

Policy measures directed at civil society, schools, universities and workplaces have sought to develop an innovation culture in view of the fact that innovation is driven by science, business, practitioners and users (Vincent-Lancrin, 2012) and pervades many spheres of human activity (Table 8.1). Such policy measures seek to improve public access to information on the future of STI and to promote society's participation in policy design. For example, the OECD Declaration on Future Policies for Science and Technology underscored the importance of raising awareness of S&T and recommended public participation in the definition of major technological orientations (OECD, 1981).

Other policy measures aim to raise awareness of and interest in S&T, especially among youth. Traditionally this has meant broad dissemination of scientific information, via the mass media, promotion of science events and other initiatives and support for the activities of science museums. The development and use of ICTs, the increasing access to digital infrastructures and the Internet, and greater interactive online communication – e.g. social media – have helped engage the public but also reduced reliance on traditional sources.

Promotion of science and innovation among youth largely takes place in classrooms. However, the evidence suggests that individuals in many countries think that schools do not make a substantial contribution to promoting entrepreneurial competencies and attitudes. Major reforms of education systems seek to add new disciplines and new learning practices to curricula. They have concerned all levels of education, from primary schools to higher education institutions, and have required building capacity in teaching and infrastructure.

Policy initiatives to build a science and innovation culture also target workplaces. They encourage a new research and innovation culture to help universities fulfil their "third" mission of transferring and co-creating relevant knowledge with the rest of

Table 8.1. **Typology of policy measures for enhancing a science and innovation culture**

Spheres	Main target populations	Key policy instruments	Some country examples
Civil society	Youth and adult population	Public dialogue (awareness workshops, conferences, standards)	Slovak Republic's Scientific Patisserie, France's Observatory of Biology
		Participation to STI policy design (public consultation)	Finland's national stakeholder conference, Great New Zealand Science Project, Turkey's technology roadmapping
		Science communication (science centre/ museum, science weeks/fairs/years/ exhibitions, science media (TV, radio, broadcasts, website and social medias), outreach programme by scientists)	Australia Questacon, Canada Science.gc.ca, Chile VA!, Korea Science Festival and Idea Festival, Start-up Expo and Start-up Fair, Germany BIOTechnikum truck, Turkey's Science Fairs Support Program (TŰBİTAK 4006)
Classrooms and education systems	Students at all educational levels	Awards/prizes and competitions in science and innovation	China innovation and entrepreneurship race, New Zealand's Future Scientist prize, Slovak Republic's Innovative Deed of the Year, Russia's Competition for Innovation in Education
		Formal education initiatives (lecture courses, new curricula)	Danish Foundation for Entrepreneurship-Young Enterprise, Norway's Action Plan for Entrepreneurship in Education, Sweden's compulsory teaching of entrepreneurship, Turkey's FATIH Project
		New pedagogical practices and networking activities (hands-on exercises, experiment labs, participatory learning, role models and mentorship)	Austria Young Science, Germany Little Scientists' House, Norway's IPRs educational scheme, Slovak Republic Scientific Patisserie
	Teachers	Capacity building for teaching, including the design of innovative teaching methods and materials	Austria's new teaching methods, Ireland's Project Maths
		Training opportunities, awareness conferences and workshops, financial incentives	Estonia's training of academic teachers on entrepreneurship, New Zealand's fellowships for S&T teachers, Young Enterprise Norway
Workplaces	Academia (researchers, doctorate students and postdocs)	Training opportunities (e.g. IPRs, start ups, etc.), awareness conferences and workshops	Technology Transfer Offices in many countries
		Support for commercialisation of public research results and industry-science linkages (remuneration schemes, performance criteria and promotion, industrial PhD)	Innova Chile CORFO, Colombia's regional alliances, Germany's VIP and EXIST grants, New Zealand's Callaghan Innovation's R&D Student Grants
	Firms	Support to industry-science linkages, and technical assistance to firms (innovation vouchers, experts detachment, industrial PhDs, extension programmes)	Technology Transfer Offices in many countries, Colombia's pilot program for training and advice in innovation management
		Training opportunities, seminars and information workshops and support, visibility	Costa Rica's CATI (IPRs) and National Portal of Innovation, New Zealand's Entrepreneurship Development Programme, South Africa's Science awareness awards, United Kingdom's Business Link

Note: IPRs = intellectual property rights.

Source: OECD (2014b), *OECD Science, Technology and Industry Outlook 2014*, OECD Publishing, Paris, *http://dx.doi.org/10.1787/sti_outlook-2014-en* based on country responses to the STI Outlook policy questionnaire 2014, and input received from national delegations.

society. Training, information workshops, and revised remuneration and promotion frameworks seek to raise awareness of IPRs and interest in the commercialisation of public research results in the research community. Researchers, especially early in their careers, are helped to launch start-ups. Firms receive technical assistance through financial and non-financial channels such as innovation vouchers, extension programmes and seconding of experts.

Managing the risks related to innovation

Climate change, loss of biodiversity, loss of topsoil, disease threats (such as multi-drug-resistant bacteria), adverse effects on health and the environment from the use of chemicals, and the consequences of population ageing are among the many complex risks facing the globe. As described in various parts of this report, breakthroughs in S&T are needed to address such global challenges in cost-effective ways. For instance, among many recent developments, digital technologies have helped to monitor disease outbreaks that threaten to become pandemics, and nucleic acid identification technologies can help to quickly identify the pathogens involved. Space technologies have unique data- and knowledge-enhancing functions with respect to climate change. Recent research has identified molecules in blood that predict, with at least 90% accuracy, whether an individual will develop mild cognitive impairment or full-blown Alzheimer's disease (Mapstone et al., 2014). Synthetic biology could allow petroleum-based products to be manufactured from sugar-based microbes (OECD, 2014f). And nano-composite materials offer the prospect of lighter vehicles, lowering fuel consumption (OECD, 2014g).

While new technologies can have diverse positive economic and social effects, potentially negative outcomes can also arise. It is unavoidable, for instance, that policy will have to manage a future where ICT allows ever more scientific information to be available to ever larger numbers of people, with some of this information being potentially dangerous. Moreover, as biotechnology advances, understanding of how to deliberately make diseases worse will also spread (for instance, adding a gene to mousepox – a version of smallpox – can make this pathogen more lethal and able to infect vaccinated individuals).[5] Indeed, the appropriateness of publishing details of the genomes of dangerous microbes is currently being debated.

Examples from biotechnology, as above, are among the more evident risks. But there are many others. For instance, some manufactured nanoparticles might have harmful effects on health. And in future, atomically precise manufacturing might be used to make weapons, in a highly proliferated way. Accidental outcomes of scientific experiments are also a potential source of risk, especially perhaps when experiments create conditions that otherwise do not exist in the natural world (Rees, 2003). And vast and critical interlinked ICT systems might behave in unpredictable and emergent ways. (In fact, interacting algorithms caused the "Flash Crash" of May 2010, when more than a trillion dollars in value were lost in minutes from global stock markets.) Indeed, scientific understanding of such complex systems is inadequate. Improved understanding is essential if governments are to protect society from potentially serious disruptions (Nesse, 2014).

More generally, uncertainty in science policy is pervasive. R&D is a probabilistic investment, and the direction of science and innovation and their ramifications involve elements of randomness and often long-term and hard-to-foresee outcomes. Even relatively short-term and prosaic policy questions can be hard to grapple with in informed ways. Accordingly, governments need to deploy a range of tools – from foresight studies and expert advice to quantitative modelling – to help steer policy and reduce unnecessary risks. But all such techniques have strengths and weaknesses.

The recent OECD Council Recommendation on the Governance of Critical Risks (OECD, 2014h) makes a range of recommendations to countries to strengthen risk governance, notably in:

1. Establishing and promoting a comprehensive, all-hazards and trans-boundary approach to country risk governance to serve as the foundation for enhancing national resilience and responsiveness, including the development of a national strategy for the governance of critical risks.

2. Building preparedness through foresight analysis, risk assessments and financing frameworks, to better anticipate complex and wide-ranging impacts.
3. Raising awareness of critical risks to mobilise households, businesses and international stakeholders and foster investment in risk prevention and mitigation.
4. Developing adaptive capacity in crisis management by co-ordinating resources across government, its agencies and broader networks to support timely decision making, communication and emergency responses.
5. Demonstrating transparency and accountability in risk-related decision making by incorporating good governance practices and continuously learning from experience and science.

The United Kingdom's Government Chief Science Adviser (GSCA, 2014) provides a useful overview of how policy can manage innovation-related risk and how science can help to assess risks in non-science domains. Key observations from this work are included in Box 8.3.

Box 8.3. **Science, innovation and the management of risk**

Innovation is essential for managing a diverse array of critical risks. Improved scientific understanding of complex systems is essential if governments are to protect society from possible disruptions to these systems. At the same time, science and innovation themselves create new risks that need to be managed. For instance, manufactured nanomaterials offer a range of possible benefits, but also have risks related to health and safety that need to be managed.

The GCSA (2014) provides useful policy guidance on innovation and risk management, in particular:

- Policy decisions on the risks and applications of a new innovation should consider the costs and benefits of not acting, as well as those of acting.
- Decision making needs to take account of the different ways of achieving the same or similar desired outcome.
- Science is one framework through which innovation and risk can be assessed, but economic, social and political considerations should also be included. Relevant social and ethical values can vary greatly among communities and countries.
- It is important to distinguish between hazard, exposure and risk. Confusing these concepts can impair communication and decision making.
- If risk is quantified, there should be transparency regarding how numbers have been generated. How numbers are presented, or framed, will also affect their interpretation.
- The incentives faced by regulators should enable innovation. Along with consumer welfare, the economic duty of regulators of infrastructure and utilities should include innovation and system resilience. Regulators should be accountable for all major decisions they take.

These issues are also core to the debate about innovation and policies for innovation today. As suggested by the UK report, several challenges put the management of risk at the core of discussions on innovation policy (GSCA, 2014):

- Governments shape the legal frameworks, institutions and policies that shape the risks and incentives faced by various actors in the innovation system.

- Innovations can cause both good and harm, leading to a debate on the risks associated with innovation. As modern societies have become safer and more free of risks, risk aversion has grown in many societies.
- Designing systems that manage the risks associated with innovation is difficult and requires a sound evidence base. However, it is also influenced heavily by social and cultural values that affect whether innovators can receive a social licence to engage in specific innovation activities.

These challenges have also been explored in OECD work on innovation, notably in work on the governance of nanotechnology and biosciences. For example, a recent report on innovative governance of biomedicine and health technologies (OECD, 2013d) notes the importance of balancing risk and benefit given limited knowledge and a context of uncertainty. In another example, OECD (2014g) examines the use of nanomaterials in tyres. This work shows that while it is straightforward to demonstrate benefits for society from this innovation, serious data gaps exist regarding risks to health and the environment.

Given the pace of discovery in the fields of biomedical and health innovation, new products or procedures may reach regulators while the available information on them may still be incomplete and insufficient to enable regulators to make entirely evidence-based decisions. Early consultation is therefore becoming an indispensable means of maximising the amount of information available to regulators. Innovators often hold more and better information than regulators in relation to at least some kinds of technical and scientific developments. The report also notes the need for timely public engagement and communication. Biomedical innovation creates challenges with respect to risk, equity, privacy, confidentiality, human dignity, right to life and freedom of research. This makes for a particularly complex environment for decision makers and the public alike, necessitating special efforts in communication and consultation.

There is another dimension of risk related to innovation that is important in the policy context, namely whether and how to address the economic risks related to innovation. Innovation is inherently risky, and firms and other innovation actors face great uncertainty in making their investments in innovation activities. For instance, the future of synthetic biology depends on achieving reliable, low-error, accurate and inexpensive DNA synthesis (i.e. writing of the genetic code). The technical difficulties involved in reaching cost parity between DNA synthesis and DNA sequencing are considerable and create high financial risks for the often small companies working in synthetic biology. Such firms are natural targets for forms of public support that help to mitigate financials risks faced by small high-technology ventures (e.g. loan guarantees, public procurement or enhanced access to equity finance).

Similarly, green nanotechnology operates in a complex landscape of fiscal and legislative policies. An important policy objective is to reduce uncertainty around the use of environmental nanotechnology. In this regard, innovative approaches to sharing risk and knowledge are being developed based on large consortia made up of companies, public laboratories and institutions (e.g. NanoNextNL, Genesis). Among other benefits, consortia can help to manage the uncertainty of bringing a product to market when no similar technologies have previously been commercialised, or when the demand for the technology is not yet clear.

Creative thinking can be brought to bear on how policy might best address risk. A relevant recent example concerns the work of Andrew Lo, professor of finance at the Massachusetts Institute of Technology (MIT). Lo and his colleagues have employed concepts from financial engineering to tackle the problem of neglected funding of drug research for certain types of disease (such as those that primarily afflict populations in poor economies). Most lines of drug research are expensive and fruitless. High costs and skewed research payoffs, combined with other features of the market for some drugs (for instance, medical authorities wish to minimise the use of new antibiotics to slow the development of resistance), have led to pharmaceutical companies abandoning some critical areas of research. But Lo and his colleagues have demonstrated that a portfolio approach to early-stage research might provide a solution (Fernandez, Stein and Lo, 2012). Above a threshold volume of research expenditure, a financial structure could be created that would pool the results of R&D projects with different risk, return and duration profiles. The portfolio entity could finance its activities by issuing debt, a critical advantage because a much larger pool of capital is available for investment in debt than equity. Based on an analysis of data from 1990 to 2011, for new molecular entities in oncology, the fund could yield rates of return sufficient to attract institutional investors such as pension funds and insurance companies.

Well-designed innovation policies, as discussed in Chapter 6, can help to mitigate some of the economic risks of innovation, in particular in areas where innovation is important to meet core public policy objectives, e.g. health or the environment. Moreover, stable and predictable policy frameworks, as discussed earlier in this report, are important to help firms manage the inherent economic risks of innovation.

8.3 Implementation and evaluation

Implementing innovation strategies

Most OECD countries and most emerging economies have adopted innovation strategies over the past decade. These typically offer a vision of the main challenges that the national research and innovation system is confronted with and the directions that should be taken to address these challenges. Innovation strategies have many possible uses: they can trigger an exchange of views among stakeholders and help achieve alignment on priorities regarding research and innovation; foster the convergence of views among stakeholders and decision makers; help in planning resources; and establish a concrete agenda for policy action.

In practice, innovation strategies often involve a range of objectives; some of these are not achieved and some are not even implemented (OECD, 2014b). Typical challenges arise from poor design, including a lack of realism in the choice of some objectives, possibly due to an inadequate design process, or from the process of implementation itself. Other barriers to implementation arise from a lack of involvement or even resistance of certain actors, whose concerns and agendas have not been sufficiently integrated into the strategy. Obstacles can also reside in institutional settings, if they are not amenable to the possible reorientation of resources required by new strategies. Moreover, as innovation becomes a tool to achieve a wide range of policy objectives, developing a coherent strategy becomes increasingly challenging, as a range of ministries and stakeholders need to be involved, as discussed in Section 8.1.

National innovation strategies are in most cases broad in scope, covering innovation, research, entrepreneurship and part of higher education (OECD, 2014b). They therefore often involve a variety of organisational actors, each with its own culture, structure, constraints and objectives. Actors involved include ministries in charge of research, higher education and the economy, as well as agencies in charge of funding, performing or evaluating research. Beyond the government, research teams, universities, various types of enterprises (multinational, start-ups, SMEs), as well as professional associations may be involved.

In addition, innovation has an important role in areas where security, social or environmental objectives are central, e.g. defence, health and the environment, and where the corresponding ministries are leading the agenda in their respective areas. Hence, although innovation is essential, the ministries directly in charge of innovation policy do not have a leading role in these areas. How to ensure coherence concerning innovation-oriented initiatives across these separate thematic areas, each led by distinct entities, is therefore a challenge. Many countries have found it difficult to mobilise the broad range of actors involved in the implementation of their national strategies (OECD, 2014b). Understanding the reasons for these difficulties, and identifying the obstacles, is often the key to ensuring a properly functioning innovation system. A particular challenge is also to include not only existing (incumbent) firms, but also young innovative firms and challengers involved in more radical innovation.

Overcoming challenges with policy reform[6]

As with other areas of policy, implementing new innovation policies can be difficult. These difficulties can emerge from many areas, including lack of funding, lack of understanding, poor institutional frameworks or poor governance, several of which have already been addressed in previous sections. Implementation of new policies can, however, also meet with resistance from specific interest groups that are affected by the policy reforms. While many new innovation policies may have only modest goals, and therefore do not necessarily give rise to such resistance, some others may, e.g. deep university reforms, or large shifts in the budgets allocated to science and/or innovation.

At the root of many of the obstacles to reform lies the large heterogeneity of citizens, firms, scientific institutions and other actors that are affected by reforms, which implies that reforms will have a differential impact on them, sometimes substantially altering the relative value of different types of human and physical capital. Even socially beneficial reforms may therefore contradict the interests of many, largely because of distributional consequences that may be unrelated to the core aims of the reform. This heterogeneity also influences the political process, not least because it structures the incentives of politicians seeking election.

When it comes to policy reform, there is thus a twofold challenge. The first is to design reforms that will enhance aggregate welfare, even allowing for the costs that reform may impose on some agents. The second is to devise strategies for securing adoption of such reforms that prevent the opponents of change from blocking reform, but that also address their legitimate concerns about its distributional consequences.

Despite these common challenges, there are no one-size-fits-all approaches to overcoming the obstacles to reform or even identifying the most urgent reform priorities. This is because the heterogeneity of institutions and economic structures across countries ensures that the challenges facing would-be reformers vary widely across both time and space. Reform design and strategies for reform adoption therefore need to reflect the specific institutional and cultural context of the country concerned. Even where common problems can be identified in different countries, the specific institutional features of a country imply that simple, unaltered "transplants" of policies and institutions from one environment to another rarely take root. Some degree of adaptation is usually required. This must be borne in mind when trying to draw lessons from reform experiences across countries.

Nevertheless, the evidence suggests that cross-country comparisons can be fruitful. First, for all their institutional, political and economic differences, countries face a large number of common challenges, also in the innovation area, e.g. in strengthening productivity, promoting the scale-up of innovative firms, addressing social challenges and strengthening inclusiveness, as has been made clear from previous sections of this report. Moreover, in many policy domains related to innovation, the recent *Science, Technology and Industry Outlook* shows that countries have increasingly adopted common approaches, even if the specific institutions and policies still vary considerably from one jurisdiction to another (OECD, 2014b).

Although the OECD's work on innovation policy has not specifically looked at these questions, experience from other areas provides some useful insights (OECD, 2010d). Notably, this work points to a number of regularities in the way reform processes unfold in different areas and across countries, which suggest that, despite the wide variety of challenges and circumstances they face, policy makers contemplating reforms need to address a certain number of basic questions in the early stages of reform design (Box 8.4). Not all of these lessons pertain also to innovation policy, but nevertheless provide important guidance for policy makers.

The discussion on the "how" of structural reform presented in Box 8.4 has not been as strongly present in the debate on innovation policies as in other areas of structural policy, such as education or environmental policy reforms. Nevertheless, it is relevant and has a bearing on the success of implementing policies in this area. For example, recent OECD work on systems innovation raises questions in regard to the political economy aspects of reform in areas such as transport innovation, green growth and cities. Policies to change such systems may negatively influence vested interests, which implies that political conflict and power struggles are likely and will need to be managed. System innovation entails not just winners, but also losers, especially when old systems are replaced by new systems. Organisations with interests linked to old systems may resist and oppose the changes. Many workers have skills that are firm- or even job-specific, and many firms are invested in capital and equipment that may be of value only for pursuing certain activities in particular locations. Such human or technological capital may be difficult to redeploy in response to a changing environment. Moreover, system innovation may require adjustments in policies, institutional frameworks, incentive structures and investment patterns.

A few other aspects of the OECD work on *Making Reform Happen* (OECD, 2010d) are relevant to innovation policy. First, a common "stylised fact" about the difficulty of structural reform is that the costs sometimes tend to be incurred up front and concentrated on a few agents, while the benefits take longer to materialise and are generally more diffuse. This may be the case with innovation policies that contribute to rapid structural change in the economy.

Box 8.4. **An initial checklist for policy reform**

While neither the OECD work nor the political economy literature in general can yield any universal formulas for reform success, the research undertaken to date and summarised in this section suggests that policy makers should bear in mind the following questions when designing both policy reforms and strategies for their adoption and implementation:

- *Do the authorities have a clear mandate for change?* The first finding of the OECD work is that, in particular for all-encompassing reforms, it is important to have an **electoral mandate**. Reform "by stealth", via the quiet adoption of a series of seemingly technical changes, can sometimes yield progress, but it has severe limits.
- *What more can be done to demonstrate the need for change and/or the desirability of the proposed solutions to the public and key stakeholders?* The second finding of the OECD work points to the importance of **effective communication**. Successful reforms have usually been accompanied by consistent co-ordinated efforts to persuade stakeholders of the need for reform and, in particular, to communicate the costs of non-reform.
- *How strong is the evidence and analysis underlying the arguments for reform?* It is important that policy design is **underpinned by solid research and analysis**. An evidence-based and analytically sound case for reform serves both to improve the quality of policy and to enhance prospects for reform adoption.
- *Are institutions in place that can manage the reform effectively, from design to implementation,* or is there a need to create/strengthen such institutions? These challenges are more likely to be met where **appropriate institutions** exist, capable of supporting reform from decision to implementation. Building such institutions can take time, as their effectiveness depends on their reputation, but where they exist, their prior analysis appears to have enhanced the prospects for reform in particular areas.
- *Does the reform have clearly identifiable "owners",* in terms of both politicians and institutions responsible for taking it forward? Virtually all of the assessments prepared in the context of OECD work point to the **importance of strong leadership** – whether by an individual policy maker or an institution charged with carrying out the reform. However, successful leadership is often about winning consent rather than securing compliance.
- *What is the expected time frame for design, adoption and implementation?* The more successful reforms examined in the OECD analyses generally took **several years to prepare and adopt**, and they often took far longer to implement. This creates challenges for innovation policies that by their nature are focused on the long term, and often go across different governments.
- *What is to be the strategy for engaging those threatened by reform?* Can they be persuaded to support it? To what extent can/should their objections be overridden? Should they be compensated for their anticipated losses – and, if so, how and to what extent? The reform experiences in OECD countries suggest that **it pays in most circumstances to engage those who will be most directly affected by reform**. Secondly, it is important to recognise that concessions to potential losers need not compromise the essentials of the reform: they may indeed be coherent with its overall logic, improving the prospects of particular groups that will be affected by the reform without contradicting its overall aims.

Source: OECD (2010d), *Making Reform Happen: Lessons from OECD Countries*, OECD Publishing, Paris, *http://dx.doi.org/10.1787/9789264086296-en*.

Scientific uncertainty often presents a further challenge, as it implies that the evidence put forward to support the case for reform will often be contested, and those threatened by policy changes will try to generate more "favourable" evidence to introduce into the debate. Consequently, the choice of analytical technique is often highly politicised. In many countries, a widespread attitude of mistrust towards government adds to the difficulty, since official claims about the need for reform are likely to meet a sceptical reception by much of the electorate. The brief discussion on trust in government in Section 8.2 and that on scientific advice in Chapter 5 of this report are relevant in this context, as is the need for evidence and indicators already highlighted in Box 8.4.

Recent policy reforms in OECD countries suggest a number of lessons about how to address these challenges. First, as already noted before, engagement of stakeholders and the public is crucial. Second, while research alone is no "fix" for politics, there is no substitute for a solid, evidence-based case for reform when dealing with stakeholders and voters. Public acceptance of a degree of scientific consensus does matter. Third, path dependence implies that the selection of policy instruments depends to some extent on existing institutions and regulatory regimes. Policy makers need to consider, for example, how easily a given legal system can cope with the introduction of new policy instruments. Efforts to ensure that policy changes are compatible with the broader institutional and regulatory environment are more likely to succeed where governments create institutions or processes that permit a whole-of-government approach to innovation policies. In many cases, these considerations also point to the need to rely on a mix of policy instruments to address multiple market failures.

Strengthening monitoring and evaluation

Another key question is how to incorporate monitoring, evaluation and policy learning in the development of innovation strategies. Effective evaluation is crucial for demonstrating accountability in public spending, establishing the legitimacy and credibility of government intervention in innovation processes, but also to support the processes of learning, prioritisation and improvement of policies over time. However, many OECD countries still put the main emphasis on accountability in their approach to evaluation, which may ignore the dynamic benefits of policy learning over time. Appropriate measurement, monitoring and evaluation are critical for the design and governance of innovation policy, allowing policy makers and analysts to:

- Assess the contribution of innovation to achieving social and economic objectives.
- Understand the determinants of and obstacles to innovation, which is crucial for designing effective innovation policies.
- Establish the impact of – often a portfolio of – policies and programmes, and whether or not policy has contributed to correcting or ameliorating the problem it set out to resolve (e.g. tackling market failures that affect the availability of finance, skills, advice and technologies).
- Evaluate the effectiveness of different policy approaches, thereby enabling government to make informed decisions about the allocation of funds. Evaluation can assist decision makers in assessing the relative effectiveness of policies and programmes and help them to make judgements about where to place their efforts in order to obtain the greatest

benefits for given costs. Thus, it can contribute to improving effectiveness, value for money, and appropriateness of policy and programme interventions *ex post* and justify future interventions.

- Continuously improve the design and administration of programmes. Evaluation is a key tool for learning about how well policies and programmes are delivering, what problems may be emerging, what practices work well and what should be done better in the future.
- Stimulate informed debate. The results of evaluations may encourage public debate that can offer opportunities to a mix of stakeholders – from programme sponsors and managers to beneficiaries – to reflect upon the appropriateness and performance of policies, programmes and institutions.
- Enhance public accountability of relevant policies.

An important consideration in evaluation is to consider the extent to which desirable outcomes would have occurred without public intervention (the "counterfactual"). Two tendencies are common here. The first is the so-called "project fallacy", whereby outcomes that are in reality cumulative and dependent upon the interaction of several factors are wholly (or mostly) attributed to the intervention under evaluation. This is also complicated by the fact that innovation policy instruments sometimes tend to work as a package, and may be difficult to evaluate in isolation. The second is the tendency to underestimate the effects of an intervention because of a narrow evaluation focus or because of the timing of an evaluation. For example, the effects might not yet have occurred or have occurred so long ago that beneficiaries fail to attribute them to the public intervention. Awareness of these tendencies is important, even if the problems they create cannot be fully solved.

Evaluation provides one source of information among many others in shaping policy and programme management processes, and appreciating this is important in informing expectations around its usefulness. In the context of evaluation, it also important to accept and understand that mistakes and errors are inevitable in the process of policy making. Part of the policy challenge is to design governance procedures to detect and correct these errors (Rodrik, 2008). It is for this reason that evaluation has become a central element in innovation policy. Fortunately, there has been a trend in the evaluation community towards greater use of more rigorous techniques that have two related characteristics: the identification of a counterfactual (what would have happened in the absence of the policy) and the inference of causality (rather than simple correlation).

For example, increased use of randomised control trials (RCTs) in countries such as the United Kingdom and the Netherlands, in areas such as R&D support, is motivated by the recognition that rather small differences in policy design can have significant implications for policy effectiveness and efficiency. (See Warwick and Nolan [2014] for a discussion.) And developments in structural econometric techniques, such as the more widespread use of instrumental variable estimation methods, have gone a long way towards increasing the robustness of evaluations of existing policy measures. Some key policy learnings on the evaluation of innovation policies are below.

Box 8.5. **Policy learnings on the evaluation of innovation policies**

The following principles – based on the recent deliberations of an OECD-mediated expert group – reflect good practice in innovation policy evaluation (and are in fact relevant to evaluation in many fields of public policy):

- **Make explicit, at the highest level, the commitment to evaluation of policy.** There should be an explicit commitment to undertake *ex post* evaluation of significant industrial and innovation policies and strategies. Overt recognition of the importance of evaluation for evidence-based policy making, by senior policy makers and agency heads, is vital in securing the necessary human and financial resources for evaluation.
- **Consider mandating evaluations when public funding is provided.** A key reason various programmes in the United States have been thoroughly evaluated – such as the Manufacturing Extension Partnerships programme – is because mandatory evaluation requirements were attached to the use of federal funds.
- **Insist on developing data and evaluation strategies as a prerequisite for the start of programmes.** A clear programme evaluation strategy should be established from the outset, with an *ex ante* evaluation plan that, to the extent possible, articulates the theory of change and shows the main expected channels of impact (from inputs and activities to outputs and outcomes). A strategy should exist to ensure that the data necessary for evaluation are collected from the outset. Governments also have a duty to make more data available so that researchers and other interested parties can also assess policy effectiveness.
- **Choose the evaluation technique in the light of the size and nature of the programme concerned.** Studies of major programmes – especially pilot schemes that could be ramped up later – should use a variety of methods: random assignment, quasi-experimental assessments, interviews with beneficiaries or participatory approaches involving stakeholders. There should be a move to more use of randomised experiments as the basis of *ex post* impact assessment.
- **At the level of innovation and industrial strategies, a mix of evaluation methods is likely to be needed.** State-of-the-art econometric methods have a role in assessing components of a strategy, but are less likely to be useful for the policy portfolio as a whole. Tracking of macro- or meso-level indicators, international benchmarking, subjective assessments via survey methods, narrative reporting, case studies and other techniques all have a role to play.
- **Insist on full disclosure in evaluation reports.** There should be a commitment to public diffusion of evaluation findings of publicly funded programmes. The choice of methods and evaluation parameters used, methodological drawbacks and areas of subjective judgement should be described in full. There should be a commitment to transparency and early publication of evaluation findings and the data on which they are based. Published evaluation findings should be accompanied by metadata that facilitate online searches.
- **Robust governance mechanisms are needed to ensure evaluation is objective.** Programmes should be evaluated by, or in collaboration with, genuinely independent experts, possibly from an audit office. Ideally, the body that implements the evaluation would work with programme managers but would not be dependent on continued contracts from the sponsor of the programme.
- **Good mechanisms for policy learning are needed to ensure that the findings of evaluation feed back into policy making.** This may require measures to increase awareness of the use of evaluation – and evidence-based decision making more generally - for policy making, and could also require measures to make evaluations mandatory for certain policy measures. Having an explicit commitment to evaluation at the highest level is particularly important.

Source: Warwick and Nolan (2014), "Evaluation of industrial policy", *http://dx.doi.org/10.1787/5jz181jh0j5k-en.*

The use of policy experimentation

The use of randomised control trials is also enabling a greater experimentation with policies (OECD/The World Bank, 2014), where governments can experiment at a small scale with specific policy initiatives, and roll them out at a larger scale when evaluations show the programme is effective in meeting its objectives. Experimental approaches partly respond to the growing understanding that innovation policies characterised by top-down government interventions are not the right approach. The reasons for the failures of such policies are well known, and include the risks of capture by vested interests, lack of information on the economy and strong information asymmetry with private actors, often combined with a lack of capability in the public sector for effective policy making and implementation.

Another, more appropriate approach to innovation (and industrial) policy involves search, experimentation, monitoring, learning and adaptation, all of which need to occur in a context of international openness to knowledge, trade, investment and competition. This new approach also rests on close co-operation with private and non-governmental actors, who are often better placed than governments to help identify barriers to innovation, and point to areas for productive investment or policy action.

This new approach also rests on a much stronger focus on (diagnostic) monitoring and evaluation, which need to be embodied in programmes and policies from the outset. This is particularly important for new and emerging areas of policy, where there is significant experimentation under way and where the scope for learning and identification of good practices is the largest. Such learning should benefit from early and periodic sharing of lessons from policy experimentation at the global level, which will require strengthened mechanisms to identify and diffuse good practices, including through specific knowledge platforms and networks.

The work on experimentation also raises questions on how policy makers – and private actors – learn from experience and mistakes that are made, how to encourage more entrepreneurial experimentation and appropriate risk-taking not only by enterprises but in policy making, how not only success but failure can be openly discussed and built upon, and how such learning can be organised, embedded and institutionalised in the policy-making process. It also points to a need to better understand systems and their behaviour, and how policy can influence the behaviour of (increasingly) complex systems to achieve more sustainable growth and shared prosperity. Some other relevant recommendations from the OECD-World Bank work on experimentation are the following (OECD/The World Bank, 2014):

- Policy makers should incorporate monitoring and evaluation plans at the policy design stage to improve the quality and efficiency of public expenditures supporting innovation policy.
- Governments can sometimes also achieve better results in the innovation area by involving agencies and actors on the periphery of policy making, which can limit capture by vested interests and may enable more creative and co-operative policies than those emerging from more central agencies. Such agencies may also be able to achieve more with less.

Improving measurement

Improving measures of innovation is essential for policy making and evaluation and for promoting innovation in businesses, the public sector and society at large. However, while progress is being made, current innovation indicators remain too focused on the inputs of the innovation process rather than on its outcomes, and aggregate numbers or indices do not adequately reflect the diversity of innovation actors and processes and the links among them. Continued efforts are needed to take this work forward and to adapt the measurement agenda to experience at the national and international level. A number of policy issues – in particular, the role of broader (beyond R&D) innovation, the growing importance of the public sector in innovation and better assessment of the economic impact of innovation to name but a few – require improved measurement. However, there is no point of having a first-class data infrastructure if it is not available to the users in a standardised and accessible fashion. And to improve policy evaluation, greater attention should be paid to the quantification of policy variables and the characteristics of their design and implementation. Priorities include:

1. **Improving the measurement of broader innovation and its link to macroeconomic performance.** This includes further work to:
   - advance the measurement of knowledge-based capital and ICT investment and its inclusion in productivity statistics
   - redesign STI surveys to take a broader view of innovation
   - invest in a high-quality and comprehensive data infrastructure to measure the determinants and impacts of innovation by linking different data sets and exploiting the potential of administrative records.
2. **Recognising the role of innovation in the public sector and promoting its measurement.** As already discussed in Section 4.5, there is need to account for the use of public funds, measure the efficiency of producing and delivering public policies and services, and improve learning outcomes and the quality of the provision of public services via innovation.
3. **Promoting the design of new statistical methods and interdisciplinary approaches to data collection.** Design of policies for innovation needs to take into account the characteristics of technologies, people and locations, as well as linkages and flows among them. New methods of analysis that are interdisciplinary in nature are necessary to understand innovative behaviour, its determinants and its impacts at the level of the individual, firm and organisation. Better use of new data sources, notably the Internet, would also help.
4. **Promoting the measurement of innovation for social goals and of social impacts of innovation.** The current measurement framework fails to measure the social impacts of innovation. The development of measures that provide an assessment of the impact of innovations on well-being, or their contributions to achieving social goals, needs to be promoted. This includes better measurement of the people dimension of innovation, including skills needs in the digital economy. It also involves measurement of important policy dimensions, e.g. security, privacy and consumer protection in relation to the digital economy.
5. **Integrating and standardising data on STI.** Significant efforts to increase the availability and accessibility of STI data are taking place across the OECD and beyond, also encouraged by the latest advances in Internet technology, big data and the drive for transparency. However, these are often disparate and unconnected developments, and access to these data is limited by a patchwork of laws, regulations and practices that are unevenly applied and interpreted. Initiatives to develop these critical infrastructures, and improve access to them while ensuring data confidentiality, should be supported and co-ordinated in order to avoid wasteful duplication of efforts, and attention should be paid to the design and regulation of emerging networking platforms, interoperability and related standards.
6. **Incorporating policy monitoring and evaluation at the design stage of policy making.** In order to strengthen policy design and enable policy learning over time, metrics also need to be designed *ex ante* and generated by policy implementation and experimentation on a smaller scale. Data also need to be generated on the design and characteristics of the

policies themselves. The breadth and complexity of the science and innovation policy framework has not been conducive so far to a systematic categorisation of its different attributes across countries, resulting in a high degree of reliance on qualitative views or on established public R&D funding indicators. Financial support, public demand, tax, regulatory and many other government measures can in principle be documented and codified in a systematic fashion.

The 2016 Blue Sky conference, to be held in Belgium, will help advance the agenda in these and other areas, with the aim to strengthen the evidence base for policy making.

Notes

1. At the same time, policy makers can learn from having a variety of approaches to innovation; alignment should therefore not be confused with harmonisation.
2. Per OECD, 14 countries out of 15 reporting countries in 2010 to the OECD (2009a), *OECD-GOV Survey on the Multi-level Governance of Science, Technology and Innovation Policy*.
3. This challenge is described in, among others, OECD (2012a), *OECD Reviews of Regional Innovation: Central and Southern Denmark*, and OECD (2009b), *OECD Reviews of Regional Innovation: 15 Mexican States*.
4. OECD has facilitated the establishment of several such distributed infrastructures, most recently in relation to Scientific Collections (SciColl, 2013), and the lessons learned, including different governance options, have been analysed (OECD, 2014e).
5. *www.theguardian.com/commentisfree/2014/jul/21/five-biggest-threats-human-existence*
6. This section draws on OECD (2010d) and briefly summarises some main lessons learned, applying them to the context of innovation policy reforms.

References

European Commission (2005), "Population innovation readiness", *Special Eurobarometer*, European Commission, Brussels, August, *http://ec.europa.eu/public_opinion/archives/ebs/ebs_236_en.pdf*.

Fernandez, J.M, R.M. Stein and A.W. Lo (2012), "Commercializing biomedical research through securitization techniques", *Journal name: Nature Biotechnology*, Vol. Volume: 30/10, pp. Pages: 964-975,Year published: doi: *http://dx.doi.org/10.1038/nbt.2374*.

Gallup, Gallup World Poll (2013), *www.gallup.com*.

GCSA (UK Government Chief Scientific Adviser) (2014), *Annual Report of the Government Chief Scientific Adviser 2014: Innovation: Managing Risk, Not Avoiding It*, The Government Office for Science, London.

Mapstone, M. et al. (2014), "Plasma phospholipids identify antecedent memory impairment in older adults", *Nature Medicine*, Vol. Volume: 20, Pages: 415-418, Year published: doi: *http://dx.doi.org/10.1038/nm.3466*.

Nesse, R. (2014), "The fragility of complex systems", in Brockman, J. (ed.), *What Should We Be Worried About?*, Harper Perennial.

OECD (2015), "Scientific advice for policy making", *OECD Science, Technology and Industry Policy Papers*, No. 12, OECD Publishing, Paris, *http://dx.doi.org/10.1787/5js33l1jcpwb-en*.

OECD (2014a), Regional Outlook *Regions and Cities: Where Policies and People Meet*, OECD Publishing, Paris, *http://dx.doi.org/10.1787/9789264201415-en*.

OECD (2014b), *OECD Science, Technology and Industry Outlook 2014*, OECD Publishing, Paris, *http://dx.doi.org/10.1787/sti_outlook-2014-en*.

OECD (2014c), *Recommendation of the Council on Effective Public Investment across Levels of Government*, OECD Publishing, Paris, *www.oecd.org/gov/regional-policy/recommendation-effective-public-investment-across-levels-of-government.htm*.

OECD (2014d), *Centre Stage – Driving Better Policies from the Centre of Government*, OECD, Paris, *www.oecd.org/gov/Centre-Stage-Report.pdf*.

OECD (2014e), *International Distributed Research Infrastructures: Issues and Options*, OECD Publishing, Paris, *www.oecd.org/sti/sci-tech/international-distributed-research-infrastructures.pdf*.

OECD (2014f), *Emerging Policy Issues in Synthetic Biology*, OECD Publishing, Paris, *http://dx.doi.org/10.1787/9789264208421-en*.

OECD (2014g), *Nanotechnology and Tyres: Greening Industry and Transport*, OECD Publishing, Paris, *http://dx.doi.org/10.1787/9789264209152-en*.

OECD (2014h), *Recommendation of the Council on the Governance of Critical Risks*, OECD Publishing, Paris, *www.oecd.org/mcm/C-MIN(2014)8-ENG.pdf*.

OECD (2013a), *Regions at a Glance*, OECD Publishing, Paris, *http://dx.doi.org/10.1787/reg_glance-2013-en*.

OECD (2013b), *Regions and Innovation: Collaborating across Borders*, OECD Publishing, Paris, *http://dx.doi.org/10.1787/9789264205307-en*.

OECD (2013c), *OECD Science, Technology and Industry Scoreboard 2013*, OECD Publishing, Paris, *http://dx.doi.org/10.1787/sti_scoreboard-2013-en*.

OECD (2013d), "Toward new models for innovative governance of biomedicine and health technologies", *OECD Science, Technology and Industry Policy Papers*, No. 11, OECD Publishing, Paris, *http://dx.doi.org/10.1787/5k3v0hljnnlr-en*.

OECD (2012a), *OECD Reviews of Regional Innovation: Central and Southern Denmark*, OECD Publishing, Paris, *http://dx.doi.org/10.1787/9789264178748-en*.

OECD (2012b), *Meeting Global Challenges through Better Governance: International Co-operation in Science, Technology and Innovation*, OECD Publishing, Paris, *http://dx.doi.org/10.1787/9789264178700-en*.

OECD (2011), *Regions and Innovation Policy*, OECD Publishing, Paris, *http://dx.doi.org/10.1787/9789264097803-en*.

OECD (2010a), *The OECD Innovation Strategy: Getting a Head Start on Tomorrow*, OECD Publishing, Paris, *http://dx.doi.org/10.1787/9789264083479-en*.

OECD (2010b), *Regional Development Policies in OECD Countries*, OECD Publishing, Paris, *http://dx.doi.org/10.1787/9789264087255-en*.

OECD (2010c), *Cities and Climate Change*, OECD Publishing, Paris, *http://dx.doi.org/10.1787/9789264091375-en*.

OECD (2010d), *Making Reform Happen: Lessons from OECD Countries*, OECD Publishing, Paris, *http://dx.doi.org/10.1787/9789264086296-en*.

OECD (2009a), "OECD-GOV Survey on the Multi-level Governance of Science, Technology and Innovation Policy", GOV/TDPC/RD(2009)9, OECD, Paris.

OECD (2009b), *OECD Reviews of Regional Innovation: 15 Mexican States*, OECD Publishing, Paris, *http://dx.doi.org/10.1787/9789264060135-en*.

OECD (2005), *Governance of Innovation: Synthesis Report*, OECD Publishing, Paris, *http://dx.doi.org/10.1787/9789264011038-en*.

OECD (1981), "Declaration on Future Policies for Science and Technology", 19 March 1981, *http://acts.oecd.org/Instruments/ShowInstrumentView.aspx?InstrumentID=154&InstrumentPID=150&Lang=en&Book*.

OECD/The World Bank (2014), *Making Innovation Policy Work – Learning from Experimentation*, OECD Publishing, Paris, *http://dx.doi.org/10.1787/9789264185739-en*.

Rees, M. (2003), *Our Final Century: Will the Human Race Survive the Twenty-First Century?*, William Heinemann, United Kingdom.

Rodrik, D. (2008), "Normalizing industrial policy", *Commission on Growth and Development Working Paper*, No. 3, Commission on Growth and Development, Washington, DC.

SciColl (Scientific Collections International) (2013), *www.scicoll.org/*, launched by OECD Global Science Forum, April 2013.

Vincent-Lancrin, S. (2012), "Towards a culture of innovation: Motors and brakes", presentation to India/OECD Initiative – *Collaborative Workshop on Education and Innovation*, New Delhi, 9-10 May 2012, *www.oecd.org/edu/ceri/50477243.pdf*.

Warwick, K. and A. Nolan (2014), "Evaluation of industrial policy: Methodological issues and policy lessons", *OECD Science, Technology and Industry Policy Papers*, No. 16, OECD Publishing, Paris, *http://dx.doi.org/10.1787/5jz181jh0j5k-en*.

ORGANISATION FOR ECONOMIC CO-OPERATION AND DEVELOPMENT

The OECD is a unique forum where governments work together to address the economic, social and environmental challenges of globalisation. The OECD is also at the forefront of efforts to understand and to help governments respond to new developments and concerns, such as corporate governance, the information economy and the challenges of an ageing population. The Organisation provides a setting where governments can compare policy experiences, seek answers to common problems, identify good practice and work to co-ordinate domestic and international policies.

The OECD member countries are: Australia, Austria, Belgium, Canada, Chile, the Czech Republic, Denmark, Finland, France, Germany, Greece, Hungary, Iceland, Ireland, Israel, Italy, Japan, Korea, Luxembourg, Mexico, the Netherlands, New Zealand, Norway, Poland, Portugal, the Slovak Republic, Slovenia, Spain, Sweden, Switzerland, Turkey, the United Kingdom and the United States. The European Commission takes part in the work of the OECD.

OECD Publishing disseminates widely the results of the Organisation's statistics gathering and research on economic, social and environmental issues, as well as the conventions, guidelines and standards agreed by its members.

OECD PUBLISHING, 2, rue André-Pascal, 75775 PARIS CEDEX 16
(92 2015 04 1P) ISBN 978-92-64-23980-7 – 2015-17

www.ingramcontent.com/pod-product-compliance
Ingram Content Group UK Ltd.
Pitfield, Milton Keynes, MK11 3LW, UK
UKHW051459290926
15471UKWH00008B/1294

9 789264 239807